Advanced Polymers for Wastewater Treatment and Toxicant Removal

Advanced Polymers for Wastewater Treatment and Toxicant Removal

Guest Editors

Seyed Borhan Mousavi
Grigorios L. Kyriakopoulos

MDPI

Basel • Beijing • Wuhan • Barcelona • Belgrade • Novi Sad • Cluj • Manchester

Guest Editors
Seyed Borhan Mousavi
J. Mike Walker '66 Mechanical Engineering Department
Texas A&M University
College Station, TX
USA

Grigorios L. Kyriakopoulos
School of Electrical and Computer Engineering,
Electric Power Division
National Technical University of Athens
Athens
Greece

Editorial Office
MDPI AG
Grosspeteranlage 5
4052 Basel, Switzerland

This is a reprint of the Special Issue, published open access by the journal *Polymers* (ISSN 2073-4360), freely accessible at: https://www.mdpi.com/journal/polymers/special_issues/LX62HZAOX8.

For citation purposes, cite each article independently as indicated on the article page online and as indicated below:

Lastname, A.A.; Lastname, B.B. Article Title. *Journal Name* **Year**, *Volume Number*, Page Range.

ISBN 978-3-7258-4491-3 (Hbk)
ISBN 978-3-7258-4492-0 (PDF)
https://doi.org/10.3390/books978-3-7258-4492-0

Contents

About the Editors

Seyed Borhan Mousavi

Seyed Borhan Mousavi earned his Bachelor of Science (BSc) with distinction in Chemical Engineering from the University of Tabriz in 2016. He continued his academic journey at the same institution, obtaining his Master of Science (MSc) with honors in Chemical Engineering and Polymer Studies in 2019. He then expanded his interdisciplinary expertise by joining the University of British Columbia (UBC) as a graduate research and teaching assistant in Mechanical Engineering from 2020 to 2022. In January 2023, he became a member of the graduate program in Mechanical Engineering at Texas A&M University.

In 2024, Dr. Mousavi joined the Department of Chemical and Biomolecular Engineering at the University of Houston, where his research has since focused on sustainability, environmental sciences, and the thermochemical conversion of solid wastes. His broader research interests span the topics of polymers, nanofluids, heat transfer, tribology, CO_2 capture, and water/wastewater treatment.

To date, he has served as an editor for five scientific journals and has conducted over 300 peer reviews. He has also authored more than 30 publications in high-impact, peer-reviewed journals.

Grigorios L. Kyriakopoulos

Dr. Dr. Grigorios L. Kyriakopoulos (PhD in Chemical Engineering, PhD in Low Carbon Economy) is a teaching and research associate and senior researcher at the School of Electrical and Computer Engineering, National Technical University of Athens (NTUA). He has consistently been among the top 2% highly cited scientists in the world for the years between 1996 and 2023, based on the Stanford University list. His main areas of research are engineering, the environment, energy, renewable energy sources, low carbon economy, waste and water management, business administration, education, urban and regional development, extroversion and internationalization of Small and Medium Enterprises (SMEs), development economics, circular economy, and behavioral ecology.

Preface

Natural water sources are frequently polluted by wastewater effluents streaming into freshwater and causing the diffusion of diverse types of contaminants that are adversely affecting the quality of drinking and irrigation water. In response to this global and controversial issue, this Special Issue explores the use of advanced polymer-based materials for wastewater treatment and toxicant removal. Emphasis is placed on synthesis methods, characterization techniques, and applications for removing pollutants such as heavy metals, dyes, pharmaceuticals, and biological contaminants from contaminated water.

Seyed Borhan Mousavi and Grigorios L. Kyriakopoulos
Guest Editors

polymers

MDPI

Article

Comparative Study of Water and Milk Kefir Grains as Biopolymeric Adsorbents for Copper(II) and Arsenic(V) Removal from Aqueous Solutions

Balázs György [1], Marek Bujdoš [1], Hana Vojtková [2], Pavel Diviš [3], Michal Slaný [4,5], Peter Matúš [1,*] and Martin Urík [1]

1 Institute of Laboratory Research on Geomaterials, Faculty of Natural Sciences, Comenius University in Bratislava, Mlynská dolina, Ilkovičova 6, 842 15 Bratislava, Slovakia; gyorgy7@uniba.sk (B.G.); marek.bujdos@uniba.sk (M.B.); martin.urik@uniba.sk (M.U.)
2 Department of Environmental Engineering, Faculty of Mining and Geology, VSB—Technical University of Ostrava, 17. listopadu 2172/15, 708 00 Ostrava, Czech Republic; hana.vojtkova@vsb.cz
3 Institute of Food Science and Biotechnology, Faculty of Chemistry, Brno University of Technology, Purkyňova 464/118, Královo Pole, 612 00 Brno, Czech Republic; divis@fch.vut.cz
4 Department of Materials Engineering and Physics, Faculty of Civil Engineering, Slovak University of Technology, Radlinského 11, 810 05 Bratislava, Slovakia; michal.slany@stuba.sk
5 Department of Materials Engineering and Chemistry, Faculty of Civil Engineering, Czech Technical University in Prague, Thákurova 7, 166 29 Prague, Czech Republic
* Correspondence: peter.matus@uniba.sk; Tel.: +421-2-9014-9280

Abstract: This study investigates the biosorption capabilities of kefir grains, a polysaccharide-based byproduct of the fermentation process, for removing copper(II) and arsenic(V) from contaminated water. Unlike traditional heavy-metal removal methods, which are typically expensive and involve environmentally harmful chemicals, biopolymeric materials such as kefir grains provide a sustainable and cost-effective alternative for adsorbing hazardous inorganic pollutants from aqueous solutions. Our experimental results revealed significant differences in the sorption capacities of two types of kefir grains. Grains of milk kefir outperformed water kefir, particularly in copper(II) removal, achieving up to 95% efficiency at low copper concentrations (0.16 mmol·L^{-1}) and demonstrating a maximum sorption capacity of 49 μmol·g^{-1}. In contrast, water kefir grains achieved only 35.5% maximum removal efficiency and exhibited lower sorption capacity. For arsenic(V) removal, milk kefir grains also showed superior performance, removing up to 56% of arsenic in diluted solution with experimental sorption capacities reaching up to 20 μmol·g^{-1}, whereas water kefir grains achieved a maximum removal efficiency of 34.5%. However, these findings also suggest that while kefir grains show potential as low-cost biosorbents, further modifications are needed to enhance their competitiveness for large-scale water treatment applications.

Keywords: biopolymer; biosorption; heavy metals; polysaccharide; remediation; waste biomass

1. Introduction

Biosorption using waste biological materials has gained significant attention in recent years due to their versatile applications in environmental remediation, particularly in the removal of toxic heavy metals, metalloids, and other pollutants from contaminated water sources [1]. Traditional methods for heavy-metal removal, such as chemical precipitation, ion exchange, membrane filtration, and electroplating, often involve high operational costs and the use of environmentally harmful chemicals. In contrast, biosorbents derived from natural and waste materials offer a more sustainable, cost-effective, and eco-friendly alternative to conventional approaches for mitigating heavy-metal contamination in water [2].

To highlight the emerging potential of biopolymeric waste products from the food industry in the bioremediation of heavy-metal-contaminated waters, this study investigates the application of kefir grains, a byproduct of the fermentation process in beverage production [3], as biosorbents. Kefir grains, traditionally associated with the fermentation of dairy and water-based beverages, have recently been considered for the removal of waterborne contaminants, particularly organic toxins [4]. However, their use as biosorbents for inorganic pollutants, such as heavy metals, remains largely underexplored [5]. Therefore, this research specifically focuses on evaluating the sorptive capabilities of both water and milk kefir grains to determine their effectiveness in adsorbing copper(II) and arsenic(V)—two hazardous inorganic contaminants frequently found in industrial effluents and drinking water supplies [6].

Copper(II), though a necessary micronutrient in trace amounts, can become highly toxic when released in excess into the environment through industrial processes, leading to severe health risks for humans, plants, and aquatic ecosystems [7]. Similarly, arsenic(V), which is often found in industrial waste and agricultural runoff, is a major public health concern, poisoning millions globally through contaminated drinking water [8].

Given the high toxicity of these elements, effective removal strategies are crucial for protecting public health and preserving wildlife and aquatic ecosystems. By comparing the sorptive performance of water and milk kefir grains in the removal of copper(II) and arsenic(V), we aim to highlight the feasibility of these natural materials as effective, low-cost alternatives for remediating contaminated water. Ultimately, this study not only highlights the potential of kefir grains as low-cost biosorbents for heavy-metal removal but also sets the stage for future research into modified biosorbents and their broader applications in sustainable water treatment technologies.

2. Materials and Methods

2.1. Kefir Grain Preparation

Kefir grains were prepared through static cultivation using a commercially obtained inoculum. Water kefir was cultivated in a 1 L solution containing 100 g of sucrose and 200 g of lemon slices, while milk kefir was cultivated in 1 L of commercially available milk. After inoculation, it has been incubated up to 4 days at room temperature. Subsequently, the resulting biomass, consisting of microbial cells and exopolymers, was separated from the solution by filtration, washed with great amounts of distilled water, and used in this form in sorption experiments.

2.2. Sorption Experiments

All sorption experiments were conducted at a temperature of 25 °C in sealable 50 mL sterile plastic centrifuge tubes. In each, the 10 g of wet biomasses in the form of grains from water or milk kefir was suspended with 20 mL of an aqueous solution of selected potentially hazardous metals, including arsenic(V) and copper(II), so that the initial concentration of the analyte ranged from 0.14 to 4.2 mmol·L^{-1}. The pH of the solutions was adjusted to 6.0 using diluted HCl or NaOH solutions.

For the preparation of stock solutions of arsenic(V) and copper(II), the reagents of p.a. quality were used: $Na_2HAsO_4 \cdot 7H_2O$ (Sigma-Aldrich, St. Galen, Switzerland) and $CuSO_4 \cdot 5H_2O$ (Centralchem, Bratislava, Slovakia).

Kinetic experiments were conducted to evaluate the sorption rate, examining the effect of contact time on the sorption capacities of kefir grains over a period of 240 min. This duration provided sufficient data to calculate rate constants and offered insights into the sorption mechanism. However, for the isotherm studies, a contact time of 100 h was selected to ensure equilibrium in the sorption system. To ensure homogeneous interaction of the selected analytes with the surfaces of the water or milk kefir grains, the plastic tubes with the sample suspensions were placed on an overhead shaker (MultiBio RS-24; Biosan, Riga, Latvia) at a rotation speed of 45 rpm. Isotherm studies were conducted in triplicate, whereas kinetic experiments were performed in duplicate.

After the required contact time, the samples were centrifuged at 10,000× g for 10 min (centrifuge 5804R; Eppendorf, Hamburg, Germany) to separate the biosorbent from the remaining solution. The obtained solution was analyzed for residual concentration of the selected analytes using atomic absorption spectroscopy [9–11]. The residual biomass was dried to a constant weight at 85 °C, and the measurements were used in the following statistical procedures.

2.3. Analytical and Statistical Procedures

To determine the concentration of arsenic and copper in collected aqueous solutions (filtrates and stock solutions), a flame atomic absorption spectrometer (PerkinElmer 1100, Waltham, MA, USA) with an air-acetylene flame (air flow rate 8.0 L·min^{-1}, acetylene flow rate 2.5 L·min^{-1}) was utilized. The limits of quantification for arsenic and copper for this method were 0.5 mg L^{-1} and 0.005 mg L^{-1}, respectively. Calibration solutions were prepared from CertiPUR ICP 1000 mg·L^{-1} single-element standard solutions (Merck, Darmstadt, Germany).

To evaluate the surface morphology of the biomasses, scanning electron microscopy (SEM) was performed using a Carl Zeiss EVO 40 HV operated at 20 kV with a Bruker energy-dispersive X-ray (EDX) silicon drift detector.

The IR spectra of the kefir grains were measured on a Nicolet 6700 Fourier transform infrared (FTIR) spectrometer (Thermo Scientific, Waltham, MA, USA). For the MIR region (4000–400 cm^{-1}), 64 scans were used, and the spectrometer was equipped with a DTGS detector, an IR source, and a KBr beam splitter. The KBr pellet preparation technique was used to prepare the samples, where 1 mg of the sample was mixed with 200 mg of KBr.

The (equilibrium) sorption capacity of the waste-biomass-based sorbent (S_{eq}) was calculated using Equation (1):

$$S_{eq} = \frac{(C_0 - C_{eq})V}{m} \tag{1}$$

where C_0 and C_{eq} refer to the initial and equilibrium concentrations (mmol·L^{-1}) of the adsorbate in the suspension, respectively. V and m indicate the volume of the solution and the dry weight of the sorbent, respectively. For the calculation of the minute sorption capacity (S_t) at reaction time t (min), the same formula was used, except that the equilibrium concentration has been substituted for the minute concentration of the adsorbate in the solution (C_t).

The calculated sorption capacities at different initial arsenic(V) and copper(II) concentrations were fitted by the standard sorption isotherms, showcased in Table 1, to evaluate the sorption performance of the water and milk kefir grains. For the kinetic modelling, pseudo-first-order and the pseudo-second-order kinetics were employed, listed in Table 2. All statistical and data analyses regarding the sorption experiments were conducted using the Analysis ToolPak version 2.0.0.0 and Solver version 3.0.0.1 add-in programs (Frontline Systems, Incline Village, NV, USA) in Microsoft 365 Excel version 2402 (Microsoft, Redmond, WA, USA).

Table 1. The isotherm models used to evaluate the sorptive performance of the water and milk kefir grains.

Isotherm	Equation	Constants
Langmuir	$S_{eq} = \frac{S_{max} K_L C_{eq}}{1 + K_L C_{eq}}$	S_{max} (mmol·g^{-1}) is the theoretical sorption capacity representing the maximum amount of adsorbate that can be adsorbed per unit mass of the sorbent, and K_L (L·mmol^{-1}) is the Langmuir sorption constant, which indicates the affinity of the adsorbate (arsenic or copper) toward the sorbent
Freundlich	$S_{eq} = K_F C_{eq}^N$	N is an empirical parameter related to the intensity of sorption, and K_F (mmol$^{1\text{-}N}$·L^{N}·g^{-1}) is the Freundlich constant that represents the sorption capacity at a unitary equilibrium concentration of adsorbate (arsenic or copper) in the solution

Table 2. The models for the reaction kinetics used to evaluate the sorptive performance of the water and milk kefir grains.

Kinetic Model	Equation	Constants
Pseudo-first order	$S_t = S_{eq}\left(1 - e^{-k_1 t}\right)$	k_1 (min^{-1}) represents the rate constants for the pseudo-first-order kinetics
Pseudo-second order	$S_t = \frac{k_2 S_{eq}^2 t}{1 + k_2 S_{eq} t}$	k_2 ($g \cdot mmol^{-1} \cdot min^{-1}$) represents the rate constants for the pseudo-second-order kinetics

To assess the appropriateness of each kinetic and isothermal model for the experimental data, Akaike weights (w_i) were utilized as follows:

$$w_i = \frac{exp\left[\frac{1}{2}(AIC_{\text{min}} - AIC_{\text{i}})\right]}{\sum exp\left[\frac{1}{2}(AIC_{\text{min}} - AIC_{\text{i}})\right]} \tag{2}$$

Here, AIC represents Akaike's information criterion. AIC_{min} signifies the lowest AIC value among the models, and AIC_{i} is the AIC value of the specific model i. Akaike's information criterion serves as an estimated measure of the quality of available models for a particular dataset, making it an ideal method for model selection [12]. This approach is advantageous as it considers the number of model parameters (k_{p}), the number of data points (n), and the statistical parameter, which, in our case, is the residual sum of squares (RSS) of the respective model:

$$AIC = 2k_{\text{p}} + n\ln(RSS/n) \tag{3}$$

3. Results

3.1. Structural and Morphological Characterization of the Kefir Grains

FTIR analysis was conducted (Figure 1) to identify the main functional groups associated with the polysaccharides and other organic components in water and milk kefir grains. As polysaccharides are expected to be the predominant components in both biomasses, a broad band observed around 3350 cm^{-1} can be attributed to the stretching of hydroxyl groups in carbohydrate structures or O–H stretching vibrations associated with the water content. Additionally, the prominent fingerprint region between 1200 and 900 cm^{-1} suggests the presence of pyranose rings and associated side groups, evidenced by C–O–C stretching vibrations and C–O–H bending vibrations in carbohydrate structures [13]. The strong absorbance at this region indicates glucose and galactose units in milk kefir grains [14], while glucose units of dextran are prominent in water kefir grains [15].

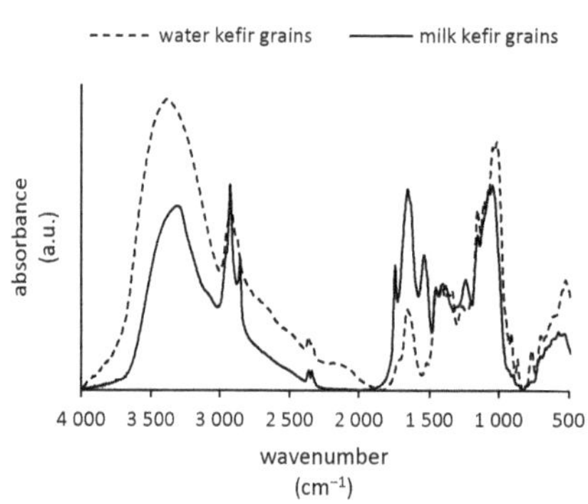

Figure 1. FTIR spectra of water and milk kefir grains.

The 3100–2800 cm^{-1} region is dominated by asymmetric and symmetric C–H stretching vibrations, with strong peaks at around 2925 and 2854 cm^{-1}, respectively, attributed to aliphatic CH_2 groups from fatty acids in membrane amphiphiles [16]. For milk kefir grains, these regions can also be linked to a carbonyl stretch near 1745 cm^{-1} suggesting the presence of milk fatty acid remnants in the biomass of milk kefir grains [17].

Notable differences between water and milk kefir grains are observed in the 1800–1200 cm^{-1} range, likely due to a higher content of proteins in milk kefir and their interpolymeric interactions with polysaccharides [14]. The 1800–1500 cm^{-1} region generally reflects the amide I and amide II bands associated with peptides and proteins. Both kefir grains display a distinct band around 1655 cm^{-1}, likely corresponding to carbonyl (C=O) stretching in peptides [18]. Additionally, a strong N–H vibration peak at 1540 cm^{-1} is unique to milk kefir grains; this peak, along with the band near 1655 cm^{-1}, may be linked to amide linkages involving aminosugars within the polysaccharide structure [19].

The external microstructure of lyophilized kefir grains was examined using scanning electron microscopy. The surface of water kefir grains (Figure 2a) appeared porous and smoother compared with the more heterogeneous surface of milk kefir grains (Figure 2b), which exhibited complex, surface-associated branched structures without evident crystalline assembly, likely consisting of precipitated peptides or other organic materials. No cells were observed directly on the surface, suggesting that they were either encapsulated within the polysaccharide biofilm or removed during sample preparation due to washing.

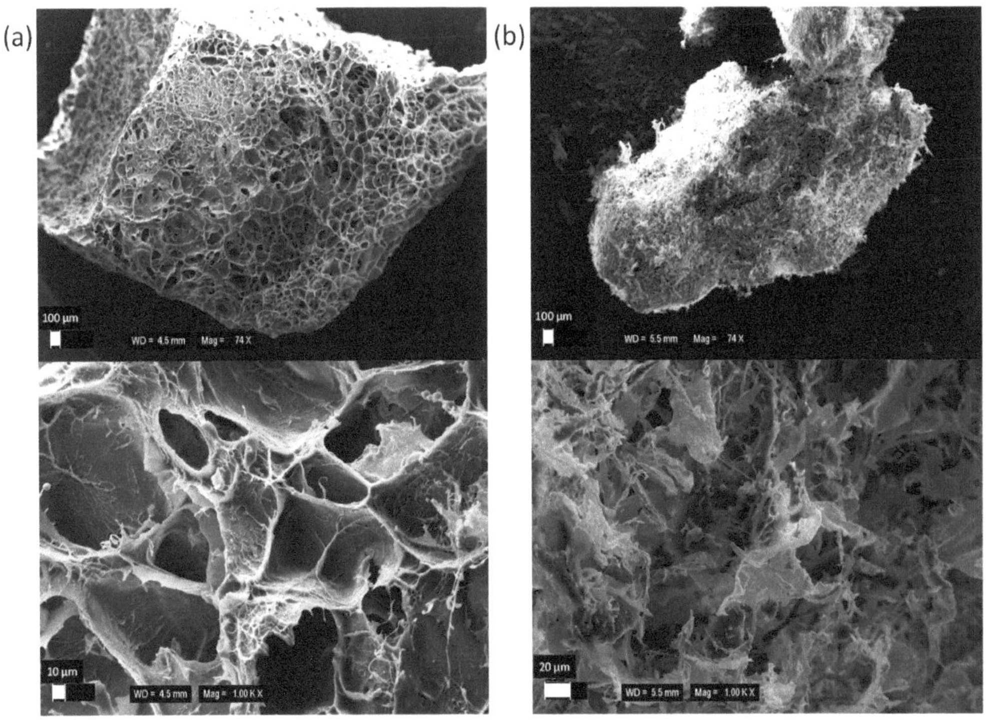

Figure 2. Scanning electron micrograph of (**a**) water and (**b**) milk kefir grains.

3.2. Removal of Copper(II) Using Water Kefir Grains

From the experimental results of copper(II) sorption onto the biomass of water kefir grains, comprising the microbial cells and their extracellular products, we determined the Freundlich constant (K_F) to be 0.008 (Table 2). Under equilibrium conditions where the sorption system reaches a concentration of 1 mmol·L^{-1} (Figure 1), this Freundlich constant can be translated into a sorption capacity of water kefir (0.008 mmol·g^{-1}). Since this value specifically applies to diluted solutions, the relatively low sorption capacity indicates a weak affinity of copper(II) for the adsorbent in such diluted systems. Notably, the highest experimentally measured sorption capacity was found to be 0.022 mmol·g^{-1} (Figure 1), further highlighting the limitations of this biosorbent in retaining copper(II) under the given conditions.

The exponent N in the Freundlich isotherm has a value of 1.03 (Table 3), indicating slight variability in the adsorption energies that bind copper(II) from the solution to the surface of the biosorbent, suggesting heterogeneity in the sorption sites.

Table 3. Langmuir and Freundlich isotherm parameters for the sorption of copper(II) using water and milk kefir grains (n = 3; temperature of 25 °C; dynamic conditions (45 rpm); initial copper(II) concentrations ranging from 0.16 to 4.2 mmol·L^{-1}; initial pH of 6.0; contact time of 100 h).

	Culture	K_L (L·mmol^{-1})	S_{max} (mmol·g^{-1})	R^2	Akaike weight
Langmuir Isotherm	Water kefir	$5.32 \cdot 10^{-6} \pm 0.03 \cdot 10^{0}$	$1.5 \cdot 10^{3} \pm 8.5 \cdot 10^{6}$	0.99	0.44
	Milk kefir	3.2 ± 1.7	0.049 ± 0.008	0.80	0.32
	Culture	**K_F (mmol$^{1\text{-}N}$·L^{N}·g^{-1})**	**N**	**R^2**	**Akaike weight**
Freundlich Isotherm	Water kefir	0.008 ± 0.0003	1.03 ± 0.05	0.99	0.56
	Milk kefir	0.035 ± 0.003	0.38 ± 0.09	0.83	0.68

The Langmuir model also provides a good fit to the experimentally obtained sorption data with R^2 reaching 0.99. However, the more statistically sensitive Akaike weight (Table 3) indicates that this model is less suitable for describing the experimental data compared with the Freundlich model. Furthermore, caution is needed when interpreting the Langmuir parameters, particularly due to the model's behavior and the lack of saturation of the biosorbent across the tested concentration range. The calculated maximum adsorption capacity (S_{max}) is unrealistic, as the sorption capacity linearly increases with rising equilibrium concentrations and shows no signs of reaching saturation (Figure 3). This discrepancy is further supported by the significantly large error associated with the S_{max} parameter, which deviates by six orders of magnitude. Therefore, the Langmuir isotherm is not suitable for analyzing copper biosorption by water kefir under the conditions of our study.

Both the low affinity of copper toward the biomass and the low sorption capacity in diluted solutions, as indicated by the Freundlich constant, contribute to the limited removal efficiency of the biosorbent. As shown in Figure 4, the overall performance of the water kefir biomass in copper(II) removal remains relatively low, with efficiency not exceeding 35.5%. This efficiency shows only minimal variation across different initial copper(II) concentrations, indicating a consistent but limited capacity for copper adsorption.

Our kinetic data (Figure 5a) indicate rapid copper(II) removal by water kefir grains, reaching equilibrium after 100 min of contact time. In comparison with the pseudo-first-order model, the experimental data fit better with the pseudo-second-order kinetic model, as suggested by a comparison of Akaike weights, with a rate constant as high as 330.3 g·mmol^{-1}·min^{-1} (Table 4).

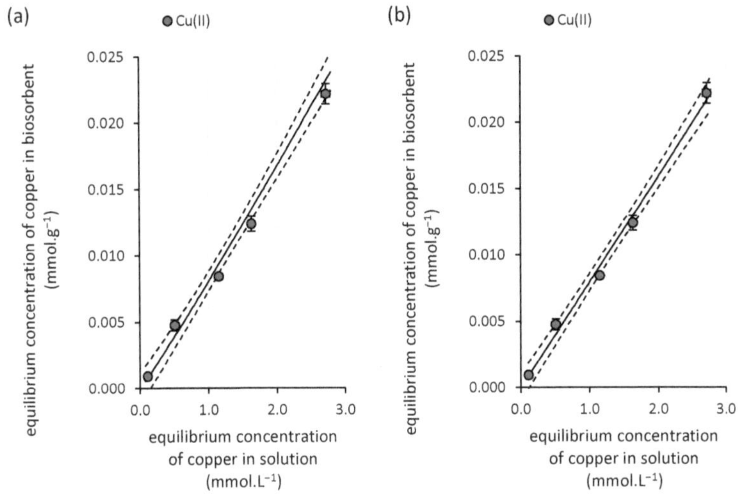

Figure 3. Isotherm fits of the experimental data for copper(II) removal from aqueous solutions by water kefir grains using the (**a**) Freundlich and (**b**) Langmuir isotherm models (solid lines). The experiments were conducted in darkness at 25 °C under dynamic conditions (45 rpm) with initial copper(II) concentrations ranging from 0.17 to 4.0 mmol·L^{-1} (initial pH of 6.0; contact time of 100 h; n = 3). The dashed line represents the 95% confidence interval for each isotherm model.

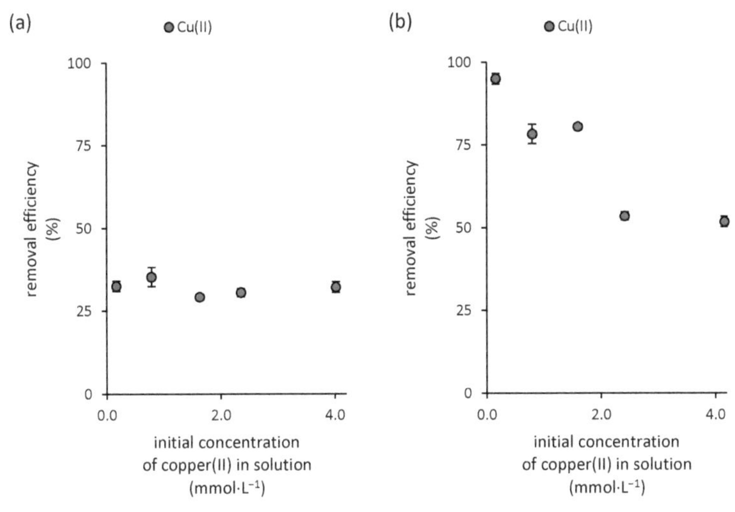

Figure 4. Removal efficiencies for copper(II) from aqueous solutions using (**a**) water and (**b**) milk kefir grains (n = 3; temperature of 25 °C; dynamic conditions (45 rpm); initial copper(II) concentrations ranging from 0.16 to 4.2 mmol·L^{-1}; initial pH of 6.0; contact time of 100 h).

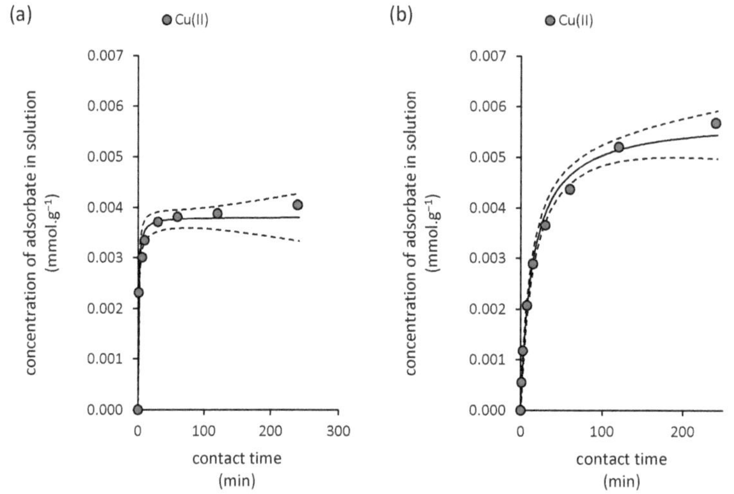

Figure 5. Kinetics of the copper(II) removal from aqueous solutions by (**a**) water kefir and (**b**) milk kefir grains fitted using pseudo-second-order kinetic model (solid line). The experiments were conducted in darkness at 25 °C under dynamic conditions (45 rpm) with initial copper(II) concentration of 0.79 mmol·L^{-1} (50 mg·L^{-1}; initial pH of 6.0; n = 2). The dashed line represents the 95% confidence interval for the applied kinetic model.

Table 4. Kinetic model parameters for the sorption of copper(II) using water and milk kefir grains (n = 2; temperature of 25 °C; dynamic conditions (45 rpm); initial copper(II) concentrations of 0.79 mmol·L^{-1}; initial pH of 6.0).

	Culture	S_{eq} **(mmol·g^{-1})**	k_1 **(min^{-1})**	**R^2**	**Akaike weight**
Pseudo-first-order kinetic model	Water kefir	0.0036 ± 0.0001	0.995 ± 0.269	0.94	0.02
	Milk kefir	0.0051 ± 0.0003	0.053 ± 0.009	0.96	0.01
	Culture	S_{eq} **(mmol·g^{-1})**	k_2 **(g·mmol^{-1}·min^{-1})**	**R^2**	**Akaike weight**
Pseudo-second-order kinetic model	Water kefir	0.0038 ± 0.0001	330.3 ± 82.1	0.98	0.98
	Milk kefir	0.0058 ± 0.0002	11.8 ± 1.83	0.99	0.99

3.3. Removal of Copper(II) Using Milk Kefir Grains

Compared with water kefir grains, the sorption kinetics of copper(II) onto the milk kefir biomass are significantly slower, as indicated by the lower pseudo-second-order kinetic rate constant of only 11.8 g·mmol^{-1}·min^{-1} (Table 4). While rapid sorption occurs within the first 100 min, equilibrium is not fully achieved within this period, suggesting a slower, time-dependent phase during which equilibrium may be reached over an extended duration (Figure 5b). Nonetheless, the data fit well with the pseudo-second-order kinetic model (Akaike weight of 0.99) in comparison with the pseudo-first-order kinetic model.

The copper sorption efficiency of the milk kefir biomass declines as the initial copper concentration in the solution increases (Figure 4b). However, at a low copper(II) concentration of 0.17 mmol·L^{-1}, milk kefir exhibited a high binding efficiency, removing up to 95% of the copper present. This result demonstrates the strong adsorption capacity of milk kefir in diluted solutions. Even at a concentration of 0.8 mmol·L^{-1}, it managed to remove 78% of copper(II). At the highest tested concentration, 4.2 mmol·L^{-1}, milk kefir still removed 51% of the copper, showcasing its superior performance compared with water kefir grains (Figure 4b).

The superior effectiveness of milk kefir as a biosorbent for copper removal from aqueous solutions is further supported by the parameters obtained from sorption isotherm analysis (Table 3). The Freundlich constant (K_F) for milk kefir is as high as 0.035, representing a fourfold increase in sorption capacity compared with water kefir at an equilibrium copper concentration of 1 mmol·L^{-1}. This higher K_F value suggests a significantly greater affinity of milk kefir for copper(II), particularly in diluted solutions.

Despite the lower determination coefficient for the Langmuir isotherm model (0.80), the maximum adsorption capacity (S_{max}) of 0.049 mmol·g^{-1} for milk kefir is more realistic than the value observed for the water kefir biomass. This is reflected in the shape of the isotherm curve, where signs of saturation of the sorbent with copper are evident at the highest tested concentration (Figure 6). These findings suggest that while milk kefir has a notable affinity for copper, its adsorption capacity likely becomes limited at equilibrium concentrations exceeding 2 mmol·L^{-1}.

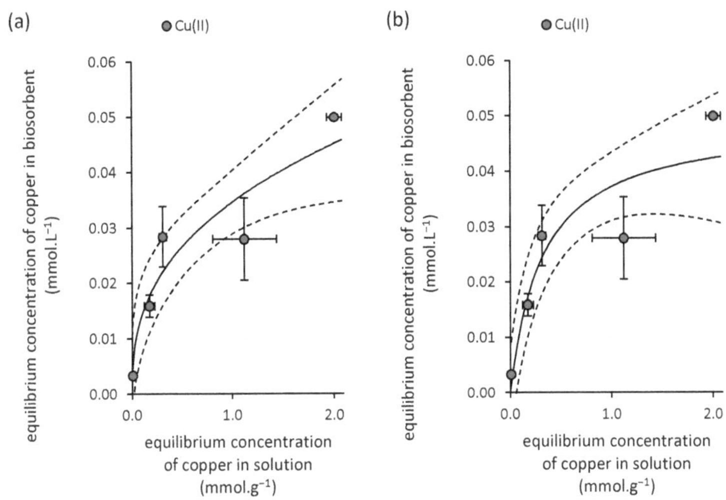

Figure 6. Isotherm fits of the experimental data for copper(II) removal from aqueous solutions by milk kefir grains using the (**a**) Freundlich and (**b**) Langmuir isotherm models (solid lines). The experiments were conducted in darkness at 25 °C under dynamic conditions (45 rpm) with initial copper(II) concentrations ranging from 0.16 to 4.2 mmol·L^{-1} (initial pH of 6.0; contact time of 100 h; n = 3). The dashed line represents the 95% confidence interval for each isotherm model.

3.4. Removal of Arsenic(V) Using Water Kefir Grains

When analyzing the experimental data on arsenic(V) sorption onto the biomass of microorganisms and exopolymers from water kefir using the Langmuir isotherm model, the results showed more realistic parameter estimates compared with copper(II) biosorption. However, the statistical uncertainty remains considerable, as evidenced by the wide error margins for the Langmuir constants: K_L (0.07 ± 0.06 L·mmol^{-1}) and S_{max} (0.13 ± 0.09 mmol·g^{-1}). These substantial errors suggest that the parameters are statistically less significant, necessitating cautious interpretation (Table 5).

The relationship between sorption capacity and equilibrium arsenic concentration (Figure 7) exhibits a linear trend similar to that observed in copper(II) sorption onto water kefir grains (Figure 6). This suggests that the Freundlich model may provide a more accurate description of the experimental data. This assumption is further supported by the significantly lower statistical error associated with the Freundlich constant K_F (0.008 ± 0.0004) (Table 5), indicating that the sorbent's affinity for arsenic(V) in diluted solutions is relatively low. Additionally, the exponent N takes a value of 0.93, which points to greater homogeneity of the adsorption sites on the surface of the water kefir biomass.

This implies a more uniform binding of arsenic(V) from aqueous solutions compared with the binding of copper(II).

Table 5. Langmuir and Freundlich isotherm parameters for the sorption of arsenic(V) using water and milk kefir biomasses (n = 3; temperature of 25 °C; dynamic conditions (45 rpm); initial arsenic(V) concentrations ranging from 0.14 to 3.5 mmol·L^{-1}; initial pH of 6.0; contact time of 100 h).

	Culture	K_L (L·mmol^{-1})	S_{max} (mmol·g^{-1})	R^2	Akaike weight
Langmuir isotherm	Water kefir	0.07 ± 0.06	0.13 ± 0.09	0.98	0.63
	Milk kefir	0.78 ± 0.22	0.03 ± 0.004	0.96	0.91
	Culture	**K_F (mmol$^{1\text{-}N}$·L^N·g^{-1})**	**N**	**R^2**	**Akaike weight**
Freundlich isotherm	Water kefir	0.008 ± 0.0004	0.93 ± 0.07	0.98	0.37
	Milk kefir	0.013 ± 0.0007	0.57 ± 0.07	0.93	0.09

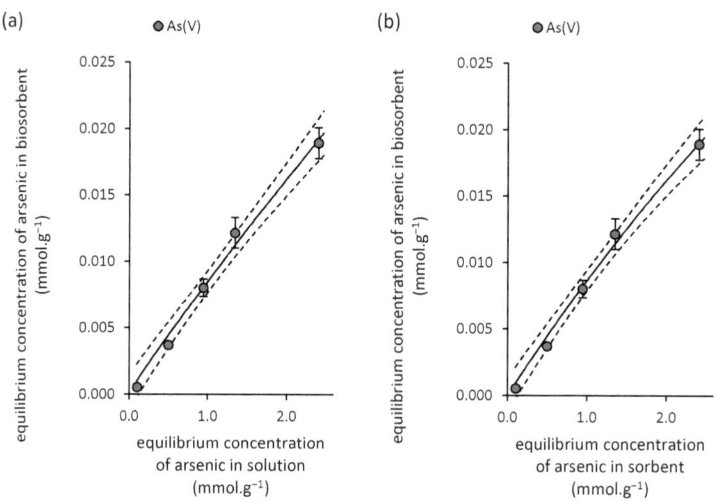

Figure 7. Isotherm fits of the experimental data for arsenic(V) removal from aqueous solutions by water kefir grains using the (**a**) Freundlich and (**b**) Langmuir isotherm models (solid lines). The experiments were conducted in darkness at 25 °C under dynamic conditions (45 rpm) with initial arsenic(V) concentrations ranging from 0.14 to 3.5 mmol·L^{-1} (initial pH of 6.0; contact time of 100 h; n = 3). The dashed line represents the 95% confidence interval for each isotherm model.

The sorption efficiency of the water kefir biomass for arsenic(V) varied depending on the concentration of arsenic in the solution, though the overall trend remained relatively stable (Figure 8a). At an initial arsenic concentration of 0.16 mmol·L^{-1}, the sorption efficiency was 22.5%. As the arsenic concentration increased to 0.7 mmol·L^{-1}, the sorption efficiency rose to 30%, indicating an increased uptake of arsenic by the water kefir biomass. This upward trend continued as the arsenic concentration increased further, reaching a maximum efficiency of 34.5% at an initial concentration of 2.1 mmol·L^{-1}. However, beyond this point, sorption efficiency declined to 31%, suggesting that the biosorbent began to approach saturation. This decline indicates that at higher concentrations, the sorbent's capacity to adsorb additional arsenic diminishes, likely due to the saturation of available adsorption sites.

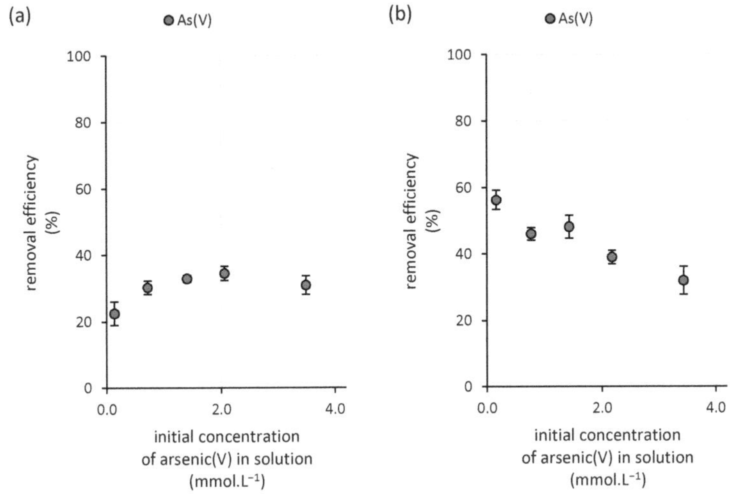

Figure 8. Removal efficiencies for arsenic(V) from aqueous solutions using (**a**) water and (**b**) milk kefir grains (n = 3; temperature of 25 °C; dynamic conditions (45 rpm); initial arsenic(V) concentrations ranging from 0.14 to 3.5 mmol·L^{-1}; initial pH of 6.0; contact time of 100 h).

The water kefir grains exhibit relatively rapid adsorption kinetic for arsenic(V) removal from aqueous solutions, achieving equilibrium in as little as 30 min (Figure 9a). As the pseudo-second-order kinetic model accurately describes the experimental data over a 240 min timeframe (Akaike weight of 0.99), the rate constant derived from this model has a relatively small margin of error, with a value of 338.0 $\pm$ 23.7 mmol^{-1}·min^{-1} (Table 6).

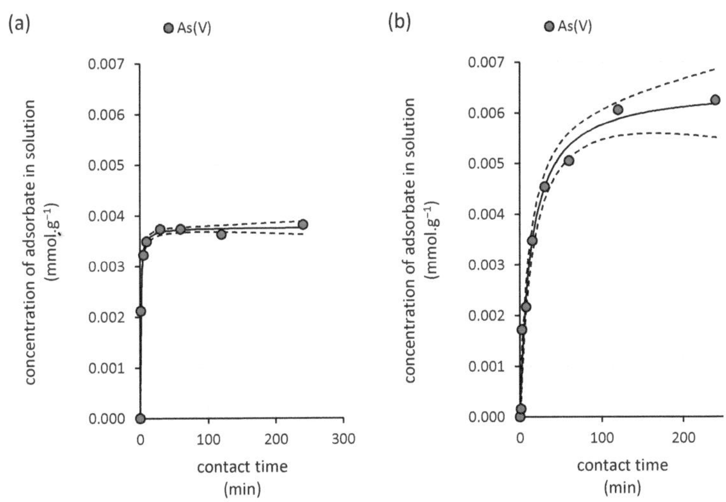

Figure 9. Kinetics of the arsenic(V) removal from aqueous solutions by (**a**) water kefir and (**b**) milk kefir grains fitted using pseudo-second-order kinetic model (solid line). The experiments were conducted in darkness at 25 °C under dynamic conditions (45 rpm) with initial arsenic(V) concentration of 0.67 mmol·L^{-1} (50 mg·L^{-1}; initial pH of 6.0; n = 2). The dashed line represents the 95% confidence interval for the applied kinetic model.

Table 6. Kinetic model parameters for the sorption of arsenic(V) using water and milk kefir grains (n = 2; temperature of 25 °C; dynamic conditions (45 rpm); initial arsenic(V) concentrations of 0.67 mmol·L^{-1}; initial pH of 6.0).

	Culture	S_{eq} (mmol·g^{-1})	k_1 (min^{-1})	R^2	Akaike weight
Pseudo-first-order kinetic model	Water kefir	0.0036 ± 0.0001	0.843 ± 0.121	0.98	0.01
	Milk kefir	0.0058 ± 0.0003	0.059 ± 0.011	0.97	0.04
	Culture	**S_{eq} (mmol·g^{-1})**	**k_2 (g·mmol^{-1}·min^{-1})**	**R^2**	**Akaike weight**
Pseudo-second-order kinetic model	Water kefir	0.0038 ± 0.0001	338.0 ± 23.7	0.99	0.99
	Milk kefir	0.0065 ± 0.0003	11.7 ± 2.29	0.98	0.96

3.5. Removal of Arsenic(V) Using Milk Kefir Grains

Although the changes in milk kefir's sorption capacity relative to the equilibrium concentration of arsenic are better described by the Langmuir isotherm according to the Akaike weight, we still recommend interpreting this model with caution. This is due to the relatively large error associated with the Langmuir constant K_L (0.78 ± 0.22), as shown in Table 5. In contrast, the Freundlich isotherm presents a much smaller statistical error, suggesting that its parameters are more reliable for describing the experimental data. The Freundlich constant K_F, which reflects the affinity of milk kefir for arsenic, indicates a moderate sorption capacity (0.013 mmol·g^{-1}), yet this value is almost twice as high as that for the water kefir biomass (Table 5), reinforcing milk kefir's superior biosorption efficiency for arsenic(V) removal.

The sorption efficiency of milk kefir exhibits a decreasing linear trend with increasing initial concentrations of arsenic(V) in the solution (Figure 8b). At lower concentrations, the sorption efficiency reached as high as 56%, whereas at the highest tested concentration of 3.4 mmol·L^{-1}, the efficiency dropped to 32%, which is only slightly higher than the efficiency observed for water kefir grains (31%) (Figure 8a). This declining trend suggests that milk kefir grain's sorption capacity is nearing saturation as arsenic concentrations increase, as indicated by the data in Figure 10.

The sorption kinetics of arsenic(V) onto the milk kefir biomass are well described by the pseudo-second-order kinetic model (Table 6), requiring a longer period to reach equilibrium compared with water kefir grains. Equilibrium is achieved more gradually, with a rate constant of 11.7 mmol^{-1}·min^{-1}, over approximately 240 min (Figure 9b).

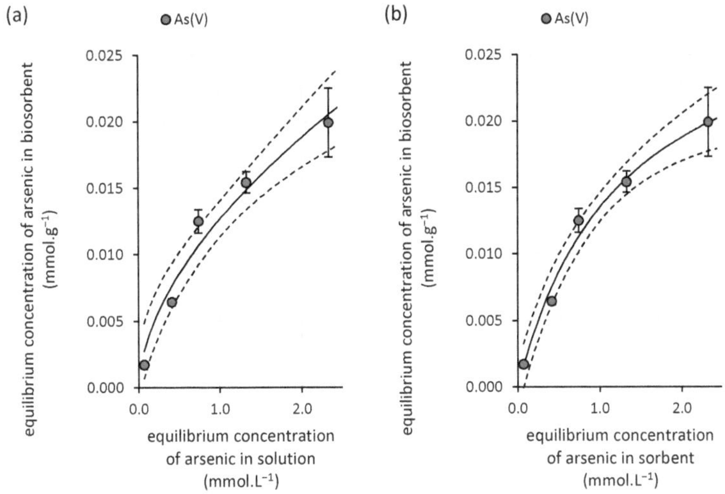

Figure 10. Fits of the experimental data for arsenic(V) removal from aqueous solutions by milk kefir grains using the (**a**) Freundlich and (**b**) Langmuir isotherm models (solid lines). The experiments were conducted in darkness at 25 °C under dynamic conditions (45 rpm) with initial arsenic(V) concentrations ranging from 0.16 to 3.4 mmol·L^{-1} (initial pH of 6.0; contact time of 100 h; n = 3). The dashed line represents the 95% confidence interval for each isotherm model.

4. Discussion

Our study demonstrates that water kefir grains generally exhibit lower sorption capacity and overall efficiency compared with the milk kefir biomass (Tables 3 and 5), particularly in diluted solutions. This difference is likely attributable to variations in the chemical composition and the availability of high-affinity sorption sites responsible for binding copper(II) and arsenic(V) from aqueous solutions.

The primary organic exopolymers in water kefir grains consist of insoluble O3- and O2-branched dextrans, along with smaller quantities of levans [15]. Both of these polysaccharides contain numerous hydroxyl (-OH) groups [20], which act as key sorption sites, enabling interactions with metals and metalloids, thereby facilitating their binding.

The results of our FTIR analysis of both kefir biomasses (Figure 1) support the presence of these heavy-metal reactive functional groups of polysaccharides. The fingerprint region between 1200 and 900 cm^{-1} highlights the presence of pyranose rings in kefir grains, with glucose present in water kefir and both glucose and galactose in milk kefir [21]. In contrast to water kefir, the matrix of milk kefir grains incorporates not only polysaccharides but also a significant amount of proteins [3], which can contribute up to 29% of the dry grain mass [22]. This composition enhances the diversity of sorption sites by introducing functional groups such as amino (-NH_2). Both arsenic(V) and copper(II) exhibit affinities for these extracellular polymeric substances [23,24]. Thus, it is likely that the protein-associated sorption sites within the milk kefir grains provide high-affinity binding locations for both arsenic(V) and copper(II), significantly enhancing the sorption performance of milk kefir grains in their removal from aqueous solutions.

The results of the FTIR analysis by Naveed, et al. [25] suggest the direct involvement of -OH, -NH-, -CO-, and C=O groups from extracellular polymeric substances in arsenic sorption, as evidenced by band shifts following arsenic(V) exposure in circumneutral solutions. In our experiment, the initial pH of the arsenic(V) solution was adjusted to 6.0, so the primary functional groups in the kefir polysaccharide matrix responsible for arsenic(V) and copper(II) removal are expected to remain negatively charged. This is consistent with findings on kefiran, which is only positively charged below pH 2 [26]—a typical trait of bacterial extracellular polymeric substances [27]. This negative charge may have impeded arsenic(V) adsorption due to repulsive interactions with the kefir grain surfaces, given that arsenic(V) predominantly exists as the negatively charged $H_2AsO_4^-$ species within our experimental pH range under oxidizing conditions, as indicated by its behavior in natural waters [28]. Conversely, copper(II) remains stable in cationic form, and observed shifts toward more acidic regions during kinetic and isotherm studies suggest a strong interaction with the negatively charged surfaces of kefir grains. This interaction likely involves the formation of inner-sphere complexes with oxygen-bearing ligands, e.g., carboxyl groups that are typically considered the primary sorption sites for copper(II) in bacterial extracellular polymeric matrices [29]. Arsenic, on the other hand, has been reported to strongly interact with amino groups in proteins, such as tyrosine- and tryptophan-like proteins within the extracellular polymeric substances of *Chlorella* sp. [30]. Thus, we hypothesize that the presence of amino groups of proteins, along with other functional groups such as hydroxyl, likely contributes to this improved efficiency, as these sites facilitate stronger interactions with the metal ions and metalloids. In contrast, water kefir's polysaccharide-based matrix offers fewer high-affinity sites, leading to its comparatively lower performance in arsenic(V) and copper(II) removal in diluted solutions (Figures 4 and 8).

The differences in sorption sites, their distribution, and their impact on the sorption mechanism can also be inferred from the isotherm analysis. For both copper(II) and arsenic(V), saturation is observed in the case of milk kefir (Figures 6 and 10), suggesting a stronger affinity but a lower density of high-affinity sorption sites within the milk grains. In contrast, water kefir grains align more closely with the Freundlich isotherm (Figures 3 and 7), indicating lower affinity but a relatively high frequency of sorption sites. This suggests that while water kefir grains possess numerous sorption sites, these are less selective for copper(II) and arsenic(V) compared with milk kefir. Furthermore, the sorption pattern observed for water kefir aligns with a C-type isotherm, indicating preferential removal through the micropores of the grains or possibly due to a more pronounced breakdown of the polymeric matrix within water kefir grains [31]. The results of our SEM analysis (Figure 2) highlighted notable structural and morphological differences between the two types of kefir grains. Water kefir grains exhibit a porous, hydrogel-like morphology, whereas milk kefir grains display a more compact surface populated with clustered, branched organic structures. It is likely that, due to more compact and structurally stable nature of milk kefir grains, sorption occurs primarily through more specific binding mechanisms rather than through porous absorption. However, differences in the sorption mechanisms are more precisely inferred from the kinetic studies. Interestingly, regardless of the type of kefir grain or element, all data fit the pseudo-second-order kinetic model exceptionally well. Conversely, the goodness of fit of the pseudo-first-order kinetic model is very low (Tables 4 and 6). This allows us to assume certain considerations that can be asserted regarding the suitability of fitting the pseudo-second-order kinetic model. These include the following: (i) the adsorption process involves chemisorption, and (ii) the rate-limiting process is the chemical reaction step [32]. Nevertheless, as indicated by the comparison of rate constants, equilibrium is achieved significantly faster with water kefir (Figures 5 and 9), suggesting that particle boundary diffusion, whether through pore or solid diffusion, may play a more prominent role in the adsorption process of both arsenic(V) and copper(II) onto the kefir biomass.

Based on the analyzed data, the milk kefir biomass emerges as a more effective biosorbent for arsenic at lower concentrations compared with the water kefir biomass. Unfortunately, to the best of our knowledge, no systematic or experimental studies have explored the potential of milk kefir grains for removing other heavy metals from solutions. Nevertheless, water kefir grains demonstrate relatively stable removal performance across various cationic heavy metals, including copper(II), nickel(II), and lead(II), as reported by Volpi, et al. [5]. This stability highlights their potential applicability for multicomponent adsorption in real industrial effluents or contaminated surface and groundwater. Still, to gain a clearer perspective, it is essential to compare water and milk kefir grains' performance with other unmodified waste biosorbents to properly assess its sorptive efficiency and potential for practical applications. Table 7 illustrates that the maximum sorption capacity derived from the Langmuir model is moderate and comparable with other non-modified waste-biomass-based biosorbents. However, the sorption capacity of chemically and physically modified biomasses far exceeds that of milk kefir, with modified materials often capable of removing hundreds of micrograms of arsenate per gram. This highlights the potential for significant improvements to enhance kefir's competitive edge. While some modification techniques still result in comparable performance, such as iron-coated cork granulates with a maximum sorption capacity of 56.07 $\mu mol \cdot g^{-1}$ [33] or wood-derived biochar with 51.9 $\mu mol \cdot g^{-1}$ [34], other approaches, particularly thermal treatments combined with nanosized inorganic compounds, yield much higher capacities. For example, zero-valent iron-composited palm-waste biochars reach a sorption capacity of 354 $\mu mol \cdot g^{-1}$ [35], and corncob-derived biochar impregnated with ZnO achieves 346 $\mu mol \cdot g^{-1}$ [36]. These modifications significantly outperform the untreated kefir grains, suggesting that targeted enhancements to kefir biosorbents can greatly improve their effectiveness for metal removal.

Table 7. Comparison of maximum sorption capacities of waste-biomass-based sorbents derived from the Langmuir isotherm (S_{max}) for arsenic(V).

Type of Sorbent	S_{max} ($\mu mol \cdot g^{-1}$)	Reference
Milk kefir grains	30	This work
Fish scale	0.38	Rahaman, et al. [37]
Wheat straw	1.5	Ebrahimi, et al. [38]
Rice polish	1.96	Ranjan, et al. [39]
Anaerobic sludge	2.19	Chowdhury and Mulligan [40]
Tea waste	28.3	Kamsonlian, et al. [41]
Sawdust	58.74	Leal, et al. [42]
Almond shell	141.7	Ali, et al. [43]

The performance of kefir grains in our study was notably lower compared with other available data on the sorptive removal of copper(II) from aqueous solutions using non-modified waste biomasses (Table 5). This discrepancy highlights the limitations of kefir grains as biosorbents in practical applications, particularly for copper removal, suggesting that kefir grains would require significant modification or enhancement to compete with other well-established biosorbents. This enhancement can be achieved through either chemical or physical modification of kefir grains or a combination of both approaches. For instance, processing the waste biomass into activated carbons can significantly improve the heavy-metal removal capacity by increasing the surface area, optimizing the pore distribution, and enhancing the chemical nature of the active sorption sites on biochar-based adsorbents [44,45].

Recently, nanobiocomposites have emerged as a promising and highly effective strategy for removing toxic metals and metalloids from contaminated aqueous environments. By combining the natural sorption properties of biological materials with the enhanced functional capabilities of nanoparticles, these composites deliver superior performance [46]. However, despite their potential, the reported effectiveness of nanobiocomposites for copper removal significantly varies. In some cases, they do not necessarily outperform untreated waste materials [47]. Nevertheless, while the removal efficiency of copper(II) by kefir grains shows potential in diluted copper solutions (Figure 4b), their overall sorption capacity remains underwhelming (Table 8). This limitation significantly reduces their practical applicability in real-world scenarios, particularly when compared with more eco-friendly alternatives like nanobiocomposites, which have demonstrated superior performance relative to the sorption capacity of kefir grains [48].

Table 8. Comparison of maximum sorption capacities of waste-biomass-based sorbents derived from the Langmuir isotherm (S_{max}) for copper(II).

Type of Sorbent	S_{max} (μmol·g^{-1})	Reference
Milk kefir grains	49	This work
Pine sawdust	35	Orozco, et al. [49]
Activated sludge biomass	138	Aslan, et al. [50]
Rice straw	196	Buasri, et al. [51]
Cone biomass of *Thuja orientalis*	303	Nuhoglu and Oguz [52]
Coffee grounds	398	Młynarczykowska and Orlof-Naturalna [53]
Sugarcane bagasse	841	Shah, et al. [54]

5. Conclusions

In this study, we investigated the efficiency of water and milk kefir grains as biosorbents for the removal of copper(II) and arsenic(V) from aqueous solutions. Our results revealed a stark contrast in performance between the two types of kefir grains. Notably, milk kefir exhibited a remarkable capacity to remove up to 95% of copper(II) at a low concentration of 0.16 mmol·L^{-1} and achieved a maximum removal efficiency of 56% for arsenic(V) at lower concentrations. This demonstrates its strong adsorption capabilities, especially in diluted solutions. In contrast, water kefir grains achieved a maximum removal efficiency of only 35.5% for copper(II) and 34.5% for arsenic(V), highlighting their limited effectiveness as a biosorbent for these contaminants.

Isotherm analyses further confirmed the superior performance of milk kefir. The results of the analysis indicated that water kefir has a lower affinity for copper and arsenic, which can be attributed to its simpler polysaccharide composition and the limited availability of high-affinity sorption sites. In contrast, milk kefir's more complex matrix, enriched with proteins and diverse functional groups, significantly enhances its binding capabilities for metal ions.

Despite the promising results, it is crucial to recognize that the overall performance of kefir grains remains lower when compared with other modified biosorbents and nanobiocomposites, which have consistently demonstrated superior metal removal efficiencies. This underscores the necessity for further modifications to the kefir biomass to enhance its sorption capacity and make it more competitive in practical applications. Future research should focus on optimizing the chemical and structural properties of kefir grains to improve their performance in real-world scenarios, thereby contributing to more sustainable and effective methods for remediating contaminated water sources.

Author Contributions: Conceptualization, B.G. and M.U.; methodology, M.U.; formal analysis, M.B., M.S. and P.M.; investigation, B.G., P.D. and H.V; writing—original draft preparation, B.G.; writing—review and editing, M.U.; supervision, M.U.; project administration, P.M.; funding acquisition, M.U., P.D. and H.V. All authors have read and agreed to the published version of the manuscript.

Funding: This work was supported by the Scientific Grant Agency of the Slovak Republic Ministry of Education and the Slovak Academy of Sciences under VEGA contract Nos. 1/0175/22 and 1/0139/22, also by SP2024/005 from the Faculty of Mining and Geology of VSB—Technical University of Ostrava, and specific research projects FCH-S-24-8526 and FCH-S-24-8597 financed by Ministry of Education, Youth and Sports of the Czech Republic.

Institutional Review Board Statement: Not applicable.

Data Availability Statement: The original contributions presented in this study are included in the article. Further inquiries can be directed to the corresponding author.

Conflicts of Interest: The authors declare no conflicts of interest.

References

1. Karić, N.; Maia, A.S.; Teodorović, A.; Atanasova, N.; Langergraber, G.; Crini, G.; Ribeiro, A.R.L.; Đolić, M. Bio-waste valorisation: Agricultural wastes as biosorbents for removal of (in)organic pollutants in wastewater treatment. *Chem. Eng. J. Adv.* **2022**, *9*, 100239. [CrossRef]
2. Karnwal, A. Unveiling the promise of biosorption for heavy metal removal from water sources. *Desalination Water Treat.* **2024**, *319*, 100523. [CrossRef]
3. Gao, X.; Li, B. Chemical and microbiological characteristics of kefir grains and their fermented dairy products: A review. *Cogent Food Agric.* **2016**, *2*, 1272152. [CrossRef]
4. Asadi Touranlou, F.; Noori, S.M.A.; Salari, A.; Afshari, A.; Hashemi, M. Application of kefir for reduction of contaminants in the food industry: A systematic review. *Int. Dairy J.* **2023**, *146*, 105748. [CrossRef]
5. Volpi, G.; Ginepro, M.; Tafur-Marinos, J.; Zelano, V. Pollution Abatement of Heavy Metals in Different Conditions by Water Kefir Grains as a Protective Tool against Toxicity. *J. Chem.* **2019**, *2019*, 8763902. [CrossRef]
6. Li, Y.; Qi, X.; Li, G.; Wang, H. Efficient removal of arsenic from copper smelting wastewater via a synergy of steel-making slag and KMnO4. *J. Clean. Prod.* **2021**, *287*, 125578. [CrossRef]
7. Rehman, M.; Liu, L.; Wang, Q.; Saleem, M.H.; Bashir, S.; Ullah, S.; Peng, D. Copper environmental toxicology, recent advances, and future outlook: A review. *Environ. Sci. Pollut. Res.* **2019**, *26*, 18003–18016. [CrossRef]
8. Hughes, M.F.; Beck, B.D.; Chen, Y.; Lewis, A.S.; Thomas, D.J. Arsenic Exposure and Toxicology: A Historical Perspective. *Toxicol. Sci.* **2011**, *123*, 305–332. [CrossRef]
9. Hagarová, I. Speciation of arsenic in waters by AAS techniques. *Chem. Listy* **2007**, *101*, 768–775.
10. Hagarová, I.; Kubová, J. Speciation of antimony in waters using separation coupled with atomic spectrometry. *Chem. Listy* **2008**, *102*, 782–790.
11. Hagarová, I.; Nemček, L. Recent advances in speciation analysis of trace antimony in environmental and biological samples based on cloud point extraction and spectrometric methods. In *Sustainable Agriculture Reviews*; Lichtfouse, E., Ed.; Springer: Berlin/Heidelberg, Germany, 2021; Volume 52, pp. 49–77.
12. Bozdogan, H. Model selection and Akaike's Information Criterion (AIC): The general theory and its analytical extensions. *Psychometrika* **1987**, *52*, 345–370. [CrossRef]
13. Piermaria, J.; Bosch, A.; Pinotti, A.; Yantorno, O.; Garcia, M.A.; Abraham, A.G. Kefiran films plasticized with sugars and polyols: Water vapor barrier and mechanical properties in relation to their microstructure analyzed by ATR/FT-IR spectroscopy. *Food Hydrocoll.* **2011**, *25*, 1261–1269. [CrossRef]
14. Ramírez Tapias, Y.A.; Rezzani, G.D.; Delgado, J.F.; Peltzer, M.A.; Salvay, A.G. New Materials from the Integral Milk Kefir Grain Biomass and the Purified Kefiran: The Role of Glycerol Content on the Film's Properties. *Polymers* **2024**, *16*, 3106. [CrossRef] [PubMed]
15. Fels, L.; Jakob, F.; Vogel, R.F.; Wefers, D. Structural characterization of the exopolysaccharides from water kefir. *Carbohydr. Polym.* **2018**, *189*, 296–303. [CrossRef] [PubMed]
16. Dziuba, B.; Babuchowski, A.; Nałęcz, D.; Niklewicz, M. Identification of lactic acid bacteria using FTIR spectroscopy and cluster analysis. *Int. Dairy J.* **2007**, *17*, 183–189. [CrossRef]
17. Rutten, M.J.M.; Bovenhuis, H.; Hettinga, K.A.; van Valenberg, H.J.F.; van Arendonk, J.A.M. Predicting bovine milk fat composition using infrared spectroscopy based on milk samples collected in winter and summer. *J. Dairy Sci.* **2009**, *92*, 6202–6209. [CrossRef]
18. Alves, E.; Ntungwe, E.N.; Gregório, J.; Rodrigues, L.M.; Pereira-Leite, C.; Caleja, C.; Pereira, E.; Barros, L.; Aguilar-Vilas, M.V.; Rosado, C.; et al. Characterization of Kefir Produced in Household Conditions: Physicochemical and Nutritional Profile, and Storage Stability. *Foods* **2021**, *10*, 1057. [CrossRef]
19. Zhang, L.; Liu, J.; Kong, S.; Chen, N.; Hung, W.-L.; Zhao, W.; Zeng, Z.; Zhang, J.; Yang, Z. Lipoteichoic acid obtained from Lactobacillus paracasei via low-temperature pasteurization alleviates the macrophage inflammatory response by downregulating the NF-κB signaling pathway. *J. Funct. Foods* **2023**, *107*, 105673. [CrossRef]
20. Coma, M.E.; Peltzer, M.A.; Delgado, J.F.; Salvay, A.G. Water kefir grains as an innovative source of materials: Study of plasticiser content on film properties. *Eur. Polym. J.* **2019**, *120*, 109234. [CrossRef]
21. Prado, M.R.; Blandón, L.M.; Vandenberghe, L.P.S.; Rodrigues, C.; Castro, G.R.; Thomaz-Soccol, V.; Soccol, C.R. Milk kefir: Composition, microbial cultures, biological activities, and related products. *Front. Microbiol.* **2015**, *6*, 1177. [CrossRef]
22. Han, X.; Yi, H.; Zhao, S.; Sun, J.; Wang, Y. Prospects of Artificial Kefir Grains Prepared by Cheese and Encapsulated Vectors to Mimic Natural Kefir Grains. *J. Food Qual.* **2020**, *2020*, 8839135. [CrossRef]
23. Naveed, S.; Li, C.; Zhang, J.; Zhang, C.; Ge, Y. Sorption and transformation of arsenic by extracellular polymeric substances extracted from *Synechocystis* sp. PCC6803. *Ecotoxicol. Environ. Saf.* **2020**, *206*, 111200. [CrossRef] [PubMed]
24. Zanetti, R.; Zecchin, S.; Colombo, M.; Borgonovo, G.; Mazzini, S.; Scaglioni, L.; Facchetti, G.; Gandolfi, R.; Rimoldi, I.; Cavalca, L. Ni^{2+} and Cu^{2+} Biosorption by EPS-Producing Serratia plymuthica Strains and Potential Bio-Catalysis of the Organo–Metal Complexes. *Water* **2022**, *14*, 3410. [CrossRef]
25. Naveed, S.; Yu, Q.; Zhang, C.; Ge, Y. Extracellular polymeric substances alter cell surface properties, toxicity, and accumulation of arsenic in Synechocystis PCC6803. *Environ. Pollut.* **2020**, *261*, 114233. [CrossRef] [PubMed]
26. Exarhopoulos, S.; Raphaelides, S.N.; Kontominas, M.G. Conformational studies and molecular characterization of the polysaccharide kefiran. *Food Hydrocoll.* **2018**, *77*, 347–356. [CrossRef]

27. Yang, L.; Chen, Z.; Zhang, Y.; Lu, F.; Liu, Y.; Cao, M.; He, N. Hyperproduction of extracellular polymeric substance in Pseudomonas fluorescens for efficient chromium (VI) absorption. *Bioresour. Bioprocess.* **2023**, *10*, 17. [CrossRef]
28. Smedley, P.L.; Kinniburgh, D.G. A review of the source, behaviour and distribution of arsenic in natural waters. *Appl. Geochem.* **2002**, *17*, 517–568. [CrossRef]
29. Fang, L.; Yang, S.; Huang, Q.; Xue, A.; Cai, P. Biosorption mechanisms of Cu(II) by extracellular polymeric substances from Bacillus subtilis. *Chem. Geol.* **2014**, *386*, 143–151. [CrossRef]
30. Zhang, J.; Zhou, F.; Liu, Y.; Huang, F.; Zhang, C. Effect of extracellular polymeric substances on arsenic accumulation in Chlorella pyrenoidosa. *Sci. Total Environ.* **2020**, *704*, 135368. [CrossRef]
31. Giles, C.H.; Smith, D.; Huitson, A. A general treatment and classification of the solute adsorption isotherm. I. Theoretical. *J. Colloid Interface Sci.* **1974**, *47*, 755–765. [CrossRef]
32. Ho, Y.S.; McKay, G. Pseudo-second order model for sorption processes. *Process Biochem.* **1999**, *34*, 451–465. [CrossRef]
33. Carneiro, M.A.; Pintor, A.M.A.; Boaventura, R.A.R.; Botelho, C.M.S. Efficient removal of arsenic from aqueous solution by continuous adsorption onto iron-coated cork granulates. *J. Hazard. Mater.* **2022**, *432*, 128657. [CrossRef] [PubMed]
34. Niazi, N.K.; Bibi, I.; Shahid, M.; Ok, Y.S.; Shaheen, S.M.; Rinklebe, J.; Wang, H.; Murtaza, B.; Islam, E.; Farrakh Nawaz, M.; et al. Arsenic removal by Japanese oak wood biochar in aqueous solutions and well water: Investigating arsenic fate using integrated spectroscopic and microscopic techniques. *Sci. Total Environ.* **2018**, *621*, 1642–1651. [CrossRef] [PubMed]
35. Ahmad, M.; Usman, A.R.A.; Hussain, Q.; Al-Farraj, A.S.F.; Tsang, Y.F.; Bundschuh, J.; Al-Wabel, M.I. Fabrication and evaluation of silica embedded and zerovalent iron composited biochars for arsenate removal from water. *Environ. Pollut.* **2020**, *266*, 115256. [CrossRef]
36. Cruz, G.J.F.; Mondal, D.; Rimaycuna, J.; Soukup, K.; Gómez, M.M.; Solis, J.L.; Lang, J. Agrowaste derived biochars impregnated with ZnO for removal of arsenic and lead in water. *J. Environ. Chem. Eng.* **2020**, *8*, 103800. [CrossRef]
37. Rahaman, M.S.; Basu, A.; Islam, M.R. The removal of As(III) and As(V) from aqueous solutions by waste materials. *Bioresour. Technol.* **2008**, *99*, 2815–2823. [CrossRef]
38. Ebrahimi, R.; Maleki, A.; Shahmoradi, B.; Daraei, H.; Mahvi, A.H.; Barati, A.H.; Eslami, A. Elimination of arsenic contamination from water using chemically modified wheat straw. *Desalination Water Treat.* **2013**, *51*, 2306–2316. [CrossRef]
39. Ranjan, D.; Talat, M.; Hasan, S.H. Biosorption of arsenic from aqueous solution using agricultural residue 'rice polish'. *J. Hazard. Mater.* **2009**, *166*, 1050–1059. [CrossRef]
40. Chowdhury, M.R.I.; Mulligan, C.N. Biosorption of arsenic from contaminated water by anaerobic biomass. *J. Hazard. Mater.* **2011**, *190*, 486–492. [CrossRef]
41. Kamsonlian, S.; Suresh, S.; Majumder, C.B.; Chand, S. Biosorption Of As(V) From Contaminated Water Onto Tea Waste Biomass: Sorption Parameters Optimization, Equilibrium and Thermodynamic Studies. *J. Future Eng. Technol.* **2011**, *7*, 34–41. [CrossRef]
42. Leal, M.A.L.; Martínez, R.C.; Villanueva, R.A.C.; Flores, H.E.M.; Penagos, C.D.C. Arsenate Biosorption by Iron-Modified Pine Sawdust in Batch Systems: Kinetics and Equilibrium Studies. *Bioresources* **2012**, *7*, 1389–1404.
43. Ali, S.; Rizwan, M.; Shakoor, M.B.; Jilani, A.; Anjum, R. High sorption efficiency for As(III) and As(V) from aqueous solutions using novel almond shell biochar. *Chemosphere* **2020**, *243*, 125330. [CrossRef]
44. Demirbas, E.; Dizge, N.; Sulak, M.T.; Kobya, M. Adsorption kinetics and equilibrium of copper from aqueous solutions using hazelnut shell activated carbon. *Chem. Eng. J.* **2009**, *148*, 480–487. [CrossRef]
45. Demiral, H.; Güngör, C. Adsorption of copper(II) from aqueous solutions on activated carbon prepared from grape bagasse. *J. Clean. Prod.* **2016**, *124*, 103–113. [CrossRef]
46. Singh, S.; Kumar, V.; Datta, S.; Dhanjal, D.S.; Sharma, K.; Samuel, J.; Singh, J. Current advancement and future prospect of biosorbents for bioremediation. *Sci. Total Environ.* **2020**, *709*, 135895. [CrossRef]
47. Kayalvizhi, K.; Alhaji, N.M.I.; Saravanakkumar, D.; Mohamed, S.B.; Kaviyarasu, K.; Ayeshamariam, A.; Al-Mohaimeed, A.M.; AbdelGawwad, M.R.; Elshikh, M.S. Adsorption of copper and nickel by using sawdust chitosan nanocomposite beads—A kinetic and thermodynamic study. *Environ. Res.* **2022**, *203*, 111814. [CrossRef] [PubMed]
48. Eldeeb, T.M.; El-Nemr, A.; Khedr, M.H.; El-Dek, S.I. Novel bio-nanocomposite for efficient copper removal. *Egypt. J. Aquat. Res.* **2021**, *47*, 261–267. [CrossRef]
49. Orozco, C.I.; Freire, M.S.; Gómez-Díaz, D.; González-Álvarez, J. Removal of copper from aqueous solutions by biosorption onto pine sawdust. *Sustain. Chem. Pharm.* **2023**, *32*, 101016. [CrossRef]
50. Aslan, S.; Yldiz, S.; Ozturk, M. Biosorption of Cu^{2+} and Ni^{2+} Ions From Aqueous Solutions Using Waste Dried Activated Sludge Biomass. *Pol. J. Chem. Technol.* **2018**, *20*, 20–28. [CrossRef]
51. Buasri, A.; Chaiyut, N.; Tapang, K.; Jaroensin, S.; Panphrom, S. Removal of Cu^{2+} from Aqueous Solution by Biosorption on Rice Straw—An Agricultural Waste Biomass. *Int. J. Environ. Sci. Dev.* **2012**, *3*, 10–14. [CrossRef]
52. Nuhoglu, Y.; Oguz, E. Removal of copper(II) from aqueous solutions by biosorption on the cone biomass of *Thuja orientalis*. *Process Biochem.* **2003**, *38*, 1627–1631. [CrossRef]

53. Młynarczykowska, A.; Orlof-Naturalna, M. Biosorption of Copper (II) Ions Using Coffee Grounds—A Case Study. *Sustainability* **2024**, *16*, 7693. [CrossRef]
54. Shah, G.M.; Nasir, M.; Imran, M.; Bakhat, H.F.; Rabbani, F.; Sajjad, M.; Umer Farooq, A.B.; Ahmad, S.; Song, L. Biosorption potential of natural, pyrolysed and acid-assisted pyrolysed sugarcane bagasse for the removal of lead from contaminated water. *PeerJ* **2018**, *6*, e5672. [CrossRef] [PubMed]

polymers

MDPI

Review

Exploring the Possibilities of Using Recovered Collagen for Contaminants Removal—A Sustainable Approach for Wastewater Treatment

Annette Madelene Dancila and Magdalena Bosomoiu *

Department of Analytical Chemistry and Environmental Engineering, Faculty of Chemical Engineering and Biotechnologies, National University of Science and Technology Politehnica Bucharest, 7 Polizu Street, 011061 Bucharest, Romania; madelene.dancila@upb.ro
* Correspondence: magdalena.bosomoiu@upb.ro

Abstract: Collagen is a non-toxic polymer that is generated as a residual product by several industries (e.g., leather manufacturing, meat and fish processing). It has been reported to be resistant to bacteria and have excellent retention capacity. However, the recovered collagen does not meet the requirements to be used for pharmaceutical and medical purposes. Due to the scarcity of water resources now affecting all continents, water pollution is a major concern. Another major field that could integrate the collagen generated as a by-product is wastewater treatment. Applications of collagen-based materials in wastewater treatment have been discussed in detail, and comparisons with already frequently used materials have been made. Over the last years, collagen-based materials have been tested for removal of both organic (e.g., pharmaceutical substances, dyes) and inorganic compounds (e.g., heavy metals, noble metals, uranium). They have also been tested for the manufacture of oil-water separation materials; therefore, they could be used for the separation of emulsified oily wastewater. Because they have been analysed for a wide range of substances, collagen-based materials could be good candidates for removing contaminants from wastewater streams that have seasonal variations in composition and concentration. The use of recovered collagen in wastewater treatment makes the method eco-friendly and cost efficient. This paper also discusses some of the challenges related to wastewater treatment: material stability, reuse and disposal. The results showed that collagen-based materials are renewable and reusable without significant loss of initial properties. In the sorption processes, the incorporation of experiments with real wastewater has demonstrated that there is a significant competition among the substances present in the sample.

Keywords: circular economy; wastewater treatment; recovered collagen

1. Introduction

Following the increasing climatic changes, ensuring the availability of necessary water has become a major global emergency. Water is an important resource that is not available in all countries; restricted access to clean water limits socio-economic development and accentuates poverty. All the activities that take place, in industry, energy production, agriculture, leisure and everyday life, lead to water consumption and implicitly to the generation of wastewater (municipal, industrial, agricultural, a mixture of those listed previously). Depending on its source, the wastewater can be contaminated not only with secondary by-products but also with leached compounds (e.g., oil refinery wastewater also contains lead, nickel, chromium and iron that are leached from the pipes, equipment or catalysts used in the refinery [1]). High adsorption capacity towards heavy metals, dyes, vitamin C and p-nitrophenol was reported for a collagen-based hydrogel, suggesting its applicability for depolluting wastewater that contains a vast range of contaminants [2]. In the textile industry, which is a major contributor to water contamination [3], dye removal is usually accompanied by the need for simultaneous removal of heavy metals [4]. Collagen is often included in the formulation of materials that are both antimicrobials and adsorbents. In this way, the concentration of contaminants is reduced, at the same time decreasing the number of microorganisms present in the wastewater [5]. Coal gasification, which is still largely used in countries from Asia and in the United States of America, is another process that generates large amounts of wastewater; this water is charged with difficult-to-remove organic compounds that have mutagenic and carcinogenic characteristics [6]. Wastewater is not charged only with chemical compounds but also with microorganisms that can be pathogenic [7].

Latest statistics and reports released by the United Nations Environment Programme warn about the significant quantities of wastewater that are discharged in water bodies without being treated: nearly two million tons of wastewater, consisting of sewage and agricultural and industrial wastewater, are released yearly [8,9]. From that, about 50% is directly discharged into the environment without preliminary treatment [10]. The United Nations have adopted a set of measures applied with the purpose of reducing the amount of untreated wastewater on one side and increasing the reuse of treated wastewater (e.g., in irrigation, as a cooling agent) [11].

Large amounts of waste rich in collagen are disposed by the leather manufacturing and meat and fish processing industries; in the context of a circular economy, collagen can be efficiently extracted to be reused [12,13]. For example, in the leather processing industry, raw hide has been reported to contain about 28% collagen; the mass percentage of collagen in trimmings is about 18%. Collagen mass balance in a leather production facility indicated that almost half of the collagen in the leather ends up disposed as waste (of 304 kg of collagen, 155 kg is found in the leather processing solid waste) [14]. Yearly, about 9 MT raw skin is processed, which in turn generates, according to the above given percentages, about 1.25 MT of collagen waste [15].

Attempts to use collagen in wastewater treatment have been made since 1995 when collagen was tested as a coagulant for sludge dewatering; the authors of the study proposed the use of collagen fibres together with alum ($(Al_2(SO_4)_3 \cdot 18H_2O)$) for liquid-solid separation processes and sludge dewatering [16]. A later study also confirmed the ability to use collagen in composite flocculants with increased flocculation ability, compared to aluminium sulphate [17]. Reusing the collagen in wastewater treatment not only will provide a sustainable solution for waste minimisation but will also contribute to obtaining a valuable resource: water. The advantages that collagen brings as a potential component of the materials used in a wastewater plant (e.g., adsorbents, membranes, filtering material, catalysts) are as follows:

- Strength and flexibility provided by its structure [18];
- Ability to interact with a large number of chemical substances due to increased number of functional groups (amino, carboxyl, hydroxyl);
- No toxicity at low concentrations [19] (in the case of leaching in the treated water, will pose no health risk);
- Resistance to bacteria [20];
- Biodegradability [21,22].

Collagen's ability to interact with different substances can be used in two directions: collagen can be easily incorporated in different materials (adsorbents, membranes, catalysts), and collagen will create bonds with contaminants present in the wastewater.

2. Collagen-Based Materials for Wastewater Treatment

In wastewater treatment, collagen-based materials have been tested in the processes of adsorption, membrane separation, column filtration, column absorption and advanced oxidation processes (photocatalysis and Fenton-like oxidation) and as components of antimicrobial materials. However, most of the studies are focusing on the use of collagen in adsorbents and separation materials (Figure 1). All these technologies are used in the secondary or tertiary stages of wastewater treatment [23]. The tertiary stage is considered the stage that brings the wastewater parameters to the levels set by the wastewater reuse directives. To achieve an optimal management of water resources and to reduce the water deficit, more and more treatment plants should be adapted by including the tertiary treatment stage.

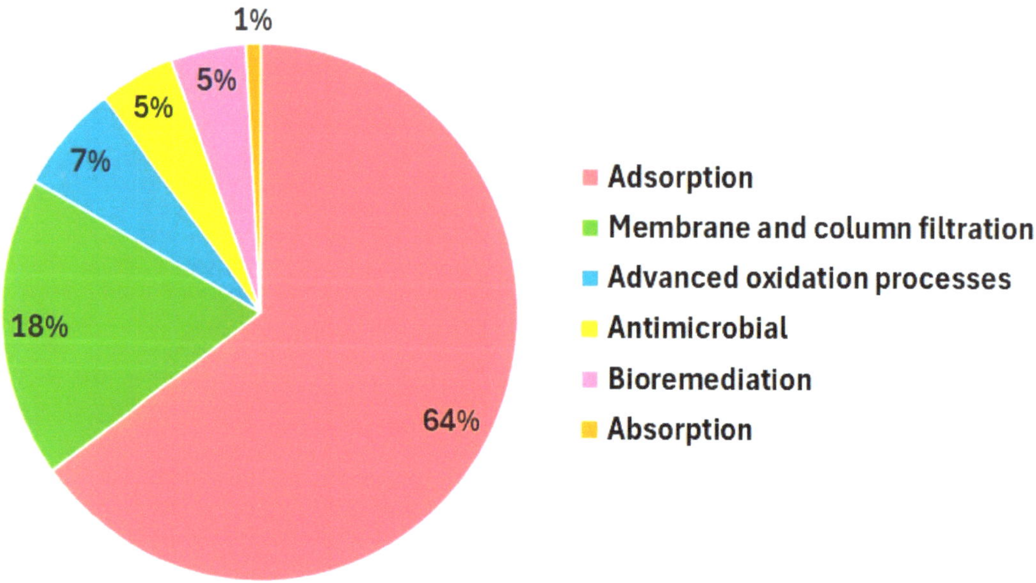

Figure 1. Studies on collagen's use in the wastewater treatment.

2.1. Adsorption Technology

Adsorption is used in wastewater treatment plants due to the ease of operation and possibility of using a wide range of materials that can be adapted according to the type of contaminants (e.g., activate carbon synthesised from different waste materials, natural materials: zeolites, clay minerals). The cost of the material remains, however, a major disadvantage. In addition, the load factor of the adsorbent is an important parameter; a low load factor results in a larger volume of adsorbent in the column [24].

To improve the efficiency of the adsorption process, different materials containing collagen have been tested for wastewater treatment. The adsorbent materials need to fulfil at least the following requirements: high adsorption capacity, non-toxic, reusable for extended period and low cost. The adsorption is currently used in the tertiary stage of wastewater treatment along with photocatalysis, catalytic oxidation, electro-chemical oxidation, ozonation or membrane separation [25].

The studies published so far have evidenced that collagen not only serves as a support for the active compounds of the adsorbent but also creates bonds with the adsorbate, enhancing the adsorption capacity. The multitude of functional groups of the collagen matrix, such as $-NH_2$, -COOH, $-C(O)NH_2$ or -OH, determines its ability to interact with different contaminants (e.g., heavy metal ions, radioactive ions, etc.) and remove them from wastewater [26–28]. It has been demonstrated that some heavy metal ions, in the presence of polypeptides, form a three-helix bundle structure [29,30]. This gives stability to the newly formed structure, explaining the good adsorption capacity of collagen-based materials towards heavy metal ions. The high adsorption of Cr(III) was attributed to the additional carboxyl groups created by modifying the collagen structure with oxazolidine [31]. Zhao et al. (2024) found that the phenolic hydroxyl groups present in the tannin extract in the composite collagen/tannin were responsible for the high affinity towards Cu(II) ions [32].

To improve the properties of the final adsorbent, some of the collagen-based adsorbents investigated so far for wastewater depollution are crosslinked with synthetic polymers [33–39].

The use of nanocomposites is preferred because they provide a higher internal surface area available for contaminants adsorption [26,39–42].

A significant number of studies have been dedicated to heavy metals, noble metals and metalloids adsorption on collagen-based adsorbents [31–34,43–59].

The enhanced removal of As(III) on Zr(IV) loaded on collagen fibres (CF) was explained by the ability of As(III) to form different configurations of complexes with zirconium-supported collagen fibres (Figure 2a): (1) bidentate, binuclear As(III) connects through oxygen with two zirconium ions; (2) bidentate, mononuclear As(III) forms two bonds with one zirconium ion via the oxygen ions; (3) monodentate, mononuclear As(III) creates one bond with one zirconium ion. This mechanism was similar to As(III) adsorption on alumina [60].

It was found that adjacent hydroxyl groups play a significant role in the retention mechanism. Removal of silver ions by persimmon tannins/CF also proceeds via a complexation between silver ions and adjacent hydroxyl groups of the adsorbent. Furthermore, the hydroxyl groups of the persimmon tannins oxidize to carbonyl groups, while the silver ions are reduced to elemental silver that is uniformly dispersed at the surface of the adsorbent [59].

The functional groups—NH_2 (Lewis base) and -COO^- enhance Pb(II) and Cd(II) removal by coordination, respectively electrostatic attractions (Figure 2b); to increase the number of functional groups, gum copal CF was cross-linked with polyacrylic acid [45]. Similarly, Cr(VI) is adsorbed from wastewater on collagen-based porous aerogel; the ions of Cr diffuse in the porous structure of the adsorbent to coordinate with amino, hydroxyl and carboxyl groups (Figure 2c); since the synthesised aerogel had fluorescent properties, it can also serve as an indicator of Cr(VI) concentration [56].

To increase the number of adjacent hydroxyl groups, Zhao et al. (2024) synthesised a composite bovine hide collagen/tannin extract; the first step in Cu(II) removal consists of deprotonation of two adjacent hydroxyl groups of the tannin, followed by the creation of bonds due to electrostatic attractions (Figure 2d); the other functional groups coordinate with Cu(II) to form stable structures [32].

For Hg(II) adsorption on CF/ZIF-7-NH_2 (ZIF—zinc imidazolate framework), it has been discovered that at low adsorbate concentration in wastewater, the removal mechanism is governed by electrostatic interactions and coordination, while at high Hg(II) concentration, the contribution of coordination is insignificant, and the Hg(II) adsorption takes place mainly by electrostatic interactions (Figure 2e) [57].

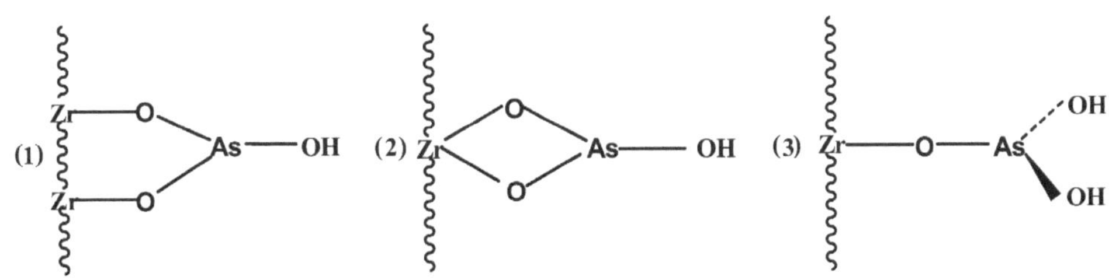

(a)

Figure 2. *Cont.*

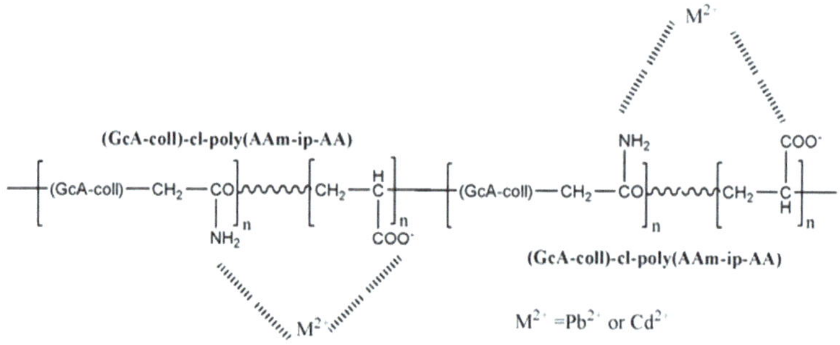

(b)

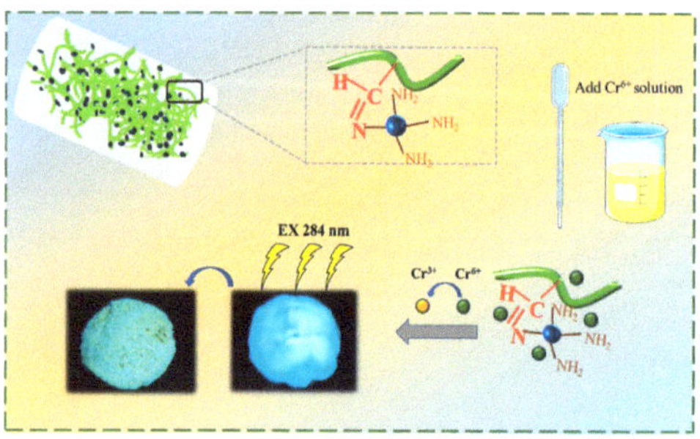

(c)

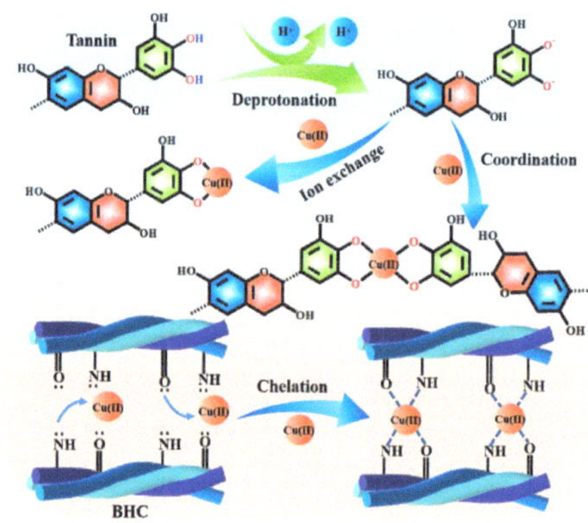

(d)

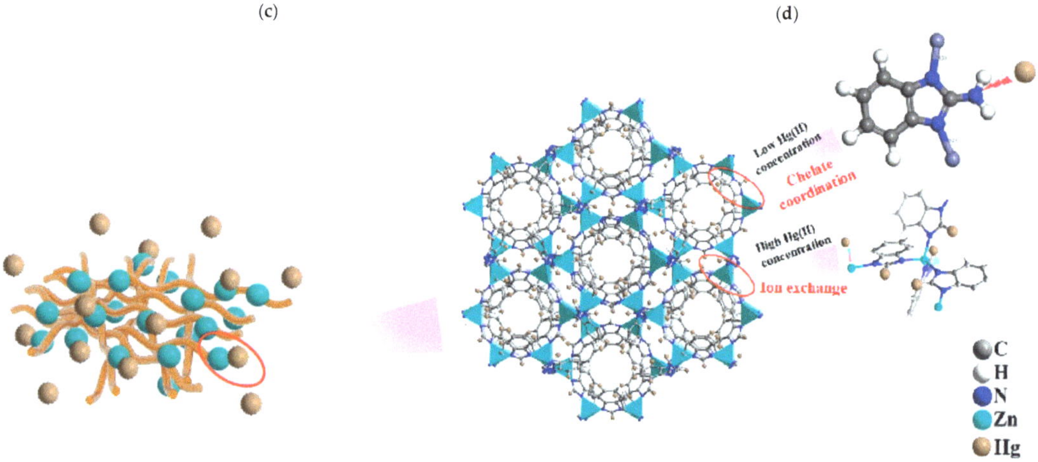

(e)

Figure 2. Mechanisms of heavy metals and metalloids ions adsorption on collagen-based materials; (**a**); As(III) removal by Zr(IV)—loaded collagen fibre [60]; (**b**) Removal of Pb(II) and Cd(II) over gum copal-collagen hybrid adsorbent [45]; (**c**) Cr (VI) adsorption by collagen-based porous fluorescent aerogel [56]; (**d**) Cu(II) retention on bovine hide collagen/tannin extract composite (BHC/TE) [32]; (**e**) Hg(II) retention on amino functionalized collagen fibre combined with metal–organic frameworks (MOFs) [57].

Collagen fibre was modified with polyethyleneimine (PEI) and used for p-arsanilic acid adsorption (Figure 3); the obtained adsorbent has an increased capacity for contaminant removal via the combined effect of collagen amphiphilicity, formation of hydrogen bonds and electrostatic interactions [35].

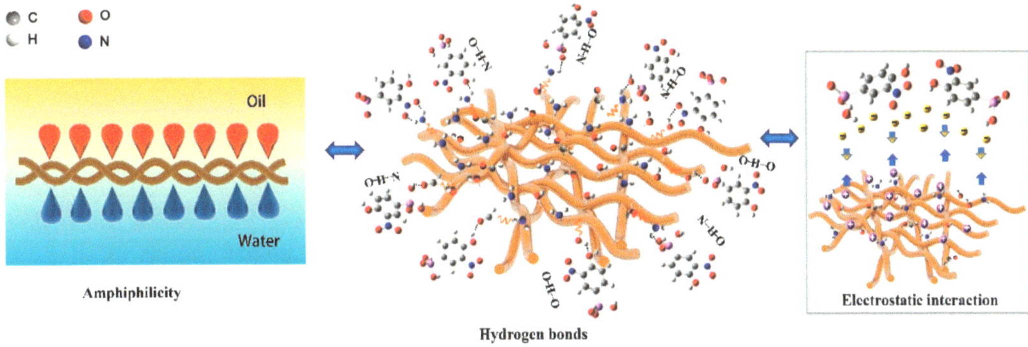

Figure 3. Mechanisms of organic pollutants adsorption on collagen-based materials—adsorption of p-arsanilic acid on amphiphilic amino modified collagen fibre [35].

Many studies have been carried out for dye removal from wastewater [36–39,42,61–75]. To increase the number of -NH_2 groups, ethylenediamine was grafted on CF; the resulting structure was combined with TpPa-I (covalent organic frameworks) to control the porosity and stability of the adsorbent [76]. The removal mechanisms of dyes reactive blue 19 and acidic fuchsin is based on electrostatic attraction forces and π-π stacking (Figure 4a) [70]. Cross-linking collagen fibres with PEI contributed to the increasing of hydrophilic swelling of the final material; this helped to improve the adsorption capacity towards anionic dye removal (acid red and soap yellow) [38]. PEI also helped to increase the number of functional groups available on the collagen support; these groups will create electrostatic attractions with negatively charged dye molecules, enhancing adsorption (Figure 4b). The collagen surface can be modified to allow the synthesis of a hydrogel with abundant negative charges; the positively charged molecule of methylene blue (cationic dye given by the quaternary amino group) reacts with the negatively charged groups present on the adsorbent surface (Figure 4c) [36]. Similarly, Sitab et al. (2023) developed a nanomaterial using collagen hydrolysate and polyvinyl alcohol; the nanomaterial was tested for methylene blue removal, and the mechanism proceeds via the interaction of the alkyl group of the dye and of the nanofibers chain; this is followed by electrostatic interactions between the cationic dye and -$COO^-$and -O^- groups (Figure 4d) [37]. Tanned bovine collagen fibres having macro (>50 nm), meso (2–50 nm) and micro (<2 nm) pores was used for acid dyes (acid brown 369, AB369, acid red 131, AR131 and acid blue 113, AB113) removal from wastewater; taking into account the configuration of the dye molecules, the bulkier AB369 molecule cannot pass through the micro pores, while the smaller molecules AR131 and AB113 can be adsorbed on pores of all sizes (Figure 4e) [73].

Adsorbents with magnetic properties have recently been synthesised; they have an affinity for removing target contaminants and are easily separated from water by applying a magnetic field [42,52]. The concept itself is not new; however, for collagen-based materials employed in wastewater treatment, it has only recently been tested. The percentage of contaminant removal can be controlled by changing the value of the electric field: an increase in the electric field can enhance the removal rate by up to 40% compared to when no electric field was applied [52]. When comparing the influence of electric and magnetic fields on the removal efficiency, it was found that Ni(II) adsorption is enhanced by applying an electric field compared to a magnetic field [52].

Oil sorption capacity of collagen-polydimethylsiloxane (PDMS) has been tested for silicone, vegetable and motor oil; compared with commercially available biomass materials, the collagen-based sorbent showed similar performances [77]. Modifying the collagen matrix with hexadecyl trimethyl ammonium bromide led to a slight increase in the oil sorption capacity [78].

The retention of uranyl ions on unmodified cellulose-collagen is low (Table 1) [79]. To increase the adsorption capacity, different metal ions or compounds must be added [28,80–84].

The removal of Cs^{+} and Sr^{2+} from radioactive wastewater by adsorption on Zr molybdopyrophosphate-functionalised collagen fibres (ZMPPCFs) consists of ion exchange of Cs^{+} with K^{+} from ZMPPCFs while Sr^{2+} is retained by electrostatic interaction with pyrophosphate anions (Figure 5a) [85].

Liao et al. (2020) developed a material (Fe/Ni bayberry tannin/CF) used for adsorption of uranyl ions and subsequent reduction of U(VI) to U(IV) that can be easily separated further by precipitation. The mechanism consists of uranyl ions adsorption facilitated by bonds formation with bayberry tannin (BT) and collagen's functional groups; U(VI) is reduced to U(IV) in redox reactions with Fe/Ni (Figure 5b) [86]. However, the stability of the material for reuse and its long-term performance were not discussed.

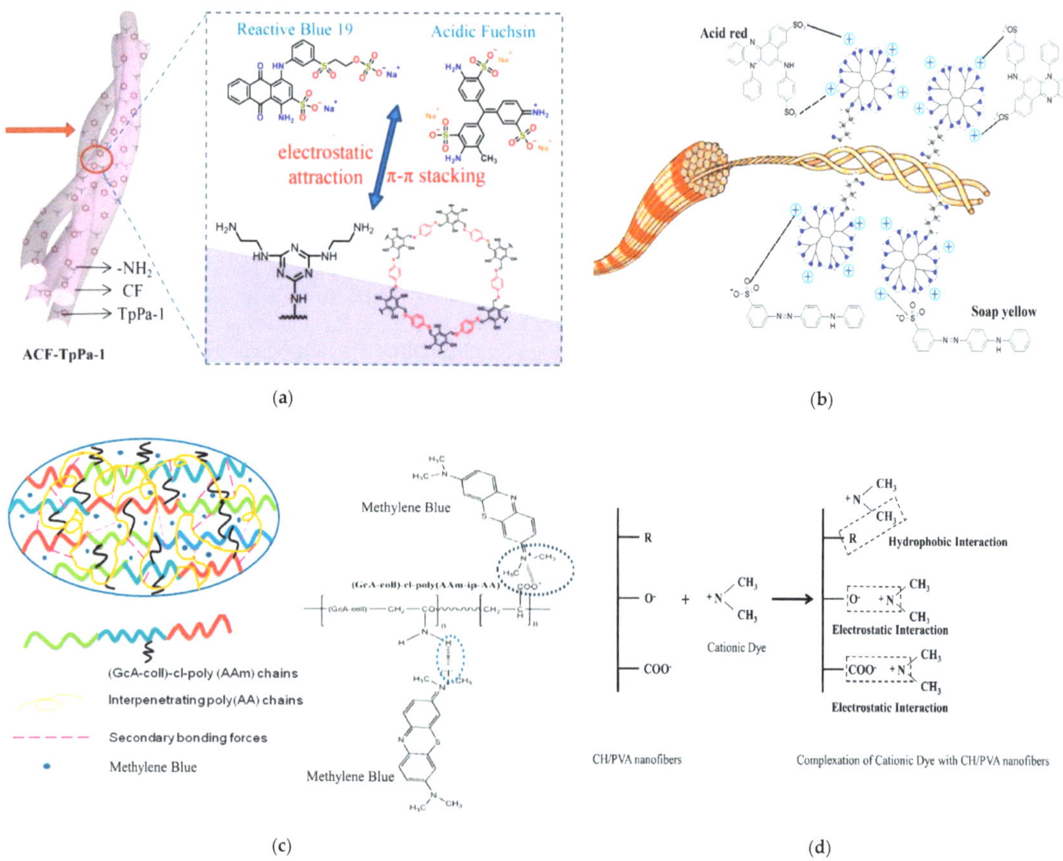

Figure 4. *Cont.*

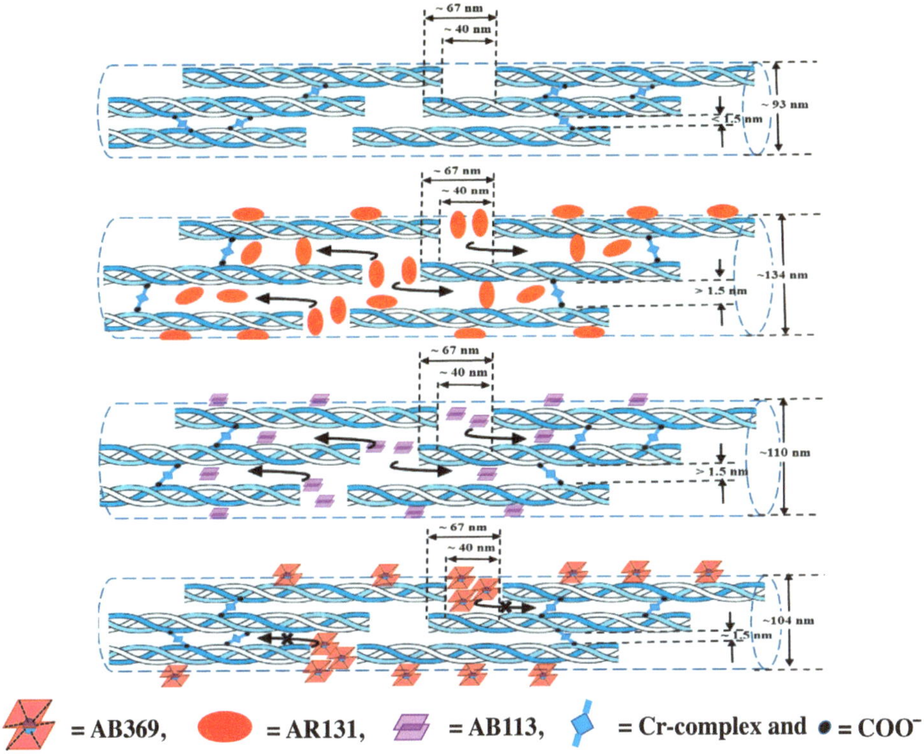

(e)

Figure 4. Mechanisms of dye adsorption over collagen-based materials: (**a**) Acid fuchsin and reactive blue 19 (anionic dyes) on composite COF (covalent organic frameworks)—aminated collagen fibres, abbreviated ACF-TpPa-1 [70]; (**b**) Anionic dyes (acid red and soap yellow) adsorption on multilayer ammoniated collagen fibres [38]; (**c**) Methylene blue adsorption on interpenetrating network hydrogel from (gum copal alcohols collagen)-co-poly(acrylamide) and acrylic acid [36]; (**d**) Cationic dye (methylene blue) adsorption on collagen hydrolysate (CH) crosslinked with polyvinyl alcohol (PVA) [37]; (**e**) Acid Brown 369 (AB369), Acid Red 131 (AR131) and Acid Blue 113 (AB113) removal by tanned bovine collagen fibres (TBCFs) [74].

The retention mechanism of uranyl ions involves the formation of bonds (electrostatic and chelation) between the uranyl ions and the hydroxyl groups of hydrated titanium oxide, HTO (electrostatic), the ortho phenolic hydroxyl groups of BT (electrostatic) and the functional groups of collagen ($-NH_2$, -COOH, $-C(O)NH_2$, -OH electrostatic and chelation) (Figure 5c) [81]. *Myrica rubra* tannin also has a large number of ortho phenolic groups; immobilisation on collagen leads to obtaining a material with a high adsorption capacity of uranyl ions [82].

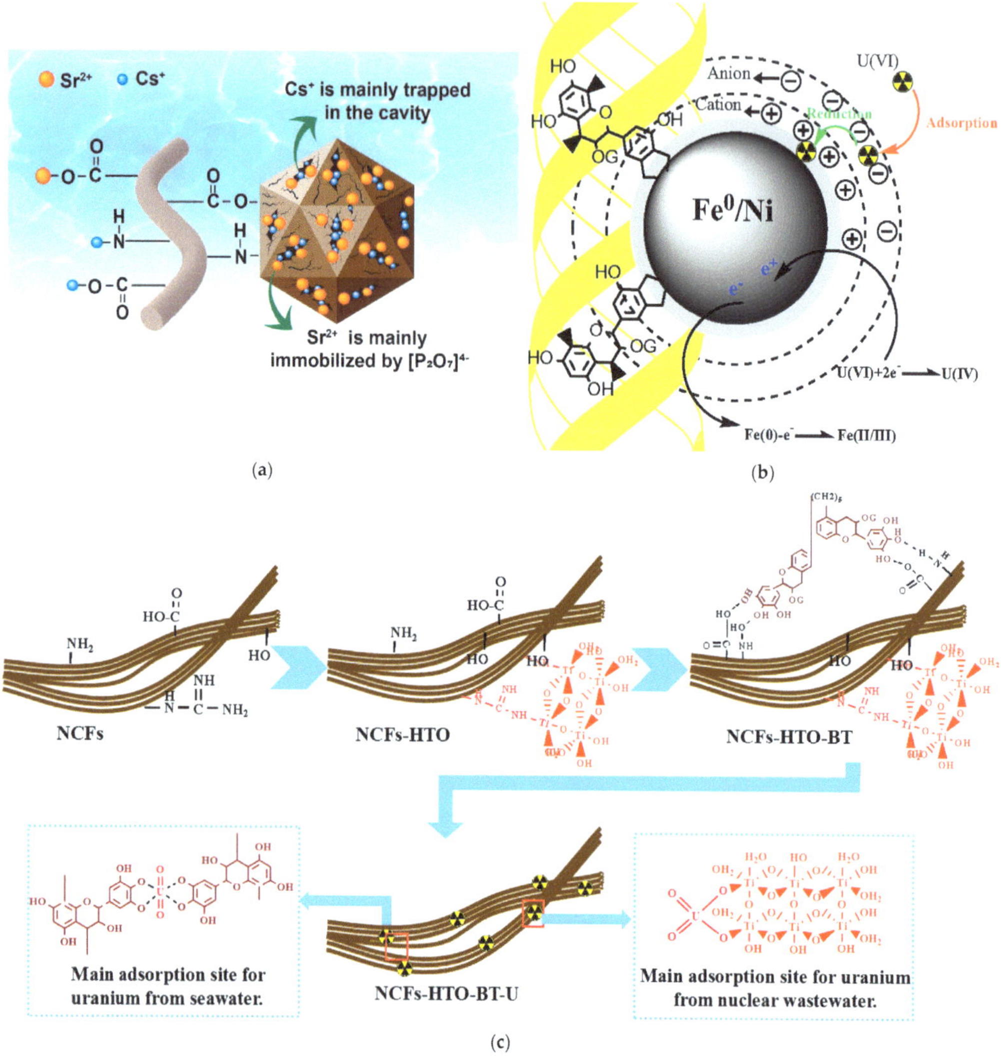

Figure 5. Mechanisms of radioactive ions adsorption: (**a**) Adsorption and subsequent separation of radioactive Cs^+ and Sr^{2+} by zirconium molybdopyrophosphate-functionalized collagen fibres [85]; (**b**) Uranyl ions adsorption over nano-zero-valent Fe/Ni particles (NZFNP) loaded on collagen fibres [86]; (**c**) Uranium extraction from seawater vs. removal from wastewater NCFs—nano collagen fibrils, HTO—hydrous titanium oxide, BT—bayberry tannin [81].

Titanium(IV) loaded collagen fibre has been tested for uranium removal from nuclear industrial wastewater [28,81,83]. It has been proven that HTO exhibits high affinity towards uranyl (UO_2^{2+}) extraction; however, its ability to store fluids is low [28]. This can be remediated by loading the HTO on collagen fibres that are well known for their properties to retain large amounts of water and ions and release only water in drought conditions [12]. Because of collagen ability to adsorb other ions too, the ion interference studies are extremely important. Fluoride anion usually present in radioactive wastewater gives the biggest interference (among F^-, HCO_3^-, Cl^-, NO_3^-, Ca^{2+}, Mg^{2+} and Cu^{2+}); the fluoride effect could be suppressed by the addition of Al^{3+} [28,83]. The selectivity of composite ZnS/alkali-activated collagen fibre for uranyl ions removal from a concentrated mixture of Fe^{3+}, Ni^{2+}, Cu^{2+}, VO^{3-} and UO_2^{2+} or K^+, Ca^{2+}, Na^+, Mg^{2+}, Cl^-, HCO_3^-, SO_4^{2-}, Br^- and UO_2^{2+} remains high; this is due to the combined effect of collagen's functional groups and nano-ZnS [84].

Only a few studies have aimed at the simultaneous adsorption of different classes of contaminants from wastewater [2,4,5,40,87]. As a result of the variable composition of the wastewater, the adsorption materials that have been tested for the retention of several types of pollutants are preferable (e.g., wastewater resulted from dyeing also contains heavy metals) [2,87]. A collagen-guar gum-MOFs (metal–organic frameworks) composite was found to be effective for dye (methylene blue and indigo carmine—acid blue 74—with anionic character because of two sulfonate groups) and heavy metals (Ni(II), Cu(II), and Zn(II)) removal but also for neutralising the wastewater pH in some cases (wastewater resulted from cotton dyeing). Results indicated that simple collagen-guar gum hydrogel had a higher adsorption capacity for methylene blue removal than the composite material and similar behaviour in the case of indigo carmine and Ni(II). The composite material was more efficient for copper and zinc removal than the hydrogel; this is probably due to the higher uniform porosity of MOFs that allow bulkier copper and zinc internal diffusion [87]. FTIR and XPS analysis evidenced the mechanism of Hg(II) and Pb(II) adsorption on a highly functionalised adsorbent (guar gum (GG)-g-(acrylic acid (AA)-co-3-acrylamido propanoic acid (AMPA)-co-acrylamide (AM))-g-cow buffing dust (CBD)); Pb(II) preferably coordinates with -COO^- functional groups, mostly in bidentate chelating, while Hg(II) preferably creates bonds with the N-donor (Figure 6). Methylene blue, methyl violet, vitamin C and nitrophenol are adsorbed on the surface via hydrogen bonds, van der Waals forces and electrostatic interactions; the adsorbent showed high adsorption capacity against all tested pollutants, compared to other materials (Table 1) [2]. A collagen-based nanocomposite was also found to be extremely efficient for metal ions and dye removal [40]. Wang et al. (2022) developed an aerogel material AgNPs/Fe@CF with simultaneous activity in antibiotics' photo degradation, antibiotic-resistant bacteria elimination and adsorption of heavy metal ions [5].

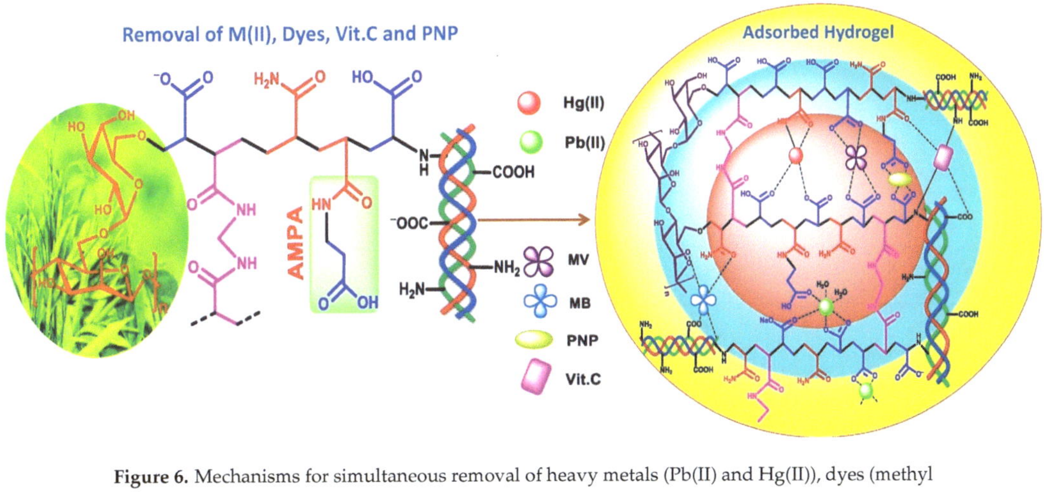

Figure 6. Mechanisms for simultaneous removal of heavy metals (Pb(II) and Hg(II)), dyes (methyl violet, MV; methylene blue, MB), vitamin C and p-nitrophenol (PNP) by carbohydrate and collagen-based doubly grafted interpenetrating terpolymer hydrogel (AMPA-3-acrylamido propanoic acid) [2].

Most of the adsorption studies are oriented only towards batch adsorption experiments (static regime) that allow the understanding of the removal mechanism and determination of the adsorption kinetics and thermodynamics. There are, however, some studies that also perform continuous column experiments (dynamic column breakthrough, DCB). This is a compulsory step before large-scale utilisation of the adsorbent and gives information about the macroscopic performance of the material.

The breakthrough point for uranyl adsorption from wastewater on HTO/CF was around 250 bed volume (BV); the material in the column was regenerated using a volume of 4 BV 0.1 M HNO_3 with a recovery rate of 98% [80].. The breakthrough point of a column packed with amidoxime/CF for uranyl removal was about 175 BV (this value was achieved in the presence of competing ions Fe^{3+}, Al^{3+}, Zn^{2+}, Mg^{2+}, Cu^{2+}, Cr^{3+}, VO^{2+}, F^- and NH_4^+) [83]. For both materials, new cycle adsorption—regeneration gave similar results, indicating excellent reusability [80,83]. It was found that one gram of PBA (Prussian blue analog)/CF could remove more than 95% of radioactive Cs^+ contained in four litres of wastewater (flow rate of 0.3 mL min^{-1}) [88]. For highly concentrated water in uranyl ions (120.5 mg·L^{-1}), ZnS/CF showed a good column adsorption performance with a breakthrough point of 75 BV [84]. The breakthrough point of a column packed with collagen fibres for thorium(IV) adsorption was 12.5 BV [89].

Compared with other materials (recycled or commercial adsorbents), collagen-based adsorbents show higher adsorption capacities for metal ions and similar adsorption capacities towards uranyl ions, dyes and oils (Table 1).

Table 1. Adsorption of different pollutants found in wastewater using collagen-based materials (1–47) and other materials (48–55).

	Adsorbent	Targeted Pollutant	Adsorption Capacity (at Equilibrium), mg·g^{-1}	Reference
1.	CF/ZIF-7-NH_2	Hg(II)	909.09	[57]
2.	Bayberry tannin-immobilized CF	Hg(II)	198.49	[47]
3.	Collagen-based porous fluorescent aerogel	Cr(VI)	103.3	[56]
4.	Collagen modified with oxazolidine	Cr(III)	143	[31]
5.	Carboxylated CF CF	Cr(III)	106.88 75.82	[51]
6.	Bovine hide collagen/tannin extract composite	Cu(II)	14.94	[32]
7.	Bovine hide collagen/tannin extract/sodium alginate, (SA/BHC)@TE	Cu(II)	140.56	[58]
8.	Collagen-tannin resin	Cu(II)	16.52	[54]
9.	Collagen/cellulose hydrogel	Cu(II)	67.36	[55]
10.	Manganite/collagen-polyurethane-chitosan hydrogel	Pb(II)	13.22	[34]
11.	Collagen fibre/carbon quantum dot	Pb(II)	183	[48]
12.	Persimmon tannins immobilized on collagen fibre	Ag(I)	1947	[59]
13.	AgNPs/Fe crosslinked CFs (NPs—nanoparticles)	Cr(VI) Ni(II) Pb(II)	90.734 73.82 78.19	[5]
14.	Zr—loaded collagen fibre	Cr(VI) V(V)	27.55 100.86	[50]
15.	Fish scales	Cu(II) Ni(II)	400 2.73	[52]
16.	Tannins immobilised on collagen	Cu(II) Pb(II) Cd(II) Cr(III) Zn(II)	13.30 18.41 6.5 10.4 0.8	[49]

Table 1. *Cont.*

	Adsorbent	Targeted Pollutant	Adsorption Capacity (at Equilibrium), $mg \cdot g^{-1}$	Reference
17.	CF-PEI (collagen fibres crosslinked with polyethyleneimine)	p-arsanilic acid	285.71	[35]
18.	Hydrogel from (gum copal alcohols collagen)-co-poly(acrylamide) and acrylic acid	methylene blue	1.70	[36]
19.	Collagen hydrolysate/polyvinyl alcohol	methylene blue	99.9	[37]
20.	Black wattle tannin-immobilised mesostructured collagen	methylene blue	46.5	[75]
21.	Collagen-based cryogel, isinglass-graphene oxide	rhodamine B	120	[72]
22.	Fish scale	acid blue 113	145.3	[62]
23.	Composite hydrogels (collagen, guar gum and metal-organic frameworks)	methylene blue indigo carmine	15.8 0.46	[87]
24.	CF-PEI	soap yellow acid red (anionic dye)	538.2 369.7	[38]
25.	Tanned bovine collagen fibres	acid brown 369 acid red 131 acid blue 113	38.29 78.14 73.25	[74]
26.	Aminated collagen fibres (ACF)	acid black dye	125.63	[68]
27.	Collagen-g-poly(acrylic acid-co-N-vinylpyrrolidone)/Fe_3O_4@SiO_2	methylene blue brilliant green rhodamine B	207.33 212.68 221.97	[39]
28.	Magnetic hematitenanoparticle@collagen nanobiocomposite	methylene blue rhodamine B	27.57 56.14	[42]
29.	CF	methylene blue reactive red	80 163	[61]
30.	ACF—TpPa-1	acid fuchsia reactive blue 19	257.98 449.54	[70]
31.	Zr—loaded collagen fibre Fe—loaded collagen fibre	phosphate	87.3 79.96	[90]
32.	Zirconium(IV)-Impregnated CF	fluoride	43.49	[91]
33.	Carbohydrate and collagen-based doubly grafted interpenetrating terpolymer hydrogel	Pb(II) Hg(II) methyl violet methylene blue vitamin C p-nitrophenol	976.64 859.23 116.80 58.52 212.91 59.01	[2]
34.	Collagenic-waste/natural rubber biocomposite	Hg(II) safranine brilliant cresyl blue	166.46 303.61 46.14	[4]
35.	Collagen-based hydrogel nanocomposite	Cd(II) Pb(II) methylene green crystal violet	~120 mg/g ~120 mg/g 179 652	[40]
36.	Zirconium molybdopyrophosphate-functionalised CF	radioactive Cs(I) and Sr(II)	149.52 38.99	[85]
37.	Ti(IV)/CF	UO_2^{2+}	167.4	[28]

Table 1. *Cont.*

	Adsorbent	Targeted Pollutant	Adsorption Capacity (at Equilibrium), $mg \cdot g^{-1}$	Reference
38.	Ti(IV)/CF	UO_2^{2+}	372.4	[80]
39.	Ti(IV)/bayberry tannin/CF	UO_2^{2+}	393.19	[81]
40.	*Myrica rubra* tannin/CF	UO_2^{2+}	321.3	[82]
41.	Amidoxime/CF	UO_2^{2+}	301.18	[83]
42.	nano-ZnS/alkali-activated CF	UO_2^{2+}	359.72	[84]
43.	Cellulose—collagen	UO_2^{2+}	$64.94\ 10^{-3}$	[79]
44.	Prussian blue analog (PBA)/CF	Cs(I)	175.4	[88]
45.	Collagen—tannin rearranged fibre	Th(IV)	114.97	[89]
46.	Collagen polydimethylsiloxane (PDMS)	Silicone oil Motor oil Vegetable oil	13.6×10^3 12.5×10^3 11.92×10^3	[77]
47.	Collagen polydimethylsiloxane (PDMS) modified with hexadecyl trimethyl ammonium bromide	Silicone oil Motor oil Vegetable oil	15.9×10^3 14.0×10^3 12.0×10^3	[78]
48.	Olive stone	Cd(II) Pb(II) Ni(II) Cu(II)	7.73 9.26 2.13 2.03	[92]
49.	Beal fruit shell	Cr(VI)	17.27	[93]
50.	Biochar	Pb(II)	46.46	[94]
51.	Biochar	Cd(II)	25	[95]
52.	Biochar	methylene blue basic fuchsin	99.11 78.01	[96]
53.	$Cu_xO/Fe_2O_3/MoC$	reactive red 195 A reactive yellow 84	61.3 93.95	[97]
54.	nano-MgO biochar	UO_2^{2+}	333.11	[98]
55.	Recycled wool	Motor oil Vegetable oil	15.8×10^3 13.16×10^3	[99]

2.2. Membrane and Column Separation

Wastewater purification by membranes has gained much interest due to the process efficiency in separating a wide range of compounds from wastewater (persistent organic compounds [100,101], microorganisms [102], microplastics [103,104], heavy metals [105,106]). The separation with membranes offers the advantages of metal recovery and ensures the removal of dissolved species. The extended application of membrane technologies for different classes of contaminants allows the production of good quality water that can be directly reused [107–109]. This combined with the ease of operation and the efforts to find new materials that are durable and cost-effective make membrane separation one of the best technologies for producing water and in some cases recovering valuable components from wastewater.

An example of industry that uses large amounts of water is petroleum processing and refining. This means that the amount of wastewater is also significant; it was reported that the processing of one barrel of crude oil generates ten times more wastewater [110] The wastewater has a complex and variable composition and includes organic matter (polycyclic aromatic hydrocarbons, benzene, toluene, xylene, ethylbenzene, phenols), heavy metals (mercury, lead, chromium Cr^{6+}, cadmium, copper Cu^{2+}, manganese, zinc and nickel), arsenic, As^{3+} and As^{5+}, inorganic salts and oil emulsions [111,112]. However, the collagen-based membranes and column filtration materials were tested almost exclusively for emulsions separation and not for more complex wastewater mixtures. There are two types of emulsions that are discharged by the industry: oil in water (O/W) and water in oil (W/O); industrial emulsions also contain substances with the role of a surfactant.

Conventional separation membranes (Figure 7) have either hydrophilic or hydrophobic surfaces and are consequently subjected to rapid pore clogging (membrane fouling). Collagen has both hydrophilic and hydrophobic functional groups (amphiphilic character); this gives different wetting affinities for water and oil, allowing the separation of the emulsion [113]. The collagen fibre membrane (CFM) consists of interwoven collagen fibres that provide good mechanical resistance to the membrane [114]. CFM fails, however, in separating micro and nanoemulsions because the size of the voids between the fibres can be larger than the particle size of the emulsion droplets. To remediate this, the CFMs undergo surface modifications that improve their separation performance [115–117]. For instance, to enhance the amphiphilic behaviour of collagen fibres, additional compounds can be added on the surface of the collagen fibres (Figure 8). Tannic acid is a natural amphiphilic compound that can be easily deposited on the collagen fibres, improving separation [117].

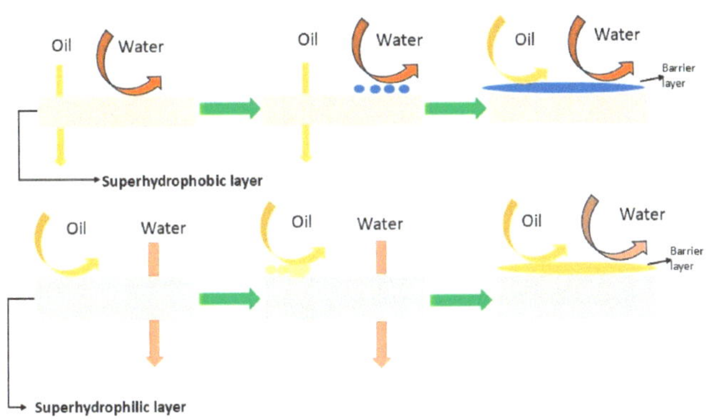

Figure 7. Pore clogging in conventional separation membrane (membrane fouling).

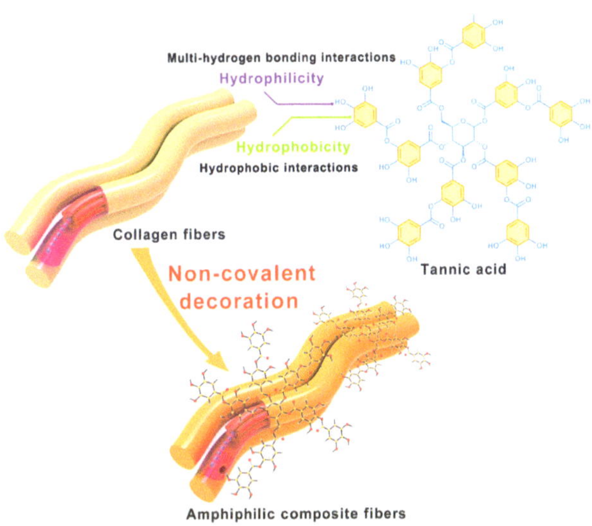

Figure 8. Amphiphilic character of collagen fibres [117].

Molecular dynamics simulations have shown that the driving forces allowing W/O emulsion separation are the electrostatic interactions and van der Waals forces. The anionic -COO^- and cationic -NH^{3+} groups of the collagen matrix generate electrostatic attraction forces towards the water molecules [113]. The oily compounds will agglomerate at the hydrophobic regions while being separated from water, which wets the surface on the hydrophilic part. The oil further permeates through the collagen fibres and is separated. This phenomenon is called underoil hydrophilicity. When the collagen fibres are wetted by O/W emulsions, the oil molecules agglomerate in the hydrophobic regions of the collagen fibres (by van der Waals forces), while the rest of the collagen fibres are wetted by the water (underwater oleophobicity) [118].

Effective separation materials consisting of blended superhydrophilic collagen fibres and superhydrophobic polypropylene fibres [119] or superhydrophilic and superhydrophobic collagen fibres [120] have been used for water recovery from O/W emulsions. Using intertwisted superhydrophobic and superhydrophilic fibres will enhance demulsification when the emulsion reaches the superhydrophilic fibres; water will be further transported along the superhydrophilic fibres while the oil will be repelled towards the hydrophobic fibres. By this method, the material fouling was delayed, and continuous separation of emulsion for 1440 min was achieved [120].

The addition of the surfactants for emulsion stabilisation also intensifies the electrostatic interactions in the separation process [107–109]. Industrial wastewater may contain emulsions with at least two surfactants [121]. The usual surfactants used in pharmaceutical, food, and cosmetics industries are Tween80 (non-ionic surfactant) and Span80 (biodegradable surfactant) [114,121]. Anionic or cationic surfactants can also be used (anionic: SDS—sodium dodecyl sulphate; anionic SDBS—sodium dodecyl benzene sulfonate; cationic CTAB—hexadecyl trimethylammonium bromide; cationic CPB—bromohexadecylpyridine) [107–109,116,117]. For an efficient reuse of water or oily compounds, the resulted aqueous filtrate should be surfactant free.

Ye et al. (2019) synthesised an amphiprotic membrane by deposition of amino-modified multi-walled carbon nanotubes (MWCNTs-NH_2) and carboxyl multi-walled carbon nanotubes (MWCNTs-COOH) on a collagen nanofibres-based membrane (MCN) [114]. The separation mechanism for this consists of W/O emulsion spreading at the membrane interface (Figure 9a) followed by charge neutralisation by oppositely charged groups—COO^- or -NH_3^+ (demulsifying). The oil is transported further through the membrane by the capillary effect along the collagen fibres.

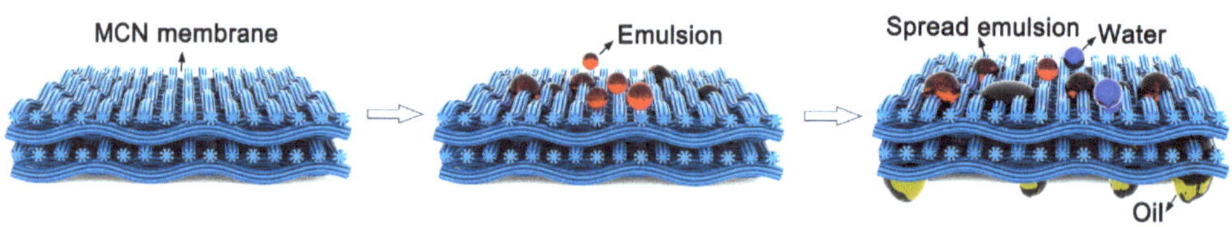

(1) (2) (3)

(a)

Figure 9. *Cont.*

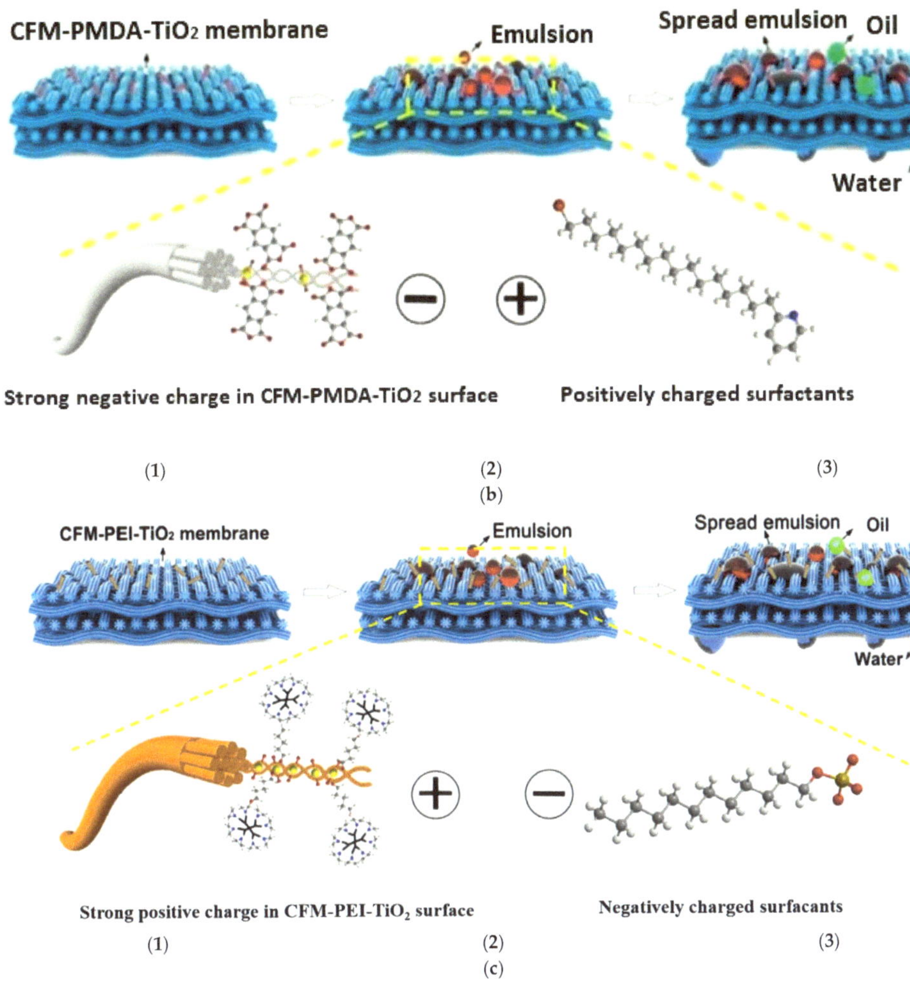

Figure 9. Separation of emulsion containing wastewater: (**a**) Oil—water emulsion, MCN—millimetre-scale collagen nanofibres: (1) fresh membrane; (2) emulsion spreading at the amphiprotic membrane surface; (3) separation [114]; (**b**) Cationic surfactant-stabilised oil-water emulsion, CFM-PMDA-TiO_2 (CFM—collagen fibre mem-brane; PMDA—pyro-mellitic dianhydride) [107]; (**c**) Anionic surfactant-stabilised oil-water emulsion, CFM-PEI-TiO_2 (CFM—collagen fibre membrane; PEI—polyethyleneimine) [108].

Grafting synthetic polymers on the surface of CFM will contribute to the creation of additional charges that will enhance the emulsion separation by electrostatic interactions (e.g., polyethyleneimine, will create positive charges) [108]. The so synthesised membrane is efficient in the separation of both surfactant-free and surfactant-stabilised O/W emulsions. Adding pyromellitic dianhydride (PMDA) will increase the number of negative charges [107]. If the aim is to obtain high purity water, TiO_2 particle deposition will amplify membrane superhydrophilicity (Figure 9b,c) [107,108]. Zr^{4+} was used to change the wettability properties of the collagen fibres matrix, obtaining a material that is superoleophobic under water (O/W emulsions) and hydrophilic under oil (W/O emulsions) [122].

The concept of size sieving membranes was used in the development of membrane that are using metal–organic framework MOFs for the emulsion separation and the collagen as a facilitator for the liquid transport along the fibres [109,115,116]. The uneven pore size distribution of CFM can be modified by incorporating MOFs with uniform pore sizes. Moreover, the pore sizes of MOFs can be customizable and can be disposed to form successive sieve-like layers to improve the separation efficiency [109]. The MOFs that have been tested consist of a zinc imidazolate framework (ZIF–8) and copper-containing MOFs (HKUST–1) [115]. Li et al. (2021) developed a double-layer collagen-based membrane CFM/UiO–66(12)/PDMS (PDMS—polydimethylsiloxane) for the separation of water in oil micro and nanoemulsions [116]. The so-called UiO 66 structure is a zirconium MOF that was set within the CFM. The principle of emulsion separation is based on the difference between the UiO-66 micropore size (which is about 6.0 Å) and the particle emulsion size, which is bigger. The superhydrophobic surface of the membrane allows the selective permeation of oil. The collagen fibres have the role of enhancing the oil transport through the membrane by the capillary effect; this has been proven by performing comparative tests on membranes with and without collagen fibres (filtrate flux of 2038 L $m^{-2}h^{-1}$ versus 866 L $m^{-2}h^{-1}$, for separation efficiencies higher than 99.99%) [114,115].

The simple collagen fibres allow a good separation of the emulsions; however, the filtrate flux is significantly lower than in the cases of modified collagen fibres or membranes. The tested membranes have proven the ability to generate high purity filtrate; they have separation efficiencies around 99% and retain the surfactant in the case of surfactant-stabilised emulsions [107–109,115,116,121]. Although the synthesis of collagen-based membranes is more complicated than commercial ones, the filtrate flux values are significantly higher for CFM (Table 2). The performances obtained in the case of wastewater filtering over collagen fibres are also superior to those obtained on commercial membranes (Table 2, lines 12–14).

Table 2. Performance of different collagen-based fibres (1–5) and membranes (6–14) for wastewater treatment.

	Filtration Material/ Membrane	Targeted Pollutant	Separation Efficiency %	Filtrate Flux $L \cdot m^{-2} \cdot h^{-1}$ *	Reference
1.	Amphiphilic composite fibres—tannic acid onto collagen fibres (ACFs)	Heptane/Water Kerosene/Water Dodecane/Water Octane/Water (SDBS) Water/Heptane (Span80) Water/Kerosene (Span80) Water/Dodecane (Span80) Water/Octane (Span80)	>99.99	1911 (W) 1745 (W) 1822 (W) 1720 (W) 2166 (O) 2471 (O) 1936 (O) 2408 (O)	[117]
2.	Collagen(I) Fibres (CF)	n-Dodecane/Water Kerosene/Water Petroleum ether/Water Water/n-Dodecane (Span80) Water/Kerosene (Span80) Water/Petroleum ether (Span80)	>99.99 99.987 99.989 >99.99	257.3 (W) 313.4 (W) 262.4 (W) 2929.9 (O) 2738.5 (O) 3337.6 (O)	[113]
3.	Zr^{4+} supported on Collagen(I) fibres	Dodecane/Water Olive oil/Water Pump oil/Water Water/Dodecane Water/Kerosene Water/Petroleum ether	>99.99	3031 (W) 2407.6 (W) 2598.7 (W) 2331.2 (O) 2751.6 (O) 2216.6 (O)	[122]
4.	CFs-PPFs (PPF—polypropylene fibres, dual-channels) Simple collagen fibres	Dodecane/Water Dodecane/Water	>99 99.4	2844 (W) 434 (W)	[119]

Table 2. *Cont.*

	Filtration Material/ Membrane	Targeted Pollutant	Separation Efficiency %	Filtrate Flux $L \cdot m^{-2} \cdot h^{-1}$ *	Reference
5.	SCFs-DC (superwetting collagen fibres with water-oil dual-channels) Simple collagen fibres	Dodecane/Water Kerosene/Water Hexadecane/Water Octane/Water Dodecane/Water	98.99 99.10 99.04 96.22 99.45	666 (W) 587 (W) 487 (W) 796 (W) 477 (W)	[120]
6.	CFM-PMDA-TiO_2 (CFM—collagen fibre membrane; PMDA—pyromellitic dianhydride)	Dodecane/Water (CTAB) Heptane/Water (CTAB) Octane/Water Dodecane/Water (CPB)	98.35 98.70 99.86 97.28	600 (W) 900 (W) 1436.40 (W) 1100 (W)	[107]
7.	CFM-PEI-TiO_2	Dodecane/Water Hexane/Water (SDS) Dodecane/Water (SDBS) Hexane/Water (SDBS) Heptane/Water (SDS) Heptane/Water (SDBS)	99.93 98.79 99.83 99.73 99.94 99.68	988.90 (W) 880.75 (W) 1351.92 (W) 1346.09 (W) 1148.58 (W) 1458.50 (W)	[108]
8.	UiO-66-NH_2 membrane incorporating MOFs	Nanoemulsions Dodecane/Water (SDS) n-Hexane/Water (SDS) n-Octane/Water (SDS) n-Hexadecane/Water (SDS) Microemulsions Dodecane/Water (SDBS) n-Hexane/Water (SDBS) n-Octane/Water (SDBS) n-Hexadecane/Water (SDBS)	>99.8 >99.7	217.25 (W) 242.12 (W) 210.86 (W) 308.46 (W) 200.66 (W) 225.01 (W) 370.05 (W) 291.87 (W)	[109]
9.	CFM/UiO-66/PDMS (PDMS—polydimethylsiloxane)	Nanoemulsions Water/Dodecane (SDBS) Water/Dodecane (CTAB) Water/n-octane (SDBS) Water/n-heptane (SDS) Microemulsions Water/Dodecane (SDS) Water/Dodecane (CTAB)	>99.99	540.4 (O) 504.6 (O) 969.8 (O) 973.3 (O) 545.4 (O) 351.1 (O)	[116]
10.	CF/ZIF-8/PDMS, (ZIF—zinc-based MOFs)	Nanoemulsions Water/Dodecane (SDBS) Water/Dodecane (CTAB)	>99.99	1982 (O) 1809 (O)	[115]
11.	MWCNTs/CFM (MWCNTs—multiple-walled carbon nanotubes)	Water/Heptane	>99.99	1051 (O)	[121]
12.	Commercial PTFE (polytetrafluoro ethylene)	Nanoemulsions Water/Dodecane (SDBS) Water/n-heptane (SDS)	>99.99	77.6 (O) 305.6 (O)	[116]
13.	Commercial double-sided polyvinylidene fluoride (PVDF) Commercial PTFE	Water/Dodecane (SDBS) Water/Dodecane (SDBS)	>99.99	51 (O) 95 (O)	[115]
14.	Commercial polyamide Commercial PTFE	Water/Heptane	>99.99	64 (O) 35 (O)	[121]

* (O)—oil; (W)—water.

Besides emulsions separation, CFM has been tested for acid recovery under vacuum. The recent study of Xiao et al. (2023) focused on obtaining an ecological membrane to be used for acid recovery [123]. The biocompounds collagen, casein and chitosan allowed the preparation of a thin film composite (TFC) membrane tested for the separation of acid from a solution of H^+/Fe^{2+} synthesised to reproduce the wastewaters discharged by several industries like mining and steel production [124,125]. The mechanism that governs the protons transfer through the membrane is based on the establishment of electrostatic interactions and hydrogen bonds facilitated by the functional groups -OH, and -COOH, while Fe^{2+} ions remain at the membrane surface due to a larger ion radius and the chelation process. The performances of the TFC membrane in acid separation were found to be comparable to those obtained on commercial membranes tested in similar conditions.

Grafting copolymer MAA-co-GMA (MAA—methacrylic acid, GMA—glycidyl methacrylate) can create a pH-responsive separation material; the pH response is obtained by the large number of carboxyl groups provided by MAA, while GMA works as a reactive polymer creating links with MAA and collagen amine groups [126]. This can be used in the separation of emulsions (surfactant free or surfactant stabilised) as a material that absorbs or releases oil under the action of pH (Figure 10). Because of pH responsive switching wettability, the material shows exceptional anti-fouling characteristics compared with a conventional collagen-based oil absorbent [127].

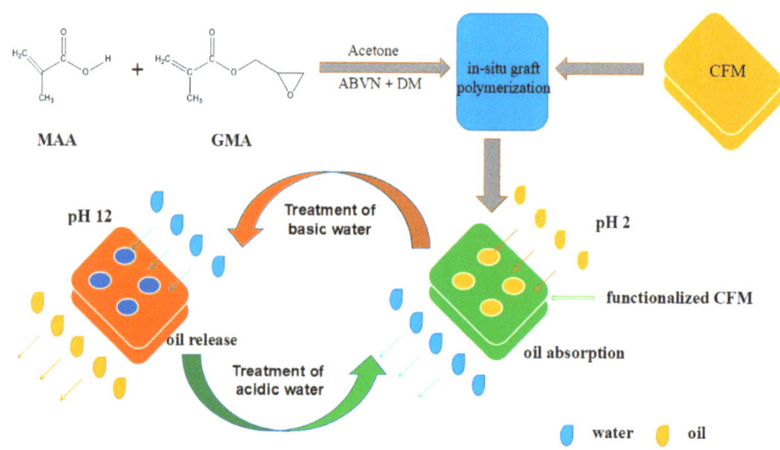

Figure 10. Separation of emulsion containing wastewater by pH-responsive collagen fibre material (MAA—methacrylic acid, GMA—glycidyl methacrylate, DM—dodecyl mercaptan, ABVN—2,2-azo bis(2,4-dimethylvaleronitrile)) [126].

2.3. *Advanced Oxidation of Chemical Contaminants*

Advanced oxidation processes (AOP) are used when the concentration of the target organic compound is low, and its recovery is not considered economically advantageous. Otherwise, alternative recovery methods such as adsorption or separation technologies are used. The main disadvantages related to AOP are the incomplete oxidation in certain cases of contaminants to compounds that are more toxic than the original ones and the accidental over-dosage of chemicals that can neutralise the effect of a subsequent biological purification step [128].

Collagen has been tested as a component of materials used in photocatalysis and Fenton oxidation. These technologies are used in the tertiary step of wastewater treatment for the removal of persistent organic compounds [129]. Photocatalysis offers the advantage of high degradation percentage [130].

Collagen is usually used as a support for catalytically active compounds due to its organised structure in fibres and fibre bundles that will lead to a catalyst with an ordered structure [131–133].

The mechanism of crystal violet dye degradation in the presence of a collagen–cellulose–Fe_3O_4/TiO_2 sponge photocatalyst consists of initial adsorption of an organic compound on the catalyst surface until the adsorption equilibrium is reached (about 12.4% of the dye is adsorbed in the first 30 min in the absence of any light source) [130]. Subsequent organic compound degradation in the presence of light contributes to the achievement of a higher removal degree (Figure 11) [130,134]. Silver chloride has been reported to be a more efficient catalyst support than TiO_2 for the photocatalytic removal of organic compounds [135].

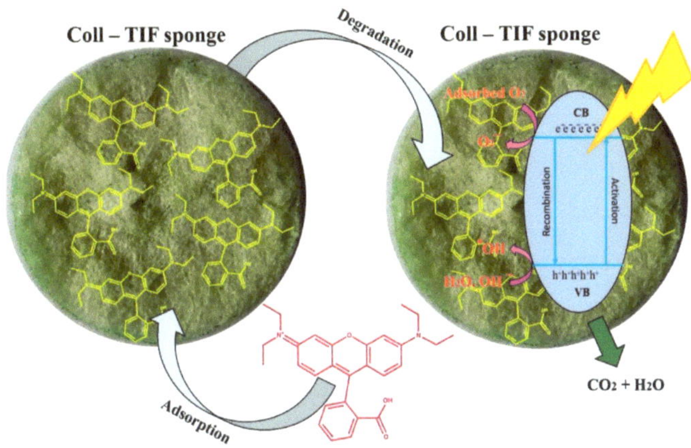

Figure 11. Mechanisms for pollutants removal by photo-oxidation: Rhodamine-B photocatalytic degradation over collagen-TiO_2 nano—sponge [134].

The high efficiency of sponge-type photocatalysts [130,134] is due to the extremely porous 3D structure that allows the easy diffusion of pollutants through the structure, determining a higher degree of adsorption of pollutants on the catalyst surface. This is adding to the material's ability of floating at the air–water interface, which facilitates the photons transfer.

The importance of H_2O_2 dosage at the optimum concentration has been highlighted, an excess of hydrogen peroxide being responsible for the photocatalytic activity decrease; this is due to the formation of a hydroperoxyl radical that has a decreased oxidative activity compared with a hydroxyl radical [136]. Several silver salts (Ag_2MoO_4, Ag_3PO_4, CH_3COOAg and AgCl) have been deposited on CFs and tested for methyl orange degradation under UV or visible light; silver chloride was found to be the most efficient under these conditions. Its efficiency was explained by the ability of AgCl to form silver nanoparticles that react with oxygen molecules to form highly reactive O_2^- species [135].

Comparing the oxidation efficiency of collagen-based materials with other materials reported in the literature, it can be concluded that they represent a viable alternative for replacing more expensive materials (Table 3).

Table 3. Performance of collagen-based (1–7) and other materials (8–11) for pollutants removal by photocatalytic oxidation.

	Photocatalyst	Targeted Pollutant	Removal Degree (%)	Degradation Time (min)	Light Source	Reference
1.	Collagen-cellulose-Fe_3O_4/TiO_2	Crystal violet	91.2 86.6	180 180	H_2O_2, Visible light (200 W Hg (Xe)) H_2O_2, direct sunlight irradiation	[130]
2.	CFs-TiO_2	Rhodamine B	95 73 100	130 130 300	H_2O_2,Visible light (200 W Hg (Xe)) H_2O_2, direct sunlight irradiation	[134]
3.	Fe(III)/CFs (CFs-collagen fibres)	Orange II p-nitrophenol	~100 95	20 20	H_2O_2, UVC irradiation (254 nm, 8 W or 4 W)	[132]
4.	Fe(III)/CFs	Orange II	73.8	90	H_2O_2, UVC irradiation (254 nm, 10 W)	[136]
5.	Fe(III)/CFs	Malachite green	55	120	H_2O_2, UVA irradiation (365 nm, 10 W)	[133]
6.	AgCl/CFs	Methyl orange	80 >90	210 30	UV light (370 nm, 36 W) visible light (500 W Xe lamp)	[135]
7.	AgNPs/Fe crosslinked CFs (NPs—nanoparticles)	Mixture of antibiotics (Tetracycline hydrochloride, sulfamethoxazole, ciprofloxacin, vancomycin and levofloxacin)	>90	30	H_2O_2, direct sunlight	[5]
8.	Cu-doped nanosized ZnO	Methylorange	85	200	Direct sunlight	[137]
9.	Zn-doped CdS	Rhodamine B	93	135	Visible light	[138]
10.	CoO/TiO_2	Rhodamine B	97	120	Infrared light irradiation	[139]
11	BiOI (bismuth oxyhalide)	Reactive blue 19	95	120	Direct sunlight	[140]

2.4. Antimicrobial Activity of Collagen-Based Materials

Each effluent that discharges wastewater containing microorganisms has a distinct bacterial community [141]. The groups of microorganisms present in wastewater include bacteria, protozoa, fungi and viruses [142]. While some of the microorganisms could have a beneficial effect in wastewater treatment [143], some microorganisms (e.g., Bacteroides, Clostridium, Enterococcus, Leptospira, Acinetobacter, Pseudomonas, Streptococcus, Mycobacterium [144]) are pathogens and must be monitored and removed.

Collagen's resistance to bacterial activity makes it a good candidate for incorporation into antibacterial materials [145]. The antibacterial activity can be enhanced by additional constituents [146–148]. For example, adding red propolis to a chitosan/collagen membrane will give a minimum inhibitory concentration value of 7.8 $\mu g \cdot mL^{-1}$ for Staphylococcus aureus and 1.9 $\mu g \cdot mL^{-1}$ for Pseudomonas aeruginosa [146]. Ions with known antibacterial activity, like silver, can be incorporated in woven-like collagen structures, or enzymes can be immobilised on collagen surface [5,135].

Aerogels synthesised from collagen (AgNPs/Fe@CF and Fe@CF) were tested for the simultaneous elimination of antibiotic-resistant bacteria (tetracycline resistant *E. coli*), removal of trace antibiotics by Fenton oxidation and heavy metals adsorption from wastewater (Figure 12); like in the case of sponge-type structures, the aerogel is characterised by an extremely porous structure and low density [5]. The silver nanoparticles incorporated in the collagen structure were found to be responsible for the bactericidal activity. It has been shown that silver ions adsorb on the surface of the bacterial membrane, accumulate and break the cell walls and lead to the death of the bacteria by interfering with glucose metabolism by inactivating thiol groups (-SH) [149].

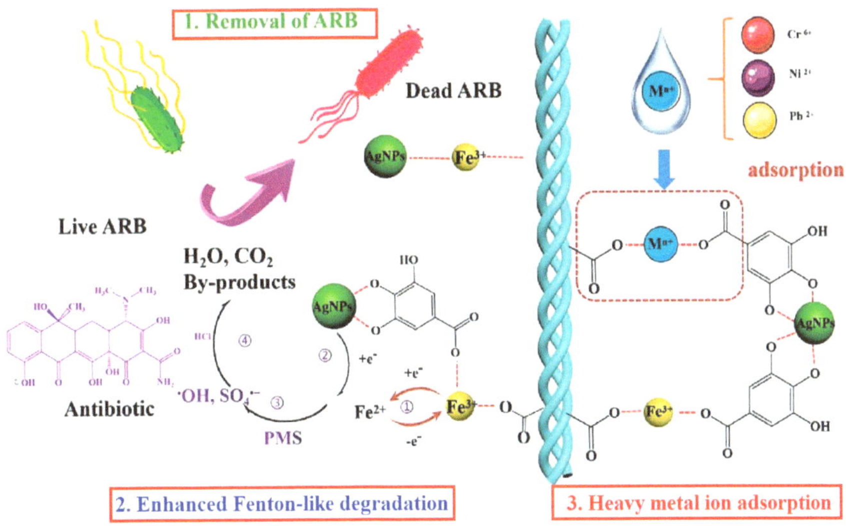

Figure 12. Hybrid Fenton oxidation—adsorption for simultaneous removal of antibiotics, antibiotic-resistant bacteria (ARB) and heavy metal ions (M^{n+}). Peroxymonosulfate (PMS) is used as an oxidant in the Fenton process to provide hydroxyl radicals (•OH) and sulphate radicals (SO_4^-•) [5].

Pure collagen polymers, although they do not degrade in the presence of bacteria, do not have antibacterial activity [20]. Ag_2MoO_4, Ag_3PO_4, CH_3COOAg and AgCl, deposited on CFs, were tested for antibacterial performance; silver chloride deposited on collagen fibres was found to be the most efficient against Escherichia coli and Staphylococcus aureus, while pure collagen showed no inhibitory activity under these conditions [135].

Subbiah et al. (2022), made comparative tests of real wastewater and synthetic wastewater photocatalytic disinfection [150]. The real water was collected from the unhairing section of a tannery in Chennai (India). The synthetic water was obtained by adding *E. coli* strains (~107 colony forming units (CFU)/mL). From Table 4, it can be seen that water containing antibiotic-resistant bacteria needs more time to achieve complete disinfection (line 1). In addition, real wastewater necessitates more time to ensure a full destruction of microorganisms (line 2); this is probably due to the presence of other chemical compounds in the wastewater that are competing for the photocatalyst active sites. However, the real wastewater was not microbiologically characterised before the experiment to know its exact composition. For comparison purposes, studies from the literature report similar or lower performances for water disinfection [151,152].

Table 4. Performance of collagen-based (1–2) and other materials (3–4) for wastewater disinfection under the action of light.

	Material	Microorganisms	Removal Degree (%)	Inactivation Time (min)	Light Source	Reference
1.	AgNPs/Fe crosslinked CFs (NPs—nanoparticles)	Tetracycline resistant *E. coli* Methicillin-resistant *Staphylococcus aureus*	99.99 99.99	60 60	H_2O_2, direct sunlight	[5]
2.	Ag-SiO_2/amino-functionalised collagen	Tannery wastewater disinfection *E. coli* containing wastewater	100 100	20 15	Visible light (200 W Hg (Xe))	[150]

Table 4. *Cont.*

	Material	Microorganisms	Removal Degree (%)	Inactivation Time (min)	Light Source	Reference
3.	Ag/CuS/carbon cloth	*Bacillus subtilis*	99.99	22.5	Visible light (500 W, Xe)	[151]
4.	Ag/polymeric carbon nitride	*E. coli*	95.5	120	Visible light	[152]

In addition to photooxidation, filtration of microorganisms has been used to reduce their loading in wastewater. Yu et al. (2022) tested a collagen-based membrane (TA@CFN-M, TA—tannic acids) for the microorganism filtration and wastewater disinfection [153]. The performances of this membrane were compared with those of a commercial product—mixed cellulose esters (MCE) membrane (from Millipore). While the modified collagen membrane showed good antifouling performance, the commercial membrane developed a thick biofilm layer, indicating severe biofouling due to bacteria adhering to the surface and decreased water flux. For TA@CFN-M, the *E. coli* counting in filtrate was lower than 10 CFU/mL, which is in agreement with the in force European directives for water reuse [154]. This membrane was also showing good separation performance for the retention of microplastic particles. The principle of antibacterial filtration is illustrated in Figure 13 [153]; the SEM images of four bacteria E.coli, S.aureus, MRSA (*Methicillin-resistant Staphylococcus aureus*) and *P. aeruginosa* before and after filtration were analysed. The phenolic groups of the tannic acid are providing the antibacterial effect of the membrane [155].

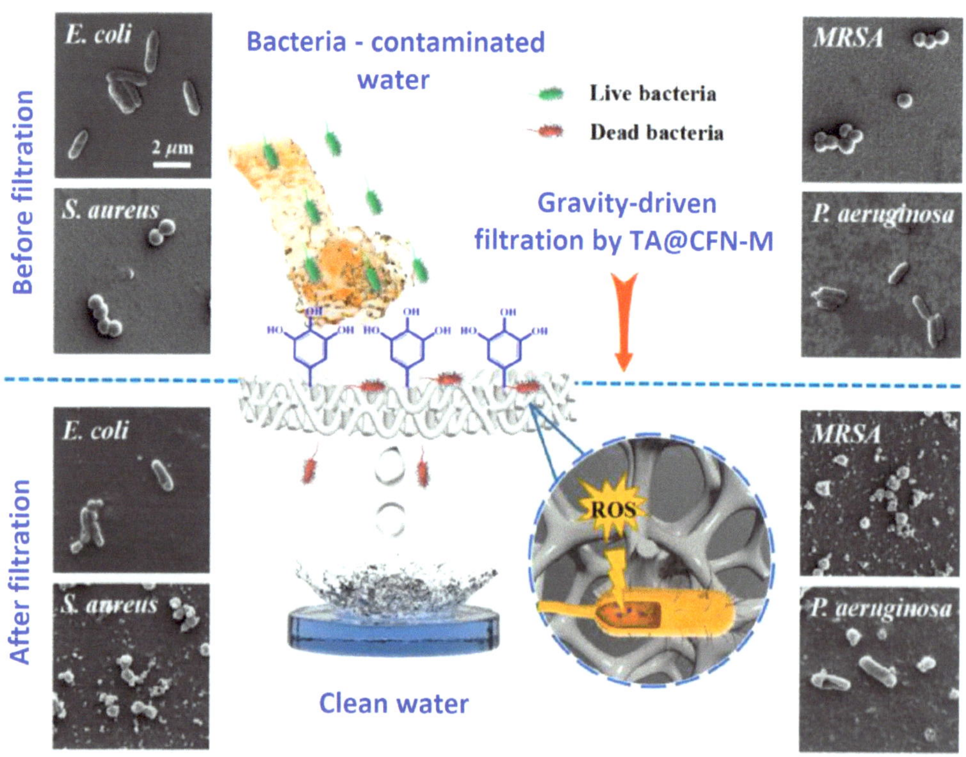

Figure 13. Antibacterial filtration principle, TA—tannic acids, CFN—collagen fibrous network, M—membrane, ROS—reactive oxygen species, MRSA—*methicillin-resistant Staphylococcus aureus* [153].

2.5. Bioremediation

Some researchers focused on the immobilisation of the enzymes on the collagen support for wastewater depollution [156–158]. Catalase, an antioxidant enzyme, is used in wastewater treatment; to reduce the cost of the method and increase its efficiency, it is preferred to immobilise the catalase on supports (alumina, natural polymers, synthetic polymers, etc.) [159,160]. Catalase has been immobilised on pure collagen or on modified support containing collagen [156]. Based on its ability to decompose about one million of hydrogen peroxide molecules per molecule of enzyme into water and oxygen [159], catalase supported on collagen can be used in the oxidation processes and in the aerobic bioremediation.

However, bioremediation can be applied only in specific locations where the environmental conditions allow the development of microorganisms.

The targeted contaminants for wastewater treatment using these materials are organic contaminants (e.g., dyes and bleaching effluents from the textile industry, persistent organic compounds), total nitrogen and total phosphorous [158,159,161]. Wastewater from the textile industry resulting from the bleaching process has a high amount of hydrogen peroxide. Instead of conventional use of hydrosulphite or sodium bisulphite, catalase immobilised on collagen could be used to degrade the hydrogen peroxide [159]. The resulting reactive oxygen species could be used to further degrade other organic compounds that are present in the wastewater.

Incorporation of Zr into collagen used for catalase immobilisation increased the catalase denaturation temperature from 37 to 75 °C [162], while catalase on pure collagen has a denaturation temperature of about 64 °C [156].

It has been reported that the activity of immobilised enzymes depends on the immobilisation method (cross-linking, adsorption or embedding); the activity of the enzyme immobilised by adsorption was the highest, while the addition of glutaraldehyde as a cross-linking agent improved the mechanical properties of collagen film (the elongation at break and tensile strength) and increased the denaturation temperature to about 84 °C [157].

Some natural polymers, including collagen, have been tested for bacteria immobilisation in the ANAMMOX process (anaerobic ammonium oxidation). Because of increased degradation of the natural polymer, the studies were redirected towards the use of synthetic polymers (like poly(vinyl alcohol) and poly(ethylene glycol) [163]. However, the good compatibility between the natural polymer and bacteria suggests that perhaps a compromise consisting of using both natural and synthetic polymers in the synthesis of a new support would be the best solution [164].

3. Regeneration and Reuse

The reusability of the material is a key factor for cost reduction and waste minimisation. In determining whether a material is feasible for large-scale use in wastewater remediation, one of the decisive factors is the ability of the material to regenerate and maintain high performance over the long term.

Most of the adsorption studies using collagen-based adsorbents were also tested to appreciate how the material behaves for pollutant desorption and for several adsorption-desorption cycles. For pollutants' desorption, acid, bases, alcohols or salts are used.

The regeneration of CF/ZIF 7-NH_2 by Hg(II) desorption in acid led to the decline of the adsorption capacity explained by the MOFs structural collapse determined by the ion exchange that took place between Hg and Zn ions (98.63% retention in the first cycle vs. 34.58% retention in the second one); this indicates that the material is not stable for multi-cycle use [57]. Covalent organic frameworks offer a better structural stability compared with MOFs, especially if the wastewater contains a large amount of metallic ions that are susceptible to being transferred in the MOFs structure via ion exchange [76]. Another collagen-based material tested for Hg(II) removal and exhibiting lower adsorption capacity for mercury compared to CF/ZIF 7-NH_2 had remarkably constant removal efficiency on the tested interval (four cycles) [47]. Jing et al. (2022) developed a collagen-based fluorescent aerogel for hexavalent chromium adsorption; the regeneration of the material was performed by Cr(VI) desorption in a solution of NaOH, and the material was reused for three cycles. The removal efficiency decreased from almost 100 before reuse to about 40% at the end of the third cycle [56]. Collagen fibre/carbon quantum dot fluorescent adsorbent had an Pb(II) removal efficiency greater than 85% after five cycles [48].

Hydrogel-based collagen used for methylene blue adsorption maintained good removal efficiency for the first three adsorption—desorption cycles followed by a slow decline in the efficiency attributed to a difficult regeneration [36]. ACF-TpPa-1 maintained constant high removal efficiency for acid fuchsin during six cycles of testing, while the adsorption efficiency decreased significantly for reactive blue 19 [70].

CF/PEI maintained almost constant adsorption capacity towards anionic dyes even after five cycles of reuse; the anionic dyes were easily desorbed using an HCl solution [38]. CF/PEI effectively removes p-arsanilic acid after five times of reuse (adsorption capacity 90.20 $mg \cdot g^{-1}$) [35]. For five cycles of adsorption-desorption, carbohydrate and collagen-based doubly grafted interpenetrating terpolymer hydrogel had a slow decay of the adsorption capacity for Pb(II), Hg(II), methyl violet and methylene blue; the adsorption capacity for these contaminants remained, however, significant [2]. This suggests that cross-linking collagen with synthetic polymers leads to adsorbents that maintain a high adsorption capacity for a greater number of cycles [35,38].

Similarly, zirconium molybdopyrophosphate-functionalised collagen fibres showed a high removal rate of radioactive Cs^+ and Sr^{2+} even after four cycles [85]. Collagen/tannin extract composite had a limited decline of the adsorption capacity towards Cu(II) during six cycles of adsorption-desorption, suggesting that the material can be reused at least six times [32].

Collagen-based bio-adsorbents showed excellent reusability behaviour. Azadi et al. (2022) focused on the synthesis of bio-adsorbents from collagen and graphene oxide; this material retained a high adsorption capacity of rhodamine B even after five cycles [72]. Collagenic-waste/natural rubber biocomposite for the removal of Hg(II), safranine and brilliant cresyl blue maintained 85% from initial adsorption capacity after five cycles, suggesting good reusability [4].

Acid use for adsorbent regeneration was found to be partially responsible for the decay of the contaminants' removal efficiency in time [40]. For a collagen-cellulose adsorbent, only the materials with a higher cellulose content were resistant in the presence of the HCl used for regeneration [55]. Another cause for adsorbent efficiency decrease is the fact that not all the adsorbate is removed during the desorption step, blocking in this way some of the adsorption centres and gradually decreasing the adsorption capacity [39].

For the adsorption of radioactive ions from nuclear wastewater (e.g., UO_2^{2+}, Cs^+, Sr^{2+}), the desorption and subsequent recovery of the adsorbate and adsorbent is even more critical from both environmental and economical points of view. However, not all the studies regarding radioactive species retention on collagen-based materials have discussed this aspect.

Most of the materials studied for radioactive wastewater treatment and reporting reusability studies demonstrated good removal of radioactive ions for at least three cycles of adsorption-desorption [28,80,83–85,88,165]. HTO/CF could be reused at least three times for the adsorption of uranyl from nuclear wastewater after adsorbent regeneration using nitric acid [28]. A cellulose-collagen-based biosorbent was having adsorption efficiency of 82% after five cycles of adsorption-desorption of uranyl ions, suggesting a high degree of reuse [79].

The materials used for oil sorption were regenerated by centrifugation; it was found that their sorption capacity towards silicone, motor or vegetable oil is slowly decreasing during five cycles but is kept above 93% of the initial sorption capacity; however, not all the oil could be removed from the sorbent by centrifugation, and another desorption method must be provided [78].

Emulsion viscosity was found to have an influence on the number of reuse cycles on membranes, filtration materials or adsorbents because of severe pore blockage by the viscous liquid [114,121,126]. CF material modified by copolymer P(MAA-co-GMA), has a decrease in its efficiency after the sixth reuse cycle in stabilised motor oil/water emulsion separation [126]. For membranes, this disadvantage can be mitigated by passing to vacuum filtration. The separation fluxes of nano-emulsions water-in-olive oil and water-in-pump oil under gravity effect were less than 10 $L \cdot m^{-2} \cdot h^{-1}$, while under vacuum, the filtrate fluxes were above 1299 $L\ m^{-2} \cdot h^{-1}\ bar^{-1}$ [121].

Most membranes and filtration materials tested for emulsion separation maintain their separation efficiency with no significant decline in the filtrate flux [107–109,115,117,120]. Ye et al. (2024) tested the filtrate flux stability of CFM-PEI-TiO_2 membrane over 120 min of continuous emulsion separation. It was found that the separation efficiency decreased from 99.9 to 99.2%, while the filtrate flux increased from 1351.92 $L \cdot m^{-2} \cdot h^{-1}$ (Table 2) to more than 1550 $L \cdot m^{-2} \cdot h^{-1}$ [108].

Collagen/casein/chitosan polymer membrane used for acid separation from wastewater showed excellent reusability for five cycles of testing [123].

Collagen fibres material containing Zr^{4+} was found to have a 5% decrease in the filtrate flux after 120 min of continuous use, which was attributed to the contamination of the separation material with oil and surfactant [122].

The regeneration of filtration material/membrane can be achieved by washing with ethanol and subsequent drying at about 40 °C; the material/membrane is then ready for the next continuous separation cycle (120 to 150 min) [107,120]. In this way, the oil is removed from the surface of the separation material. It has been reported that a membrane maintained its high-water filtrate flux even in the sixth cycle of reutilisation (separation efficiency up to 99.98, water filtrate flux 1804 $L \cdot m^{-2} \cdot h^{-1}$) [117].

Liu et al. (2009, 2010), discussed the reutilisation of a Fe(III) immobilized on CFs photocatalyst, used for malachite green and orange II removal [133,136]. A slight decrease in the photocatalytic activity after each cycle of reutilisation has been reported. This was attributed to the decrease in the adsorption capacity of the catalyst and to the loss of iron. To reactivate the catalyst, the Fe(III) reimmobilisation onto the catalyst has been considered.

The efficiency of tetracycline removal by photocatalytic degradation in the presence of AgNPs/Fe@CFs after five cycles was higher than 90%; the same material was tested for the elimination of tetracycline resistant *E. coli*, and its elimination rate after five cycles was very high (99.99%) [5].

When retaining oil on collagen-cellulose-Fe_3O_4/TiO_2 (sponge-type structure), it was proven that this material can be used up to at least nine cycles without a significant decrease in the degree of oil removal from water [130].

Catalase immobilised on collagen was tested for hydrogen peroxide decomposition, and it was found that the material can be reused up to 45 times when working in optimal conditions (temperature 30 °C, pH = 7) [156].

4. Environmental Implications

Although collagen toxicity is small and high quantities of collagen must be ingested before some health effects can be observed [19], the final material that is used in wastewater treatment may contain substances that are not harmless (e.g., silver, iron, synthetic polymers, etc.). Therefore, advanced studies of stability over time and degradation under the action of water pH and temperature are required before using these materials on a large scale. Collagen-based materials can be susceptible to proteolytic degradation (protein breakdown into smaller polypeptides or the corresponding amino acids); the addition of, e.g., magnetite improves the resistance of the material to acidic and proteolytic degradation [34].

In addition, the material must have good mechanical stability, so that it is not entrained, blocking the equipment and the pipes through which the water passes in the subsequent stages of decontamination. For membrane materials, some authors report performing abrasion stability tests [107–109,116,121]. These tests consist of membrane abrasion with sandpaper followed by contact angle checking at various intervals. The membrane is then reused for emulsion separation experiments, and its performance is reevaluated. For example, it was found that CFM-PEI-TiO_2 surface superhydrophilicity has been kept even after 500 cycles of abrasion [104]. CFM/UiO-66(12)/PDMS membrane maintained its high performance in the separation of six types of micro and nano-emulsions, after 200 abrasion cycles (oil separation flux up to 1012.2 L m^{-2} h^{-1}, efficiency up to 99.993%) [116].

Generally, the materials that are used in wastewater treatment must have a good resistance to degradation in a harsh chemical environment given by the presence of organic solvents, strong acids or strong alkali. The chemical stability can be tested by membrane immersion for 2 h in strong acid (pH = 2) and alkali (pH = 12) solutions [109,116]. The resistance towards the corrosive environment was attributed in the case of UiO-66-NH_2 membrane to the supplementary membrane functionalisation with -NH_2 groups [109]. The stability of a thin film composite (collagen, casein, chitosan) membrane used for acid recovery was evaluated by performing acid immersion tests for extended time (10 days); although some mesopores have been formed in the membrane, it has been reported that the membrane overall performance was not impacted significantly [123]. Li et al. (2021) also evaluated the resistance towards organic solvents (e.g., by immersion in dodecane, n-heptane or n-octane solution); the tested materials were reassembled, and emulsion separation experiments indicated that the materials kept their separating performances [116].

The stability of AgNPs/Fe@CFs aerogel used for antibiotic degradation, antibiotic-resistant bacteria elimination and heavy metals adsorption was evaluated by estimating silver leaching corresponding to each cycle; after the fifth cycle, the cumulative silver leaching was about 0.13%, which corresponds to a concentration of silver in water significantly lower than the maximum concentration allowed in the drinking water standards [5].

Collagen crosslinked with synthetic polymers was found to have better performance in acidic conditions or at high temperatures [33,38].

When foreseeing adsorption utilisation for wastewater treatment, the stability of the material to irradiation must be evaluated. This is done by performing morphological analysis and identification of the functional groups (SEM and FTIR) of the adsorbent before and after irradiation at different doses. In this regard, Tang et al. (2021) found that there is a slight decrease in the adsorption capacity of HTO/CF after the first ^{60}Co γ-ray irradiation dose, followed by constant adsorption capacity when the irradiation dose increases [80]. PBA/CF showed also good irradiation stability after exposure at different doses of ^{60}Co γ-ray irradiation (10–350 kGy) [88]. Fe/Ni loaded on bayberry tannin/CF showed adsorption capacity decrease with increasing irradiation dose, but the removal rate remains high (above 98%) [86].

Some adsorbents are intended to be biodegradable [36,45,56]. The adsorbent material synthesised by Kaur and Jindal (2020) is ecofriendly; this adsorbent biodegrades under the action of microorganisms by metabolic and enzymatic processes specific to soil burial or composting methods [36,45]. Du et al. (2018) compared the biodegradability of CF-PDMS with that of simple CF during a soil burial test performed for 75 days and found that the PDMS modified material has a weight loss above 52% after 75 days, while CF loses about 67% of its initial weight [78].

Ye at al. (2023) report biosafety tests for CFM-PMDA-TiO_2 [107]. The toxicity of the membrane was evaluated by studying zebrafish evolution in the filtrate or in the water that contains the membrane; the zebrafish population remained unaffected after 14 days of testing, both tests indicating that the membrane is not toxic.

5. Challenges and Conclusions

Experiments with real wastewater are tremendously useful to appreciate the behaviour of the materials when competitive adsorption takes place and if the affinity of the materials towards the tested pollutants shifts to water components that are not necessary to be removed (e.g., Ca^{2+}, Mg^{2+}, K^+, Na^+, Cl^-, etc.). There are a limited number of studies with real wastewater [35,58,87,107,150]. To overcome this aspect, some researchers have performed ion interference studies [32,56,58,80,81,84].

The Hg(II) retention on CF/ZIF 7-NH_2 from water contaminated with multi ions (potassium, magnesium, calcium, cadmium, manganese) evidenced the high affinity of the adsorbent towards Hg ions (315.89 vs. 332.1 mg·g^{-1} in the absence of competing ions) [57]. In the presence of other ions (trivalent iron, bivalent copper, barium and lead), it was found that the Cr(VI) removal performance of a collagen-based aerogel decreases from almost 100% to about 65%, hence, emphasising the importance to study the performances of the materials in conditions that are similar to real ones [56]. In general, for retention mechanisms that are based on coordination between metal ions and organic functional groups (little electrostatic influence), the competition for retention will be given by metal ions that have a similar hydrated radius [88]; when electrostatic interaction is predominant, the interference in the retention will be given by the ions that have the same valence as the target ion [32,58].

The removal efficiency of As(III) using Zr/CF was highly influenced by coexisting anions like Cl^-, NO_3^-, SO_4^{2-} and HPO_4^{2-}, dividing the removal efficiency by 2. On the contrary, low concentrations of HCO_3^- do not affect the As(III) adsorption [60].

p-arsanilic acid was added to river or lake water samples to study the removal of this contaminant on amphiphilic amino-modified CF; the contaminant concentration in the treated water was less than the maximum allowable concentration [35].

Antibiotics and other metal ions have also been found to compete for adsorption sites when removing heavy metals from water; the efficiency of Cu(II) removal of (SA/BHC)@TE decreased from 90.3% to 80.07% in the presence of tetracycline (Figure 14) [58]. To better understand, the authors did comparative experiments with tap water, sea water and lake water; the lowest Cu(II) removal efficiency (76.32%) was recorded for sea water due to increased water mineralisation [58].

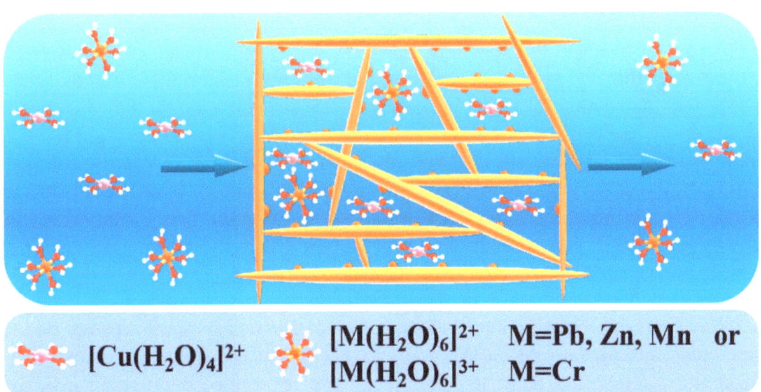

Figure 14. Influence of other species present in the wastewater [58].

Some researchers preferred to perform the adsorption kinetics studies using methylene blue and indigo carmine, while the reusability of the adsorbent was tested for 10 cycles on real wastewater resulting from the dyeing stage (cotton dyeing—containing direct red and mordant brown; acrylic fibre dyeing—basic pink dye); the adsorbent was regenerated after each adsorption cycle using NaCl and had a removal efficiency greater than 70% for the first eight cycles [87].

Collagen-graphene oxide nanocomposite was tested for dye and phenolic compounds removal from tannery effluents; the material proved to be efficient, reducing the COD (chemical oxygen demand) and BOD (biochemical oxygen demand) values below the limits imposed by the wastewater directive [73].

The efficiency of nano-ZnS/CF for uranyl removal out of lake and river water samples was found to remain high (86.64% and 92.89%) [84]. The experiences with synthetic nuclear wastewater showed a good selectivity of adsorbents towards uranyl ions; the common compound of the tested materials was collagen, and the selectivity for uranyl ions was attributed to the strong bonds that uranyl ions create with the collagen's functional groups [28,80,81,83,84,165].

Separation on membranes can be complicated by the presence of a mixture of several oily compounds instead of an emulsion formed of one oily compound in water. Ye at al. (2023, 2024) performed separation studies using emulsions prepared from commercially available cosmetics (facial creams, shampoos, hand creams, hair conditioners) to simulate real wastewater resulting from the cosmetics industry [107,108]. The results can be compared with the separation efficiencies and filtrate fluxes obtained from tests with a single oil/water emulsion on the same membrane. For CFM-PMDA-TiO_2 membrane, separation efficiency remains higher than 94%; the filtrate flux decreases due to the complex mixture of oily compounds present in the cosmetic wastewater. The filtrate flux remains high anyway (579.9 $L \cdot m^{-2} \cdot h^{-1}$ in the worst case [107]). The CFM-PEI-TiO_2 membrane gives, however, remarkably better performances then the CFM-PMDA-TiO_2 membrane (the efficiency when separating cosmetic emulsions was above 98.5%, minimum filtrate flux 985.0 $L \cdot m^{-2} \cdot h^{-1}$, maximum filtrate flux 1486.4 $L \cdot m^{-2} \cdot h^{-1}$) [108]. This highlights the importance of additional compounds that are included in the collagen membrane (PEI vs. PMDA).

An aspect that was neglected is the calculation of the cost of the material. Although collagen is recovered from waste, this generates a cost, which is added to the production cost of the final material. Peng et al. (2023) reported a small cost compared with other materials (0.027 $/gram for zirconium molybdopyrophosphate-functionalised collagen fibres vs. other industrial adsorbents for nuclear wastewater that are in the range of 0.133–18.821 $/gram) [85]. In addition, Yang et al. (2024), reported smaller manufacturing costs for their adsorbent compared with the price of industrial adsorbents for nuclear industry [84].

The purpose of this review was to explore the state of research on the use of recovered collagen as a sustainable alternative for wastewater treatment. It is rarely that wastewater is loaded with only one type of pollutant. In general, it is a mixture of different types of contaminants such as heavy metals, organic substances (e.g., cosmetics, pharmaceuticals, surfactants, oils, dyes, etc.) and acids. The collagen-based materials have proven their effectiveness in retaining a wide range of organic and inorganic compounds from the classes above mentioned, as well as in retaining isotypes from the wastewater resulting from nuclear power plants.

The current methods that have been tested for wastewater purification using collagen-based materials are adsorption, photooxidation, separation using membranes or bed filtration and coagulation-flocculation. Without a trace of doubt, collagen has proven its effectiveness in the depollution of wastewater; it can be incorporated into various materials used in many of the technologies specific to wastewater treatment plants.

Most of the adsorption studies also address the issues of material stability and reusability. The positive results of the tests performed to evaluate the irradiation stability of collagen-based materials together with the experiments in a dynamic regime suggest that collagen adsorbents are good candidates for the treatment of radioactive wastewater. The ability of the membranes to withstand mechanical stress was evaluated, and it was found that the collagen membranes generally maintain their wetting properties even after a large number of abrasion cycles. Since these materials are intended for use in wastewater treatment plants, it means that they are exposed to a harsh chemical environment. Therefore, their reusability and stability must be properly evaluated before implementing the use of these materials on a larger scale.

Despite its proven antibacterial activity, few studies have been dedicated so far to collagen-based materials used for microorganisms' removal from wastewater. These aspects must be studied more in depth considering the excellent antimicrobial properties of collagen-modified materials and the possibility of coupling good adsorption capacity with microorganism reduction.

The removal of microplastic by filtration on collagen-based materials requires advanced deepening. So far, only one study addresses the issue of microplastic retention by filtration on collagen-containing membranes.

Another aspect that needs to be highlighted is that with a few exceptions, most of the studies have been performed on synthetic wastewater samples. Some authors have studied, however, the interference of other contaminants or species in the removal process of the targeted compound.

Author Contributions: Conceptualisation, A.M.D. and M.B.; formal analysis, A.M.D. and M.B.; writing—original draft preparation, A.M.D.; writing—review and editing, M.B. All authors have read and agreed to the published version of the manuscript.

Funding: This research received no external funding.

Acknowledgments: The authors would like to thank Cristina Costache for critical reading of the manuscript.

Conflicts of Interest: The authors declare no conflicts of interest.

References

1. Thorat, B.N.; Sonwani, R.K. Current technologies and future perspectives for the treatment of complex petroleum refinery wastewater: A review. *Bioresour. Technol.* **2022**, *355*, 127263.
2. Mitra, M.; Mahapatra, M.; Dutta, A.; Roy, J.S.D.; Karmakar, M.; Deb, M.; Mondal, H.; Chattopadhyaya, P.K.; Bandyopadhyay, A.; Singha, N.R. Carbohydrate and collagen-based doubly-grafted interpenetrating terpolymer hydrogel via N–H activated in situ allocation of monomer for superadsorption of Pb(II), Hg(II), dyes, vitamin-C, and p-nitrophenol. *J. Hazard. Mater.* **2019**, *369*, 746–762. [CrossRef]
3. Singha, K.; Pandit, P.; Maity, S.; Sharma, S.R. Harmful Environmental Effects for Textile Chemical Dyeing Practice. In *Green Chemistry for Sustainable Textiles Modern Design and Approaches, The Textile Institute Book Series*; Ibrahim, N., Hussain, C.M., Eds.; Woodhead Publishing: Duxford, UK, 2021; pp. 153–164.
4. Singha, N.R.; Roy, C.; Mahapatra, M.; Dutta, A.; Roy, J.S.D.; Mitra, M.; Chattopadhyaya, P.K. Scalable Synthesis of Collagenic-Waste and Natural Rubber-Based Biocomposite for Removal of Hg(II) and Dyes: Approach for Cost-Friendly Waste Management. *ACS Omega* **2019**, *4*, 421–436. [CrossRef]
5. Wang, R.; Zhao, P.; Yu, R.; Jiang, J.; Liang, R.; Liu, G. Cost-efficient collagen fibrous aerogel cross-linked by Fe(III)/silver nanoparticle complexes for simultaneously degrading antibiotics, eliminating antibiotic-resistant bacteria, and adsorbing heavy metal ions from wastewater. *Sep. Purif. Technol.* **2022**, *303*, 122209. [CrossRef]
6. Kong, W.; Li, Y.; Zhang, Y.; Mei, Y.; Tabassum, S. Biological treatment of refractory organic compounds in coal gasification wastewater: A review. *J. Water Process. Eng.* **2024**, *60*, 105255. [CrossRef]
7. Chakraborty, P.; Mohamed, N.; Rajan, S.; Urbaniak, M.; Suresh, A.; Saha, S.; Pasupuleti, M.; Islam, E.; Vinod, P.G.; Hashmi, M.Z. Post-COVID-19 pandemic perspectives on the need for wastewater surveillance of pathogenic microorganisms, antibiotics, and antimicrobial resistance in South Asia. *Curr. Opin. Environ. Sci. Health* **2024**, *39*, 100553. [CrossRef]
8. Jones, E.R.; Bierkens, M.F.P.; Wanders, N.; Sutanudjaja, E.H.; van Beek, L.P.H.; van Vliet, M.T.H. Current wastewater treatment targets are insufficient to protect surface water quality. *Commun. Earth Environ.* **2022**, *3*, 221. [CrossRef]
9. Sick Water? The Central Role of Wastewater Management in Sustainable Development. Available online: https://wedocs.unep.org/bitstream/handle/20.500.11822/9156/Sick%20Water.pdf?sequence=1&isAllowed=y (accessed on 15 May 2024).
10. Wastewater—Turning Problem to Solution. Available online: https://www.unep.org/resources/report/wastewater-turning-problem-solution (accessed on 15 May 2024).
11. Ensure Availability and Sustainable Management of Water and Sanitation for All. Available online: https://sdgs.un.org/goals/goal6#targets_and_indicators (accessed on 15 May 2024).
12. Stefan, D.S.; Bosomoiu, M.; Constantinescu, R.R.; Ignat, M. Composite Polymers from Leather Waste to Produce Smart Fertilizers. *Polymers* **2021**, *13*, 4351. [CrossRef] [PubMed]
13. Stefan, D.S.; Bosomoiu, M.; Dancila, A.M.; Stefan, M. Review of Soil Quality Improvement Using Biopolymers from Leather Waste. *Polymers* **2022**, *14*, 1928. [CrossRef]
14. Buljan, J.; Reich, G.; Ludvik, J. *Mass Balance in Leather Processing*; US/RAS/92/120; United Nations Industrial Development Organization (UNIDO): Vienna, Austria, 2000.
15. Ayele, M.; Limeneh, D.Y.; Tesfaye, T.; Mengie, W.; Abuhay, A.; Haile, A.; Gebino, G. A Review on Utilization Routes of the Leather Industry Biomass. *Adv. Mater. Sci. Eng.* **2021**, *2021*, 1503524. [CrossRef]

16. Davis, J.F.; Maffia, G.J. Collagen dispersions for liquid-solid separations in water treatment and sludge dewatering. *Sep. Technol.* **1995**, *5*, 147–152. [CrossRef]
17. Li, R.; Zhou, J.; Liao, X.; Zhang, W.H.; Shi, B. Flocculation Behaviors of Collagen Protein—Al(III) Composite Flocculant. *JALCA* **2014**, *109*, 56–62. [CrossRef]
18. Masrouri, M.; Qin, Z. Effects of terminal tripeptide units on mechanical properties of collagen triple helices. *Extrem. Mech. Lett.* **2023**, *64*, 102075. [CrossRef]
19. Seo, H.S.; Kim, J.H.; Kim, S.H.; Park, M.K.; Seong, N.W.; Kang, G.H.; Kim, J.S.; Kim, S.H.; Kim, J.C.; Moon, C. Toxicity of a 90-day repeated oral dose of a collagen peptide derived from skate (*Raja kenojei*) skin: A rat model study. *Toxicol. Res.* **2023**, *39*, 383–398. [CrossRef] [PubMed]
20. Ersanli, C.; Tzora, A.; Skoufos, I.; Voidarou, C.; Zeugolis, D.I. Recent Advances in Collagen Antimicrobial Biomaterials for Tissue Engineering Applications: A Review. *Int. J. Mol. Sci.* **2023**, *24*, 7808. [CrossRef]
21. Kaing, V.; Guo, Z.; Sok, T.; Kodikara, D.; Breider, F.; Yoshimura, C. Photodegradation of biodegradable plastics in aquatic environments: Current understanding and challenges. *Sci. Total Environ.* **2024**, *911*, 168539. [CrossRef]
22. Mukherjee, C.; Varghese, D.; Krishna, J.S.; Boominathan, T.; Rakeshkumar, R.; Dineshkumar, S.; Brahmananda Rao, C.V.S.; Sivaramakrishna, A. Recent advances in biodegradable polymers—Properties, applications and future prospects. *Eur. Polym. J.* **2023**, *192*, 112068. [CrossRef]
23. Jafarinejad, S. Some Possible Process Configurations for Modern Wastewater Treatment Plants for Per- and Polyfluoroalkyl Substances (PFASs) Removal. *Sustainability* **2024**, *16*, 8109. [CrossRef]
24. Adsorption Techniques. Available online: https://emis.vito.be/en/bat/tools-overview/sheets/adsorption-techniques (accessed on 10 October 2024).
25. Jain, M.; Majumder, A.; Ghosal, P.S.; Gupta, A.K. A review on treatment of petroleum refinery and petrochemical plant wastewater: A special emphasis on constructed wetlands. *J. Environ. Manag.* **2020**, *272*, 111057. [CrossRef]
26. Sun, Y.; Zhou, Y.; Wang, Z.; Ye, X. Fabrication of nanocomposites by collagen templated synthesis of layered double hydroxides assisted by an acrylic silane coupling agent. *Appl. Surf. Sci.* **2009**, *255*, 4497–4502. [CrossRef]
27. Deng, D.; Liao, X.; Shi, B. Synthesis of Porous Carbon Fibers from Collagen Fiber. *Chemsuschem* **2008**, *1*, 298–301. [CrossRef] [PubMed]
28. Cheng, Y.M.; Sun, X.; Liao, X.; Shi, B. Adsorptive Recovery of Uranium from Nuclear Fuel Industrial Wastewater by Titanium Loaded Collagen Fiber. *Sep. Sci. Eng. Chin. J. Chem. Eng.* **2011**, *19*, 592–597. [CrossRef]
29. Lieberman, M.; Sasaki, T. Iron(II) organizes a synthetic peptide into three-helix bundles. *J. Am. Chem. Soc.* **1991**, *113*, 1470–1471. [CrossRef]
30. Koide, T.; Yuguchi, M.; Kawakita, M.; Konno, H. Metal-Assisted Stabilization and Probing of Collagenous Triple Helices. *J. Am. Chem. Soc.* **2002**, *124*, 9388–9389. [CrossRef]
31. Luo, Z.; Zhang, X.; Fan, H.; Liu, Y.; Shi, B. Modification of Collagen for High Cr(III) Adsorption. *JALCA* **2009**, *104*, 149–155.
32. Zhao, F.; Yang, S.; Xiao, X.; Chen, L.; Wang, Y.; Li, Y.; Tao, E. Bovine hide collagen/tannin extract composite: A revelation of the selective structure-activity relationship between phenolic hydroxyls and Cu(II) and study on adsorption properties. *Colloids Surf. A Physicochem. Eng. Asp.* **2024**, *682*, 132886. [CrossRef]
33. Reyes-Ruiz, I.; Castañeda-Calzoncit, C.E.; Claudio Rizo, J.A.; Flores-Guía, T.E.; Cabrera-Munguía, D.A.; Cano-Salazar, L.F.; Becerra-Rodriguez, J.J. Evaluation of Collagen-Polyurethane-Chitosan Hydrogels for Lead Ions Removal from Water. *Mediterr. J. Basic Appl. Sci.* **2020**, *4*, 93–104.
34. Claudio-Rizo, J.A.; Burciaga-Montemayor, N.G.; Cano-Salazar, L.F.; Flores-Guía, T.E.; Cabrera-Munguía, D.A.; Herrera-Guerrero, A.; Soriano-Corral, F. Novel Collagen-Chitosan Based Hydrogels Reinforced with Manganite as Potential Adsorbents of Pb^{2+} Ions. *J. Polym. Environ.* **2020**, *28*, 2864–2879. [CrossRef]
35. Ye, X.; Lin, H.; Chi, R.; Guo, Z.; Lv, Y.; Lin, C.; Liu, Y.; Luo, W. Effectively remove p-arsanilic acid from water over amphiphilic amino modified collagen fiber. *Chemosphere* **2022**, *288*, 132542. [CrossRef]
36. Kaur, S.; Jindal, R. Synthesis of interpenetrating network hydrogel from (gum copal alcohols-collagen)-co-poly(acrylamide) and acrylic acid: Isotherms and kinetics study for removal of methylene blue dye from aqueous solution. *Mater. Chem. Phys.* **2018**, *220*, 75–86. [CrossRef]
37. Sitab, A.A.; Tujjohra, F.; Rashid, T.U.; Rahman, M.M. Thermally crosslinked electrospun nanofibrous mat from chrome-tanned solid wastes for cationic dye adsorption in wastewater treatment. *Clean. Eng. Technol.* **2023**, *13*, 100621. [CrossRef]
38. Luo, W.; Chi, R.; Zeng, F.; Wu, Y.; Chen, Y.; Liu, S.; Lin, W.; Lin, H.; Ye, X.; Chen, J. Multilayer Structure Ammoniated Collagen Fibers for Fast Adsorption of Anionic Dyes. *ACS Omega* **2021**, *6*, 27070–27079. [CrossRef] [PubMed]
39. Nakhjiri, M.T.; Marandi, G.B.; Kurdtabar, M. Adsorption of Methylene Blue, Brilliant Green and Rhodamine B from Aqueous Solution Using Collagen-g-p(AA-co-NVP)/Fe_3O_4@SiO_2 Nanocomposite Hydrogel. *J. Polym.Environ.* **2019**, *27*, 581–599. [CrossRef]
40. Kurdtabar, M.; Peyvand Kermani, Z.; Bagheri Marandi, G. Synthesis and characterization of collagen-based hydrogel nanocomposites for adsorption of Cd^{2+}, Pb^{2+}, methylene green and crystal violet. *Iran Polym. J.* **2015**, *24*, 791–803. [CrossRef]
41. Qiang, T.; Zhu, R. Bio-templated synthesis of porous silica nano adsorbents to wastewater treatment inspired by a circular economy. *Sci. Total Environ.* **2022**, *819*, 152929. [CrossRef]

42. Roy, C.; Chowdhury, D.; Sanfui, M.D.H.; Roy, J.S.D.; Mitra, M.; Dutta, A.; Chattopadhyay, P.K.; Singha, N.R. Solid waste collagen-associated fabrication of magnetic hematite nanoparticle@collagen nanobiocomposite for emission-adsorption of dyes. *Int. J. Biol. Macromol.* **2023**, *242*, 124774. [CrossRef]
43. Liao, X.; Lu, Z.; Zhang, M.; Liu, X.; Shi, B. Adsorption of Cu(II) from aqueous solutions by tannins immobilized on collagen. *J. Chem. Technol. Biotechnol.* **2004**, *79*, 335–342. [CrossRef]
44. Wang, R.; Liao, X.; Shi, B. Adsorption behaviors of Pt(II) and Pd(II) on collagen fiber immobilized bayberry tannin. *Ind. Eng. Chem. Res.* **2005**, *44*, 4221–4226. [CrossRef]
45. Kaur, S.; Jindal, R. Exploring the heavy metal ion sequestration ability of gum copal-collagen hybrid based interpenetrating polymer network: Kinetics, isotherms, and biodegradation studies. *SPE Polym.* **2020**, *1*, 26–35. [CrossRef]
46. Liao, X.; Zhang, M.; Shi, B. Collagen-Fiber-Immobilized Tannins and Their Adsorption of Au(III). *Ind. Eng. Chem. Res.* **2004**, *43*, 2222–2227. [CrossRef]
47. Huang, X.; Liao, X.; Shi, B. Hg(II) removal from aqueous solution by bayberry tannin-immobilized collagen fiber. *J. Hazard. Mater.* **2009**, *170*, 1141–1148. [CrossRef] [PubMed]
48. Jing, L.; Ding, Q.; Li, X.; Lou, J.; Liu, Z.; Jiang, Y.; Han, W.; Cheng, Z. Bifunctional collagen fiber/carbon quantum dot fluorescent adsorbent for efficient adsorption and detection of Pb^{2+}. *Sci. Total Environ.* **2023**, *871*, 161989. [CrossRef] [PubMed]
49. Kim, H.U.; Kim, K.H.; Chang, Y.Y.; Lee, S.M.; Yang, J.K. Heavy metal removal from aqueous solution by tannins immobilized on collagen. *Desalin. Water Treat.* **2012**, *48*, 1–8. [CrossRef]
50. Liao, X.P.; Tang, W.; Zhou, R.Q.; Shi, B. Adsorption of metal anions of vanadium(V) and chromium(VI) on Zr(IV)-impregnated collagen fiber. *Adsorption* **2008**, *14*, 55–64. [CrossRef]
51. Qiang, T.; Bu, Q.; Ren, L.; Wang, X. Adsorption Behaviors of Cr(III) on Carboxylated Collagen Fiber. *J. Appl. Polym. Sci.* **2014**, *131*, 40285. [CrossRef]
52. Shourije, S.M.J.S.; Dehghan, P.; Bahrololoom, M.E.; Cobley, A.J.; Vitry, V.; Azar, G.T.P.; Kamyab, H.; Mesbah, M. Using fish scales as a new biosorbent for adsorption of nickel and copper ions from wastewater and investigating the effects of electric and magnetic fields on the adsorption process. *Chemosphere* **2023**, *317*, 137829. [CrossRef] [PubMed]
53. Sreeram, K.J.; Saravanabhavan, S.; Rao, J.R.; Nair, B.U. Use of Chromium-Collagen Wastes for the Removal of Tannins from Wastewaters. *Ind. Eng. Chem. Res.* **2004**, *43*, 5310–5317. [CrossRef]
54. Sun, X.; Huang, X.; Liao, X.P.; Shi, B. Adsorptive removal of Cu(II) from aqueous solutions using collagen-tannin resin. *J. Hazard. Mater.* **2011**, *186*, 1058–1063. [CrossRef]
55. Wang, J.; Wei, L.; Ma, Y.; Li, K.; Li, M.; Yu, Y.; Wang, L.; Qiu, H. Collagen/cellulose hydrogel beads reconstituted from ionic liquid solution for Cu(II) adsorption. *Carbohydr. Polym.* **2013**, *98*, 736–743. [CrossRef]
56. Jing, L.; Yang, S.; Li, X.; Jiang, Y.; Lou, J.; Liu, Z.; Ding, Q.; Han, W. Effective adsorption and sensitive detection of Cr^{6+} by degradable collagen-based porous fluorescent aerogel. *Ind. Crops Prod.* **2022**, *182*, 114882. [CrossRef]
57. Ye, X.; Zheng, Z.; Zhu, J.; Chen, J.; Zhou, J.; Shi, B. ZIF-7-NH_2 functionalized collagen fibers for effective and selective mercury ion capture. *Sep. Purif. Technol.* **2025**, *353*, 128354. [CrossRef]
58. Zhao, F.; Yang, S.; Wang, Y.; Peng, C.; Chen, L.; Li, Y.; Tao, E. Synthesis of novel collagen-based aerogel with slit-shaped pore structure: Study on its adsorption mechanism on copper ions. *Sep. Purif. Technol.* **2025**, *353*, 128456. [CrossRef]
59. Liu, F.; Wang, Z.; Li, G. Adsorption of Ag+ by persimmon tannins immobilized on collagen fiber. *Desalin. Water Treat.* **2014**, *52*, 7172–7179. [CrossRef]
60. Huang, X.; Jiao, L.; Liao, X.; Shi, B. Adsorptive removal of As(III) from aqueous solution by Zr(IV)-loaded collagen fiber. *Ind. Eng. Chem. Res.* **2008**, *47*, 5623–5628. [CrossRef]
61. Oliveira, L.; Goncalves, M.; Oliveira, D.; Guerreiro, M.C.; Guilherme, L.; Dallago, R.M. Solid waste from leather industry as adsorbent of organic dyes in aqueous medium. *J. Hazard. Mater.* **2007**, *141*, 344–347. [CrossRef] [PubMed]
62. Ooi, J.; Lee, L.Y.; Hiew, B.Y.Z.; Thangalazhy-Gopakumar, S.; Lim, S.S.; Gan, S. Assessment of fish scales waste as a low cost and eco-friendly adsorbent for removal of an azo dye: Equilibrium, kinetic and thermodynamic studies. *Bioresour. Technol.* **2017**, *245A*, 656–664. [CrossRef]
63. Pinheiro, N.S.C.; Perez-Lopez, O.W.; Gutterres, M. Solid leather wastes as adsorbents for cationic and anionic dye removal. *Environ. Technol.* **2020**, *2*, 1–12.
64. Qiang, T.; Luo, M.Q.; Bu, Q.; Wang, X.C. The preparation of aminated collagen fiber and its study on adsorption of dyes. *Adv. Mater. Res.* **2012**, *347–353*, 1259–1263. [CrossRef]
65. Vieira, E.F.S.; Cestari, A.R.; Carvalho, W.A.; dos Oliveira, S.C.; Chagas, R.A. The use of freshwater fish scale of the species *Leporinus elongatus* as adsorbent for anionic dyes. *J. Therm. Anal. Calorim.* **2012**, *109*, 1407–1412. [CrossRef]
66. Chowdhury, S.; Das Saha, P.; Ghosh, U.K. Fish (*Labeo rohita*) scales as potential low-cost biosorbent for removal of malachite green from aqueous solutions. *Biorem. J.* **2012**, *16*, 235–242. [CrossRef]
67. Gu, Y.C.; Liao, X.P.; Huang, Y.J.; Shi, B. Adsorption of anionic dyes on Fe(III)-loaded collagen fibre from aqueous solution. *Int. J. Environ. Pollut.* **2008**, *34*, 111–121. [CrossRef]
68. Qiang, T.; Luo, M.; Bu, Q.; Wang, X. Adsorption of an acid dye on hyperbranched aminated collagen fibers. *Chem. Eng. J.* **2012**, *197*, 343–349. [CrossRef]
69. Wang, X.; Bu, Q.; Qiang, T.; Min, L. Study on the adsorption behavior of acid dyes on collagen fiber. *Mater. Sci. Forum* **2013**, *744*, 759–767. [CrossRef]

70. Han, Y.-X.; Li, J.; He, B.; Li, L. Preparation and characterization of a novel ACFTpPa-1 composite for dye adsorption. *J. Eng. Fibers Fabr.* **2021**, *16*, 3.
71. Tang, Y.; Zhao, J.; Zhang, Y.; Zhou, J.; Shi, B. Conversion of tannery solid waste to an adsorbent for high-efficiency dye removal from tannery wastewater: A road to circular utilization. *Chemosphere* **2021**, *263*, 127987. [CrossRef]
72. Azadi, S.; Esmkhani, M.; Javanshir, S. New collagen-based cryogel as bio-sorbent materials for Rhodamine B removal from aqueous environments. *J. Sol Gel Sci. Technol.* **2022**, *103*, 405–415. [CrossRef]
73. Chowdhury, S.; Showrov, S.I.; Santo, H.R.; Uddin, E. A breakthrough approach with collagen-graphene oxide nanocomposite: Sustainable solution for dye wastewater. In Proceedings of the 7th International Conference on Civil Engineering for Sustainable Development, Khulna, Bangladesh, 7–9 February 2024.
74. Karmakar, M.; Mahapatra, M.; Dutta, A.; Chattopadhyay, P.K.; Singha, N.R. Fabrication of semisynthetic collagenic materials for mere/synergistic adsorption: A model approach of determining dye allocation by systematic characterization and optimization. *Int. J. Biol. Macromol.* **2017**, *102*, 438–456. [CrossRef] [PubMed]
75. Li, A.; Liu, J.; Qin, Z.; Wang, L.; Li, L.; Tang, K.; Pei, Y. Black wattle tannin immobilized mesostructured collagen as a promising adsorbent for cationic organic dyes (methylene blue) removal in batch and continuous fixed-bed systems. *J. Appl. Polym. Sci.* **2022**, *139*, e52452. [CrossRef]
76. Geng, K.; He, T.; Liu, R.; Dalapati, S.; Tan, K.T.; Li, Z.; Tao, S.; Gong, Y.; Jiang, Q.; Jiang, D. Covalent Organic Frameworks: Design, Synthesis, and Functions. *Chem. Rev.* **2020**, *120*, 8814–8933. [CrossRef]
77. Du, W.; Han, X.; Li, Z.; Li, Y.; Li, L.; Wang, K. Oil sorption behaviors of porous polydimethylsiloxane modified collagen fiber matrix. *J. Appl. Polym. Sci.* **2015**, *132*, 42727. [CrossRef]
78. Du, W.; Dai, G.; Wang, B.; Li, Z.; Li, L. Biodegradable porous organosilicone modified collagen fiber matrix: Synthesis and high oil absorbency. *J. Appl. Polym. Sci.* **2018**, *135*, 46264. [CrossRef]
79. Sharma, H.; Kaith, B.S.; Kumar, R.; Mehra, R.; Bhatti, M.S. Fabrication of cellulose-collagen based biosorbent as eco-friendly scavengers for uranyl ions. *Int. J. Biol. Macromol.* **2024**, *266*, 131400. [CrossRef]
80. Tang, Y.; Zhou, J.; Guo, J.; Liao, X.; Shi, B. Irradiation-stable hydrous titanium oxide-immobilized collagen fibers for uranium removal from radioactive wastewater. *J. Environ. Manag.* **2021**, *283*, 112001. [CrossRef] [PubMed]
81. Yu, J.; Yu, C.; Zhu, W.; He, G.; Wei, Y.; Zhou, J. Hydrous titanium oxide and bayberry tannin co-immobilized nano collagen fibrils for uranium extraction from seawater and recovery from nuclear wastewater. *Chemosphere* **2022**, *286*, 131626. [CrossRef]
82. Liao, X.; Lu, Z.; Du, X.; Liu, X.; Shi, B. Collagen Fiber Immobilized *Myrica rubra* Tannin and Its Adsorption to $UO_2{}^{2+}$. *Environ. Sci. Technol.* **2004**, *38*, 324–328. [CrossRef]
83. Tang, Y.; Zhou, J.; Luo, W.; Liao, X.; Shi, B. In situ chemical oxidation-grafted amidoxime-based collagen fibers for rapid uranium extraction from radioactive wastewater. *Sep. Purif. Technol.* **2023**, *307*, 122826. [CrossRef]
84. Yang, L.; Zeng, X.; Tang, J.H.; Sun, H.Y.; Liu, J.T.; Yan-Ling Guo, Y.L.; Wei-Zhen Liu, W.Z.; Feng, M.L.; Huang, X.Y. Rapid and selective uranium adsorption by a low-cost, eco-friendly, and in-situ prepared nano-ZnS/alkali-activated collagen fiber composite. *Sep. Purif. Technol.* **2024**, *333*, 125856. [CrossRef]
85. Peng, L.; Zhang, X.; Guo, L.; Li, J.; Zhang, W.; Shi, B. Rapid removal and easy separation recovery of Cs^+ and Sr^{2+} by zirconium molybdopyrophosphate-functionalized collagen fibers. *J. Clean. Prod.* **2023**, *414*, 137653. [CrossRef]
86. Liao, H.; Yu, J.; Zhu, W.; Kuang, M.; Duan, T.; Zhang, Y.; Lin, X.; Luo, X.; Zhou, J. Nano-zero-valent Fe/Ni particles loaded on collagen fibers immobilized by bayberry tannin as an effective reductant for uranyl in aqueous solutions. *Appl. Surf. Sci.* **2020**, *507*, 145075. [CrossRef]
87. Caldera-Villalobos, M.; Cabrera-Munguía, D.A.; Flores-Guía, T.E.; Viramontes-Gamboa, G.; Vargas-Correa, J.A.; Cano-Salazar, L.F.; Claudio-Rizo, J.A. Removal of water pollutants using composite hydrogels comprised of collagen, guar gum, and metal-organic frameworks. *J. Polym. Res.* **2021**, *28*, 395. [CrossRef]
88. Peng, L.; Guo, L.; Li, J.; Zhang, W.; Shi, B.; Liao, X. Rapid and highly selective removal of cesium by Prussian blue analog anchored on porous collagen fibers. *Sep. Purif. Technol.* **2023**, *307*, 122858. [CrossRef]
89. Zeng, Y.; Liao, X.; He, Q.; Shi, B. Recovery of Th(IV) from aqueous solution by reassembled collagen-tannin fiber adsorbent. *J. Radioanal. Nucl. Chem.* **2009**, *280*, 91–98. [CrossRef]
90. Liao, X.P.; Ding, Y.; Wang, B.; Shi, B. Adsorption Behavior of Phosphate on Metal-Ions-Loaded Collagen Fiber. *Ind. Eng. Chem. Res.* **2006**, *45*, 3896–3901. [CrossRef]
91. Liao, X.-P.; Shi, B. Adsorption of fluoride on zirconium(IV)-impregnated collagen fiber. *Environ. Sci. Technol.* **2005**, *39*, 4628–4632. [CrossRef]
92. Fiol, N.; Villaescusa, I.; Martínez, M.; Miralles, N.; Poch, J.; Serarols, J. Sorption of Pb(II), Ni(II), Cu(II) and Cd(II) from aqueous solution by olive stone waste. *Sep. Purif. Technol.* **2006**, *50*, 132–140. [CrossRef]
93. Anandkumar, J.; Mandal, B. Removal of Cr (VI) from aqueous solution using Bael fruit (*Aegle marmelos* Correa) shell as an adsorbent. *J. Hazard. Mater.* **2009**, *168*, 633–640. [CrossRef]
94. Park, J.-H.; Ok, Y.S.; Kim, S.-H.; Cho, J.-S.; Heo, J.-S.; Delaune, R.D.; Seo, D.-C. Competitive adsorption of heavy metals onto sesame straw biochar in aqueous solutions. *Chemosphere* **2016**, *142*, 77–83. [CrossRef]
95. Yap, M.W.; Mubarak, N.M.; Sahu, J.N.; Abdullah, E.C. Microwave induced synthesis of magnetic biochar from agricultural biomass for removal of lead and cadmium from wastewater. *J. Ind. Eng. Chem.* **2017**, *45*, 287–295. [CrossRef]

96. El-Shafie, A.S.; Rahman, E.; GadelHak, Y.; Mahmoud, R.; El-Azazy, M. Techno-economic assessment of waste mandarin biochar as a green adsorbent for binary dye wastewater effluents of methylene blue and basic fuchsin: Lab- and large-scale investigations. *Spectrochim. Acta A Mol. Biomol. Spectrosc.* **2024**, *306*, 123621. [CrossRef]
97. Bastami, T.R.; Khaknahad, S.; Malekshahi, M. Sonochemical versus reverse-precipitation synthesis of Cu_xO/Fe_2O_3/MoC nano-hybrid: Removal of reactive dyes and evaluation of smartphone for colorimetric detection of organic dyes in water media. *Environ. Sci. Pollut. Res.* **2020**, *27*, 9364–9381. [CrossRef]
98. Hu, H.; Gao, M.; Wang, T.; Jiang, L. Efficient uranium adsorption and mineralization recycle by nano-MgO biochar with super-hydrophilic surface. *J. Environ. Chem. Eng.* **2023**, *11*, 110542. [CrossRef]
99. Radetic, M.; Ilic, V.; Radojevic, D.; Miladinovic, R.; Jocic, D.; Jovancic, P. Efficiency of recycled wool-based nonwoven material for the removal of oils from water. *Chemosphere* **2008**, *70*, 525–530. [CrossRef] [PubMed]
100. Wintgens, T.; Melin, T. Removing endocrine-disrupting compounds from wastewater. *Membr. Technol.* **2001**, *2001*, 11–12. [CrossRef]
101. Li, J.; Zhu, Y.; Zhou, Y.; Zhu, J.; Chen, G. Fe^{3+}-modified polyamide membrane with a porous network structure: Rapid permeation and efficient interception of trace organic compounds. *Sep. Purif. Technol.* **2025**, *353*, 128294. [CrossRef]
102. Sadare, O.O.; Oke, D.; Olawuni, O.A.; Olayiwola, I.A.; Moothi, K. Modelling and optimization of membrane process for removal of biologics (pathogens) from water and wastewater: Current perspectives and challenges. *Heliyon* **2024**, *10*, e29864. [CrossRef]
103. Oluwoye, I.; Tanaka, S.; Okuda, K. Pilot-scale performance of gravity-driven ultra-high flux fabric membrane systems for removing small-sized microplastics in wastewater treatment plant effluents. *J. Environ. Manag.* **2024**, *363*, 121438. [CrossRef]
104. Hube, S.; Veronelli, S.; Li, T.; Burkhardt, M.; Brynjolfsson, S.; Wu, B. Microplastics affect membrane biofouling and microbial communities during gravity-driven membrane filtration of primary wastewater. *Chemosphere* **2024**, *353*, 141650. [CrossRef]
105. Liu, Y.; Liu, Y.; Wang, H.; Chi, M.; Matsuyama, H.; Guo, Y.; Li, Y.; Liu, R.; Wouters, C.; Chen, Y.; et al. One-pot fabrication of highly permeable P84/UiO-66-NH_2-PEI membrane for the efficient removal of heavy metals from wastewater. *Sep. Purif. Technol.* **2025**, *354*, 129022. [CrossRef]
106. Padrón-Páez, J.I.; De-León Almaraz, S.; Román-Martínez, A. Sustainable wastewater treatment plants design through multiobjective optimization. *Comput. Chem. Eng.* **2020**, *140*, 106850. [CrossRef]
107. Ye, X.; Zheng, Z.; Chi, R.; Liu, J.; Chen, J.; Luo, W. Waste for Waste: Interface-Intensified Durable Superhydrophilic−Superoleophobic Collagen Fiber Membrane for Efficient Separation of Cationic Surfactant-Stabilized Oil-in-Water Emulsions. *Langmuir* **2023**, *39*, 18815–18824. [CrossRef]
108. Ye, X.; Lin, X.; Chi, R.; Liu, J.; Huang, J.; Chen, J. Aminated superhydrophilic/underwater superoleophobic collagen fiber membrane for efficient separation of anionic surfactant-stabilized oil-in-water emulsions. *Process Saf. Environ. Prot.* **2024**, *183*, 1186–1197. [CrossRef]
109. Ye, X.; Huang, R.; Zheng, Z.; Juan Liu, J.; Chen, J.; Lv, Y. Anti-abrasion collagen fiber-based membrane functionalized by UiO-66-NH_2 with ultra-high efficiency and stability for oil-in-water emulsions separation. *Chin. J. Chem. Eng.* **2024**, *66*, 285–297. [CrossRef]
110. Obotey Ezugbe, E.; Rathilal, S. Membrane Technologies in Wastewater Treatment: A Review. *Membranes* **2020**, *10*, 89. [CrossRef]
111. Alomar, T.; Hameed, B.H.; Al-Ghouti, M.A.; Almomani, F.A.; Han, D.S. A review on recent developments and future prospects in the treatment of oily petroleum refinery wastewater by adsorption. *J. Water Process Eng.* **2024**, *64*, 105616. [CrossRef]
112. Singha, W.J.; Deka, H. Ecological and human health risk associated with heavy metals (HMs) contaminant sourced from petroleum refinery oily sludge. *J. Hazard. Mater.* **2024**, *476*, 135077. [CrossRef] [PubMed]
113. Chen, G.; Hao, B.; Wang, Y.; Wang, Y.; Xiao, H.; Li, H.; Huang, X.; Shi, B. Insights into Regional Wetting Behaviors of Amphiphilic Collagen for Dual Separation of Emulsions. *ACS Appl. Mater. Interfaces* **2021**, *13*, 18209–18217. [CrossRef] [PubMed]
114. Ye, X.; Xiao, H.; Wang, Y.; Ke, L.; Luo, W.; Huang, X.; Shi, B. Efficient separation of viscous emulsion through amphiprotic collagen nanofibers-based membrane. *J. Membr. Sci.* **2019**, *588*, 117209. [CrossRef]
115. Xiao, H.; Cui, Y.; Wang, Y.; Li, H.; Chen, G.; Huang, X.; Shi, B. Synergistic combination of the capillary effect of collagen fibers and size-sieving merits of metal–organic frameworks for emulsion separation with high flux. *Ind. Eng. Chem. Res.* **2020**, *59*, 14925–14934. [CrossRef]
116. Li, H.; Zheng, W.; Xiao, H.; Hao, B.; Wang, Y.; Huang, X.; Shi, B. Collagen fiber membrane-derived chemically and mechanically durable superhydrophobic membrane for high-performance emulsion separation. *J. Leather Sci. Eng.* **2021**, *3*, 20. [CrossRef]
117. Cao, Y.; Xiao, H.; Wan Zheng, W.; Wang, Y.; Wang, Y.; Huang, X.; Shi, B. Green synthesis of environmentally benign collagen fibers-derived hierarchically structured amphiphilic composite fibers for high-flux dual separation of emulsion. *J. Environ. Chem. Eng.* **2022**, *10*, 107067. [CrossRef]
118. Lu, X.; Shen, L.; Chen, C.; Yu, W.; Wang, B.; Kong, N.; Zeng, Q.; Chen, S.; Huang, X.; Wang, Y.; et al. Advance of self-cleaning separation membranes for oil-containing wastewater treatment. *Environ. Func. Mater.* **2024**, *in press*. [CrossRef]
119. Cui, Y.; Wang, Y.; Hao, B.; Xiao, H.; Huang, X.; Shi, B. Water-oil dual-channels enabled exceptional anti-fouling performances for separation of emulsified oil pollutant. *J. Hazard. Mater.* **2023**, *449*, 131012. [CrossRef] [PubMed]
120. Cui, Y.; Zheng, W.; Pu, H.; Xiong, J.; Liu, H.; Shi, Y.; Huang, X. Intertwisted superhydrophilic and superhydrophobic collagen fibers enabled anti-fouling high-performance separation of emulsion wastewater. *J. Hazard. Mater.* **2024**, *473*, 134653. [CrossRef] [PubMed]

121. Ye, X.; Wang, Y.; Ke, L.; Cui, Y.; Luo, W.; Wang, X.; Huang, X.; Shi, B. Competitive adsorption for simultaneous removal of emulsified water and surfactants from mixed surfactant-stabilized emulsions with high flux. *J. Mater. Chem. A* **2018**, *6*, 14058–14064. [CrossRef]
122. Chen, G.; Wang, Y.; Wang, Y.; Xiao, H.; Huang, X.; Shi, B. Collagen fibers with tuned wetting properties for dual separation of oil-in-water and water-in-oil emulsion. *J. Mater. Chem. A* **2020**, *8*, 24388–24392. [CrossRef]
123. Xiao, P.; Li, X.; Li, L.; Liu, F.; Wang, Y.; Li, J. Bio-functional collagen/casein/chitosan scaffolds regulated porous TFC membrane for acid recovery. *Desalination* **2023**, *558*, 116627. [CrossRef]
124. Matebese, F.; Mosai, A.K.; Tutu, H.; Tshentu, Z.R. Mining wastewater treatment technologies and resource recovery techniques: A review. *Heliyon* **2024**, *10*, e24730. [CrossRef]
125. Rawat, A.; Srivastava, A.; Bhatnagar, A.; Gupta, A.K. Technological advancements for the treatment of steel industry wastewater: Effluent management and sustainable treatment strategies. *J. Clean. Prod.* **2023**, *383*, 135382. [CrossRef]
126. Dai, G.H.; Zhang, Z.; Du, W.; Li, Z.; Gao, W.; Li, L. Conversion of skin collagen fibrous material waste to an oil sorbent with pH-responsive switchable wettability for high-efficiency separation of oil/water emulsions. *J. Clean. Prod.* **2019**, *226*, 18–27. [CrossRef]
127. Thanikaivelan, P.; Narayanan, N.T.; Pradhan, B.K.; Ajayan, P.M. Collagen based magnetic nanocomposites for oil removal applications. *Sci. Rep.* **2012**, *2*, 230. [CrossRef]
128. Chemical Oxidation Techniques. Available online: https://emis.vito.be/en/bat/tools-overview/sheets/chemical-oxidation-techniques (accessed on 10 October 2024).
129. Iyyappan, J.; Gaddala, B.; Gnanasekaran, R.; Gopinath, M.; Yuvaraj, D.; Kumar, V. Critical review on wastewater treatment using photo catalytic advanced oxidation process: Role of photocatalytic materials, reactor design and kinetics. *CSCEE* **2024**, *9*, 100599. [CrossRef]
130. Assanvo, E.F.; Nagaraj, S.; Boa, D.; Thanikaivelan, P. Hybrid collagen–cellulose–Fe_3O_4@TiO_2 magnetic bio-sponges derived from animal skin waste and Kenaf fibers for wastewater remediation. *Nat. Sci. Rep.* **2023**, *13*, 13365. [CrossRef] [PubMed]
131. Deng, D.; Tang, R.; Liao, X.; Shi, B. Using collagen fiber as a template to synthesize hierarchical mesoporous alumina fiber. *Langmuir* **2008**, *24*, 368–370. [CrossRef] [PubMed]
132. Tang, R.; Liao, X.P.; Liu, X.; Shi, B. Collagen fiber immobilized Fe(III): A novel catalyst for photo-assisted degradation of dyes. *Chem. Commun.* **2005**, *47*, 5882–5884. [CrossRef] [PubMed]
133. Liu, X.; Tang, R.; He, Q.; Liao, X.; Shi, B. Fe(III)-loaded collagen fiber as a heterogeneous catalyst for the photo-assisted decomposition of Malachite Green. *J. Hazard. Mater.* **2010**, *174*, 687–693. [CrossRef]
134. Nagaraj, S.; Cheirmadurai, K.; Thanikaivelan, P. Visible-light active collagen-TiO_2 nanobio-sponge for water remediation: A sustainable approach. *Clean. Mater.* **2021**, *1*, 100011. [CrossRef]
135. Zhou, H.; Zhou, J.; Wang, T.; Zeng, J.; Liu, L.; Jian, J.; Zhou, Z.; Zeng, L.; Liu, Q.; Liu, G. In-situ preparation of silver salts/collagen fiber hybrid composites and their photocatalytic and antibacterial activities. *J. Hazard. Mater.* **2018**, *359*, 274–280. [CrossRef]
136. Liu, X.; Tang, R.; He, Q.; Liao, X.; Shi, B. Fe(III)-Immobilized Collagen Fiber: A Renewable Heterogeneous Catalyst for the Photoassisted Decomposition of Orange II. *Ind. Eng. Chem. Res.* **2009**, *48*, 1458–1463. [CrossRef]
137. Alatawi, N.M.; Ben Saad, L.; Soltane, L.; Moulahi, A.; Mjejri, I.; Sediri, F. Enhanced solar photocatalytic performance of Cu-doped nanosized ZnO. *Polyhedron* **2021**, *197*, 115022. [CrossRef]
138. Singh, A.; Singh, D.; Ahmed, B.; Ojha, A.K. Sun/UV-light driven photocatalytic degradation of rhodamine B dye by Zn doped CdS nanostructures as photocatalyst. *Mater. Chem. Phys.* **2022**, *277*, 125531. [CrossRef]
139. Zhang, Z.; Liu, N.; Wu, J.; Liu, J.; Wang, H.; Pang, D.; Zheng, J. Constructing p-n heterojunction in CoO@TiO_2 photocatalytic material to enhance the performance of catalytic degradation of Rhodamine, B. *Chem. Phys. Lett.* **2024**, *855*, 141568. [CrossRef]
140. Mahmoodi, V.; Ahmadpour, A.; Bastami, T.R.; Mousavian, M.T.H. Facile Synthesis of BiOI Nanoparticles at Room Temperature and Evaluation of their Photoactivity Under Sunlight Irradiation. *Photochem. Photobiol.* **2018**, *94*, 4–16. [CrossRef]
141. Wang, K.; Yan, D.; Chen, X.; Xu, Z.; Cao, W.; Li, H. New insight to the enriched microorganisms driven by pollutant concentrations and types for industrial and domestic wastewater via distinguishing the municipal wastewater treatment plants. *Environ. Pollut.* **2024**, *361*, 124789. [CrossRef]
142. Kaur, B.; Choudhary, R.; Sharma, G.; Kaur Brar, L. Sustainable and effective microorganisms method for wastewater treatment. *Desalin. Water Treat.* **2024**, *319*, 100419. [CrossRef]
143. Rendón-Castrillón, L.; Ramírez-Carmona, M.; Ocampo-López, C.; González-López, F.; Cuartas-Uribe, B.; Mendoza-Roca, J.A. Efficient bioremediation of indigo-dye contaminated textile wastewater using native microorganisms and combined bioaugmentation-biostimulation techniques. *Chemosphere* **2024**, *353*, 141538. [CrossRef] [PubMed]
144. Zhan, J.; Xu, S.; Zhu, Y.; Han, Y.; Li, L.; Liu, J.; Guo, X. Potential pathogenic microorganisms in rural wastewater treatment process: Succession characteristics, concentration variation, source exploration, and risk assessment. *Water Res.* **2024**, *254*, 121359. [CrossRef] [PubMed]
145. Ahmed, M.; Kumar Verma, A.; Patel, R. Collagen extraction and recent biological activities of collagen peptides derived from sea-food waste: A review. *Sustain. Chem. Pharm.* **2020**, *18*, 100315. [CrossRef]
146. Cavalcante Loureiro, K.; Carvalho Barbosa, T.; Nery, M.; Vinicius Chaud, M.; Ferreira da Silva, C.; Nalone Andrade, L.; Bani Corrêa, C.; Jaguer, A.; Ferreira Padilha, F.; Cordeiro Cardoso, J.; et al. Antibacterial activity of chitosan/collagen membranes containing red propolis extract. *Pharmazie* **2020**, *75*, 75–81.

147. Xu, Z.; Chang, J.; Zhang, P.; Guan, X.; Chen, Y.; Fan, H. Collagen modified with epoxidized safrole for improving antibacterial activity. *RSC Adv.* **2017**, *7*, 50300. [CrossRef]
148. Zhou, T.; Sui, B.; Mo, X.; Sun, J. Multifunctional and biomimetic fish collagen/bioactive glass nanofibers: Fabrication, antibacterial activity and inducing skin regeneration in vitro and in vivo. *Int. J. Nanomed.* **2017**, *12*, 3495–3507. [CrossRef]
149. Sun, Z.; Fan, C.; Tang, X.; Zhao, J.; Song, Y.; Shao, Z.; Xu, L. Characterization and antibacterial properties of porous fibers containing silver ions. *Appl. Surf. Sci.* **2016**, *387*, 828–838. [CrossRef]
150. Subbiah, N.; Palanisamy, T. Collagen-Supported Amino-Functionalized Ag@SiO_2 Core−Shell Nanoparticles for Visible-Light-Driven Water Remediation. *ACS Appl. Nano Mater.* **2022**, *5*, 14408–14424. [CrossRef]
151. Zou, L.; Li, Z.; Sun, T.; Zhu, X.; Li, J.; Cui, Y.; Zhao, B.; Zhang, Q.; Song, Y. Inactivating microorganisms in water by using a novel Ag/CuS/carbon cloth composite for photo-electrocatalytic performance. *Chem. Phys. Impact* **2024**, *8*, 100426. [CrossRef]
152. Zhang, M.; Sun, T.; Wang, X.Y.; Xue, B. Photo-crosslinked composite hydrogels with silver-deposited polymeric carbon nitride for boosting antibacterial activity. *Colloids Surf. A Physicochem. Eng. Asp.* **2024**, *680*, 132668. [CrossRef]
153. Yu, R.; Wang, H.; Wang, R.; Zhao, P.; Chen, Y.; Liu, G.; Liao, X. Polyphenol modified natural collagen fibrous network towards sustainable and antibacterial microfiltration membrane for efficient water disinfection. *Water Res.* **2022**, *218*, 118469. [CrossRef]
154. Truchado, P.; Gil, M.I.; Lopez, C.; Garre, A.; Lopez-Aragon, R.F.; Bohme, K.; Allende, A. New standards at European Union level on water reuse for agricultural irrigation: Are the Spanish wastewater treatment plants ready to produce and distribute reclaimed water within the minimum quality requirements? *Int. J. Food Microbiol.* **2021**, *356*, 109352. [CrossRef]
155. Khosravi, Z.; Kharaziha, M.; Goli, R.; Karimzadeh, F. Antibacterial adhesive based on oxidized tannic acid-chitosan for rapid hemostasis. *Carbohydr. Polym.* **2024**, *333*, 121973. [CrossRef]
156. He, L.; Lan, W.; Cen, L.; Chen, S.; Liu, S.; Liu, Y.; Ao, X.; Yang, Y. Improving catalase stability by its immobilization on grass carp (*Ctenopharyngodon idella*) scale collagen self-assembly films. *Mater. Sci. Eng. C* **2019**, *105*, 110024. [CrossRef] [PubMed]
157. He, L.; Lan, W.; Zhao, Y.; Chen, S.; Liu, S.; Cen, L.; Cao, S.; Dong, L.; Jin, R.; Liu, Y. Characterization of biocompatible pig skin collagen and application of collagen-based films for enzyme immobilization. *RSC Adv.* **2020**, *10*, 7170. [CrossRef]
158. Lim, J.; Mohamad, Z. Enzymes immobilized polymeric supports for wastewater treatment application: A short review. *Mater. Today Proc.* **2022**, *65*, 2946–2952. [CrossRef]
159. Kaushal, J.; Mehandia, S.; Singh, G.; Raina, A.; Arya, S.K. Catalase enzyme: Application in bioremediation and food industry. *Biocatal. Agric. Biotechnol.* **2018**, *16*, 192–199. [CrossRef]
160. Abdalbagemohammedabdalsadeg, S.; Xiao, B.L.; Ma, X.X.; Li, Y.Y.; Wei, J.S.; Moosavi-Movahedi, A.A.; Yousefi, R.; Hong, J. Catalase immobilization: Current knowledge, key insights, applications, and future prospects—A review. *Int. J. Biol. Macromol.* **2024**, *276*, 133941. [CrossRef] [PubMed]
161. Gan, J.S.; Ashraf, S.S.; Muhammad Bilal, M.; Iqbal, H.M.N. Biodegradation of environmental pollutants using catalase-based biocatalytic systems. *Environ. Res.* **2022**, *214*, 113914. [CrossRef] [PubMed]
162. Song, N.; Chen, S.; Huang, X.; Liao, X.; Shi, B. Immobilization of catalase by using Zr(IV)-modified collagen fiber as the supporting matrix. *Process Biochem.* **2011**, *46*, 2187–2193. [CrossRef]
163. Bae, H.; Choi, M.; Lee, C.; Chung, Y.C.; Yo, Y.J.; Lee, S. Enrichment of ANAMMOX bacteria from conventional activated sludge entrapped in poly(vinyl alcohol)/sodium alginate gel. *Chem. Eng. J.* **2015**, *28*, 531–540. [CrossRef]
164. Quan, L.M.; Khanh, D.P.; Hira, D.; Fujii, T.; Furukawa, K. Reject water treatment by improvement of whole cell anammox entrapment using polyvinyl alcohol/alginate gel. *Biodegradation* **2011**, *22*, 1155–1167. [CrossRef]
165. Ye, X.; Liu, J.; Chen, X.; Chen, D.; Qian, Z.; Chen, J.; Lin, C. Effective uranium extraction from seawater through immobilization of conjugated microporous polymers on collagen fiber membrane. *Chem. Eng. Sci.* **2024**, *295*, 120160. [CrossRef]

polymers

MDPI

Article

Synthesis and Dye Adsorption Dynamics of Chitosan–Polyvinylpolypyrrolidone (PVPP) Composite

Hilda Dinah Kyomuhimbo [1], Wandile McHunu [1], Marco Arnold [1], Usisipho Feleni [2], Nils H. Haneklaus [3] and Hendrik Gideon Brink [1,*]

[1] Department of Chemical Engineering, University of Pretoria, Pretoria 0028, South Africa; u21830658@tuks.co.za (H.D.K.); u20623594@tuks.co.za (W.M.); u20648210@tuks.co.za (M.A.)
[2] Institute for Nanotechnology and Water Sustainability (iNanoWS), College of Science, Engineering and Technology, University of South Africa, Johannesburg 1709, South Africa; felenu@unisa.ac.za
[3] Td Lab Sustainable Mineral Resources, University for Continuing Education Krems, 3500 Krems an der Donau, Austria
* Correspondence: deon.brink@up.ac.za

Abstract: One major environmental issue responsible for water pollution is the presence of dyes in the aquatic environment as a result of human activity, particularly the textile industry. Chitosan–Polyvinylpolypyrrolidone (PVPP) polymer composite beads were synthesized and explored for the adsorption of dyes (Bismarck brown (BB), orange G (OG), brilliant blue G (BBG), and indigo carmine (IC)) from dye solution. The CS-PVPP beads demonstrated high removal efficiency of BB (87%), OG (58%), BBG (42%), and IC (49%). The beads demonstrated a reasonable surface area of 2.203 m^2/g and were negatively charged in the applicable operating pH ranges. TGA analysis showed that the polymer composite can withstand decomposition up to 400 °C, proving high stability in harsh conditions. FTIR analysis highlighted the presence of N-H amine, O-H alcohol, and S=O sulfo groups responsible for electrostatic interaction and hydrogen bonding with the dye molecules. A shift in the FTIR bands was observed on N-H and C-N stretching for the beads after dye adsorption, implying that adsorption was facilitated by hydrogen bonding and Van der Waals forces of attraction between the hydroxyl, amine, and carbonyl groups on the surface of the beads and the dye molecules. An increase in pH increased the adsorption capacity of the beads for BB while decreasing OG, BBG, and IC due to their cationic and anionic nature, respectively. While an increase in temperature did not affect the adsorption capacity of OG and BBG, it significantly improved the removal of BB and IC from the dye solution and the adsorption was thermodynamically favoured, as demonstrated by the negative Gibbs free energy at all temperatures. Adsorption of dye mixtures followed the characteristic adsorption nature of the individual dyes. The beads show great potential for applications in the treatment of dye wastewater.

Keywords: chitosan; polyvinylpolypyrrolidone; dyes; adsorption

1. Introduction

Environmental pollution, particularly from the textile and paper industries, is a major concern due to the discharge of synthetic organic dyes in industrial effluents, which can contain concentrations of hazardous dyes of up to 10% residual dyes [1,2]. The discharge of synthetic organic dyes from industrial effluents poses significant dangers to water bodies, aquatic life, and humans alike [3]. These dyes, along with other chemicals present in textile dyeing effluents, can severely degrade water quality and disrupt aquatic ecosystems, reducing oxygen levels, blocking sunlight penetration, and altering the chemical composition of the water [4–6]. This creates inhospitable conditions for aquatic organisms, leading to reduced biodiversity and increased mortality among species [7]. Moreover, the discharge of untreated or inadequately treated effluents into water bodies can impact human health, leading to skin irritations, respiratory problems, and gastrointestinal illnesses in communities living near contaminated water sources [8–10].

Various physical, chemical, and biological approaches have been employed for the removal of dyes from industrial effluents, including coagulation and flocculation [11], adsorption [12,13], ion exchange [14], reverse osmosis [15], membrane filtration [16], biological treatment [17], and irradiation [18]. These methods offer efficient colour removal but are costly, generate concentrated sludge as secondary waste, and require high amounts of energy [19,20]. For instance, coagulation–flocculation using ferric chloride leads to the production of significant amounts of sludge [21]. Fenton reactions generate iron sludge as a by-product [22,23] and electrochemical oxidation consumes significant amounts of electricity [24]. Biological treatment methods encounter technical constraints such as sensitivity towards temperature and time constraints, and they often require large land space for optimal effects [25]. Adsorption using sustainable materials like chitosan [26], alginate [27], waste fruit peels [28], zirconium [29], and tannin composites offers a favourable approach to wastewater treatment [30] due to their flexibility, ease of operation, low energy consumption, cost-effectiveness, and insensitivity to toxic pollutants [31,32]. Moreover, adsorption does not involve the generation of harmful bi-products since the dyes simply transfer from the bulk solution to the solvent [25,33]. Different materials have been explored for dye removal from wastewater including activated carbon [34,35], biomass [36,37], clay [38,39], and polymers [40,41]. Polymers have gained attention in dye adsorption due to their high efficiency in surfactant adsorption, cost-effectiveness, biocompatibility, non-toxicity, and suitability for various applications [42,43].

Chitosan, a derivative of chitin found in the exoskeletons of crustaceans and cell walls of fungi, boasts versatile applications across industries owing to its biodegradability and biocompatibility [44,45]. Chitosan has several promising characteristics in wastewater treatment such as biodegradability, low toxicity, high adsorption capacity, and antimicrobial properties [46,47]. Its unique cationic character in the acid medium gives it a chelate effect on dyes and metal ions due to ion exchange and electrostatic attraction [26,48]. Chitosan has been combined with other polymers such as polyvinyl alcohol [49], alginate [50], and polyvinylpyrrolidone [51] to enhance its performance in water treatment, especially in dye removal. For instance, Khorshidi and Khalaji [52] reported 98% removal of eosin Y dye from an aqueous solution using chitosan–polyvinylpyrrolidone composite (CS-PVP) powder. The high dye removal was attributed to electrostatic attractions, hydrogen bonding, and n–π interactions between the composite and dye molecules.

Polyvinylpolypyrrolidone (PVPP), a synthetic polymer known for its cost-effectiveness, biocompatibility, and high hydrophilicity, is adept at adsorbing polyphenolic compounds from solutions [53]. It owes its high effectiveness to the strong electrostatic forces and formation of hydrogen bonds to form a phenol–PVPP complex [54]. In this study, we focused on the synthesis of chitosan–PVPP beads and their application in the removal of four representative dyes, that is, brilliant blue G (a triphenylmethane dye), indigo carmine (a disulfoninc acid sodium salt), Bismarck brown, and orange G (azo dyes). The effect of initial pH, temperature, contact time, and dye concentrations were investigated. Although PVPP has proved proficient in the removal of phenolic compounds, it has not been explored for the removal of dyes from aqueous solution. The beads formed in this study are porous with firm physical and chemical characteristics, demonstrate acceptable reusability, and can be used in various processes from batch to continuous in different reactor systems.

2. Materials and Methods

Chitosan (medium molecular weight, $C_{12}H_{24}N_2O_9$, 99.9%), PVPP (crystalline powder, circa 110 µm, $(C_6H_9NO)_x$, 99.99%), acetic acid (CH_3COOH, 99.9%), sodium hydroxide (NaOH, 98%), bismarck Brown ($((H_2N)_2C_6H_3N_2]_2C_6H_4$, 50%), orange G ($C_{16}H_{10}N_2Na_2O_7S_2$, 80%) indigo carmine ($C_{16}H_8N_2Na_2O_8S_2$, 80%), brilliant blue G ($C_{37}H_{34}N_2Na_2O_9S_3$, 99.9%), and congo red ($C_{32}H_{22}N_6Na_2O_6S_2$, 35%) were obtained from Sigma-Aldrich, Saint-Louis, MO, USA. All other reagents used in the experiment were also obtained from Sigma-

Aldrich, Johannesburg, South Africa, and used without any further purification. The structures and physiochemical properties of dyes and polymers used in this study are presented in Table 1.

Table 1. Physiochemical properties of dyes and polymers in this study.

Dye	Structure	Log P	pKa	Solubility (g/L)	BOD (mg/g)
Bismarck brown	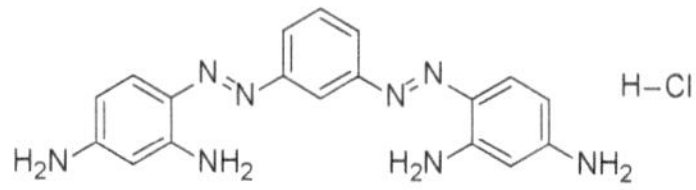	2.26	3.34	10	1.456
Orange G	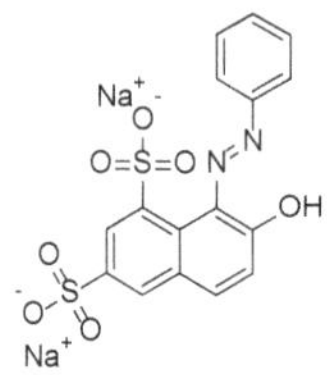	4.931	11.5	80	1.176
Indigo Carmine	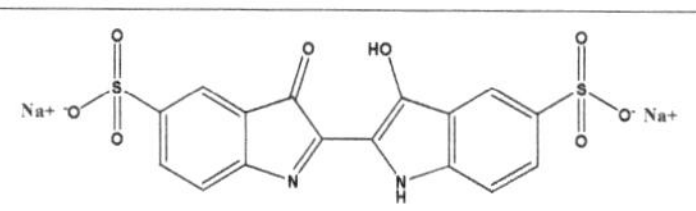	3.06	12.2	10	1.072
Brilliant blue G	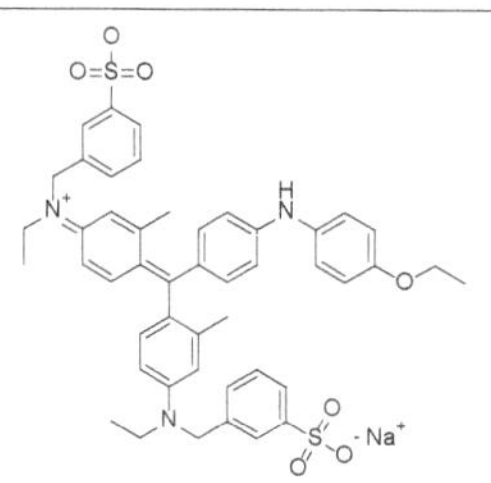	−0.35	12.4	40	0.43 g/g
Chitosan	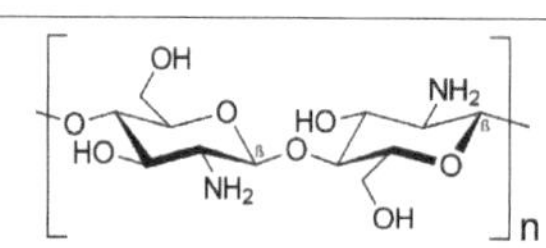	−2.3	6.5	Insoluble	-
PVPP	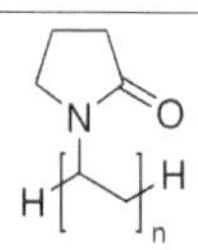	-	-	Insoluble	-

2.1. Synthesis and Characterization of Polymer Beads

Chitosan powder was added to 1.5% acetic acid to make a 2% chitosan solution while stirring at 60 °C followed by the addition of PVPP powder to make a paste of 3% PVPP. The beads were formed by dropping the paste into a 2% NaOH solution at a constant rate of 8 mL/min from a distance of roughly 10 cm using a syringe pump (Figure 1). They were allowed to cure for 4 h washed with deionised water, dried with a paper towel, and stored for future use.

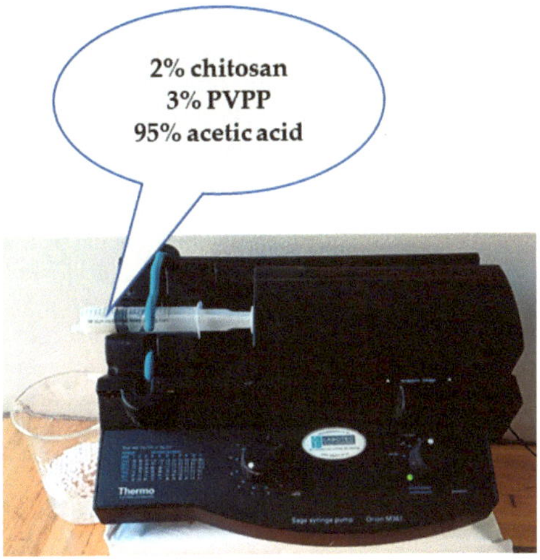

Figure 1. An illustration of the synthesis of beads in the lab using a syringe pump.

The beads were characterized by Fourier transform infrared (FTIR) to identify the functional groups responsible for dye adsorption and the surface area was determined using the dye adsorption experiment [55].

For determination of surface area, 3 g of the beads were added to 15 mL of different concentrations of congo red dye (0–400 mg/L) in phosphate buffer saline (PBS) solution at pH 6 with 0.004 wt.% NaCl in 50 mL Erlenmeyer flasks [56]. The flasks were incubated at 60 °C for 24 h while stirring at 180 rpm. The concentration of the residual Congo red was analyzed using a UV-Vis spectrophotometer (UV-1600PC; Avantor, Lutterworth, UK) at a wavelength of 500 nm. With a plot of Q_e vs. C_e, the maximum Langmuir monolayer adsorption capacity (Q_m) was calculated using Equation (1):

$$Q_e = \frac{Q_m K_L C_e}{1 + K_L C_e} \tag{1}$$

where Q_e is the equilibrium congo red concentration on the adsorbent (mg/g), C_e is the concentration of congo red in the solution at equilibrium (mg/L), and K_L is the Langmuir adsorption constant (L/mg).

The specific surface area of the beads was calculated from Equation (2):

$$Specific\ surface\ area = \frac{Q_m \times N \times S_a}{M_w \times 10^{21}} \tag{2}$$

where N is the Avogadro constant, and S_a (1.73 nm^2) and Mw (696.7 g/mol) are the surface area and molecular weight of the Congo red molecule, respectively.

Zeta potential measurement was performed on a Zetasizer Nano ZS90 (Malvern Instruments Ltd., Malvern, UK) using deionized water as the dispersant at 24.9 °C, a count rate of 18.0 kcps, and a measurement position of 2.00 mm in a clear disposable zeta cell.

Thermogravimetric analysis (TGA) was carried out using a TGA analyzer (SDT Q600 V20.9 Build 20, New Castle, DE, USA). The measurement was performed under nitrogen protection at a continuous flow of 50 mL/min and the temperature increased from 21.5 °C to 900 °C with a heating rate of 10 °C/min.

2.2. Adsorption of Dyes Using CS/PVPP Beads

Batch degradation experiments were carried out in triplicate on four representative dyes, that is, Bismarck brown (BB), orange G (OG), brilliant blue G (BBG), and indigo carmine (IC) dissolved in deionized water (DIW). A 1000 mg/L stock solution of the dyes was prepared in DIW and diluted to the required concentrations during degradation using DIW. The experiments were carried out at room temperature by adding 1 g of beads to 15 mL of 50 mg/L dye solution in 250 mL Erlenmeyer flasks while shaking at 150 rpm. The concentrations of the dyes over the degradation times were measured using a VWR 1600PC spectrophotometer at wavelengths 465 nm, 485 nm, 595 nm, and 610 nm for BB, OG, BBG, and IC, respectively. The adsorption kinetics were studied at room temperature (25 °C) by contacting 1 g of beads with 2.5 mg/L, 5 mg/L, 10 mg/L, 20 mg/L, 50 mg/L, and 100 mg/L dye solutions with the solution pH set at 9. For comparison purposes, identical experiments were carried out at varying temperatures (25, 35, and 45 °C) and initial pH (7, 9, and 12.5). The initial solution pH was adjusted using conc. NaOH or conc. HCl before the addition of beads. The equilibrium concentration, C_e (mg/L), was recorded after 24 h and the amount of dye adsorbed onto the beads at equilibrium, Q_e (mg/g) was calculated using Equation (3):

$$Q_e = (C_o - C_e) \times \frac{V}{m} \tag{3}$$

where C_o is the initial dye concentration, V is the experimental solution volume, and m is the mass of the beads.

Binary and quaternary mixtures of the four dyes were also degraded using the beads in deionized and their concentrations at different degradation times were determined. All the dye mixtures were prepared to give a total dye concentration of 50 mg/L and the dye composition was measured using UV-Vis spectroscopy.

3. Results and Discussions

3.1. Synthesis and Characterization of Polymer Beads

The white chitosan–PVPP beads formed were spherical with a slightly translucent quality, a glossy surface, and an average diameter of 4.48 ± 0.156 mm (Figure 2).

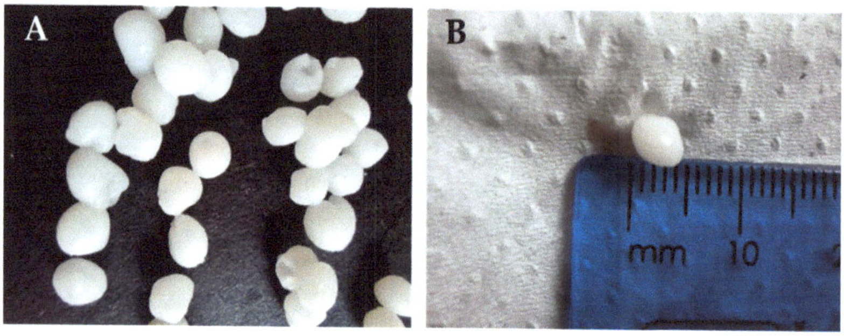

Figure 2. CS-PVPP beads (**A**) synthesized in the lab and (**B**) a demonstration of their size.

3.1.1. Surface Area and Zeta Potential of the Beads

The specific surface area (SSA) of the beads was calculated by a congo red adsorption experiment according to the procedure described by Spence et al. [56]. The adsorption capacity of the beads using different concentrations of congo red dye was fitted using a non-linear Langmuir monolayer adsorption model [55]. The beads showed higher adsorption capacity with increasing congo red concentration (Figure 3A) and the Q_m, K_L, and SSA of the beads were obtained as 1.473 mg/g, 0.057 L/mg, and 2.203 m^2/g, respectively, using Equations (1) and (2).

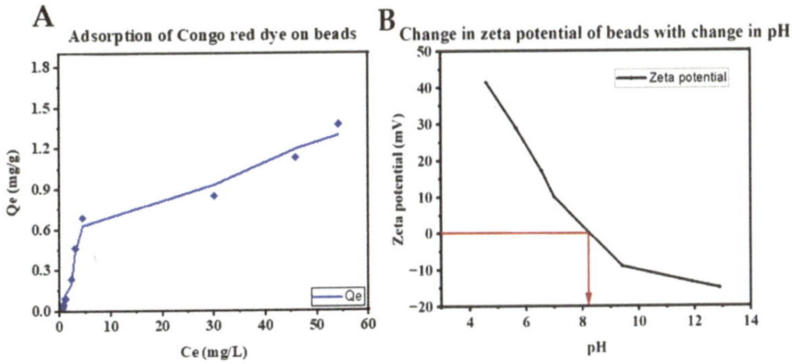

Figure 3. A graph of (**A**) the amount of dye adsorbed on beads vs concentration of Congo red dye at equilibrium and (**B**) the change of zeta potential of the beads' surface with pH.

As observed in Figure 3B, the isoelectric point for the surface of the bead is around pH 8.26. The surface of the beads possesses a positive charge at lower pH values and favours the attraction of negatively charged or anionic dye molecules. An increase in pH results in negative zeta potential causing electrostatic repulsion between the anionic adsorbate/dye molecules and the bead surface [57].

3.1.2. TGA Analysis

TGA was performed on the beads to determine their thermal stability and decomposition behaviour, as shown in Figure 4. The beads underwent three stages of thermal degradation with the first stage of 90.80% occurring from 25 to 150 °C. This loss was attributed to the loss of free water since the beads comprise 95% water and 5% polymer material. From 180 to 300 °C, another 1.313% mass loss was registered, which could account for the thermal decomposition of unpolymerized monomers and monomers with lower degree of polymerization as well as loss of bound water [58]. After 400 °C, a further 4.976% mass is lost due to drying of the sugar rings, depolymerization, breakdown of intra and intermolecular hydrogen bonds, and decomposition of acetylated and deacetylated units [59,60]. The third transition of the weight loss could be due to the breakdown of the oligosaccharide backbone of the polymer [60]. The stability of the polymer up to 400 °C can be attributed to the strong intramolecular and intermolecular hydrogen bonds formed between chitosan and PVPP [61].

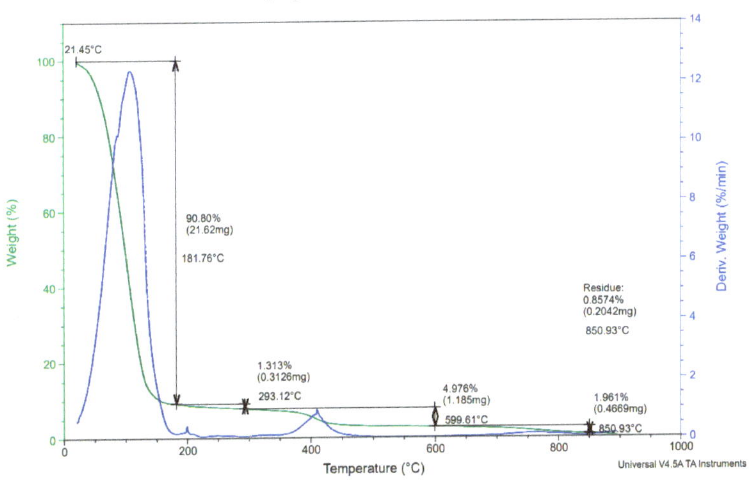

Figure 4. Thermogravimetric curves of the polymer beads.

3.1.3. FTIR Analysis of Composite Beads before and after Dye Adsorption

FTIR was performed on the dyes and the beads before and after dye adsorption. It was obtained using a Bruker Alpha II Platinum ATR over the range of 400–4000 cm^{-1} and 25 scans. BB (Figure 5A) showed absorption bands around 3310 and 3140 cm^{-1} due to N-H (primary and secondary amine), 1631 cm^{-1} due to C=C stretching (conjugated alkene), 1514 cm^{-1} due to N-H bending (amine), 1409 and 1143 due to C-N stretching (aromatic amine), and 1030, 880, and 707 cm^{-1} due to C=C bending (alkene). OG (Figure 5B) bands were displayed on 3452 cm^{-1} for O-H stretching (alcohol), 1629 cm^{-1} for C=C stretching for trans-alkene (disubstituted), 1479, 1199, and 1036 cm^{-1} for S=O for sulfo groups, and 977, 915, 831, 759, and 660 cm^{-1} for C=C bending (alkene). BBG (Figure 5C), on the other hand, displayed bands at 1571 cm^{-1} for N-H bending (amine), 1501 cm^{-1} for C-H bending alkene, 1329 and 1154 cm^{-1} for S=O stretching sulfo group, 1024 cm^{-1} for C-O stretching ether, and 909, 685, and 617 cm^{-1} for C=C bending alkene. IC (Figure 5D) produces bands at 3361 cm^{-1} and 1479 cm^{-1} for O-H alcohol, 1629 cm^{-1} for C=N stretching, 1154 cm^{-1} for and 1024 cm^{-1} for S=O stretching (sulfo group), 1099 cm^{-1} for C-N stretch, and 820, 730, and 678 cm^{-1} for C=C bending (alkene).

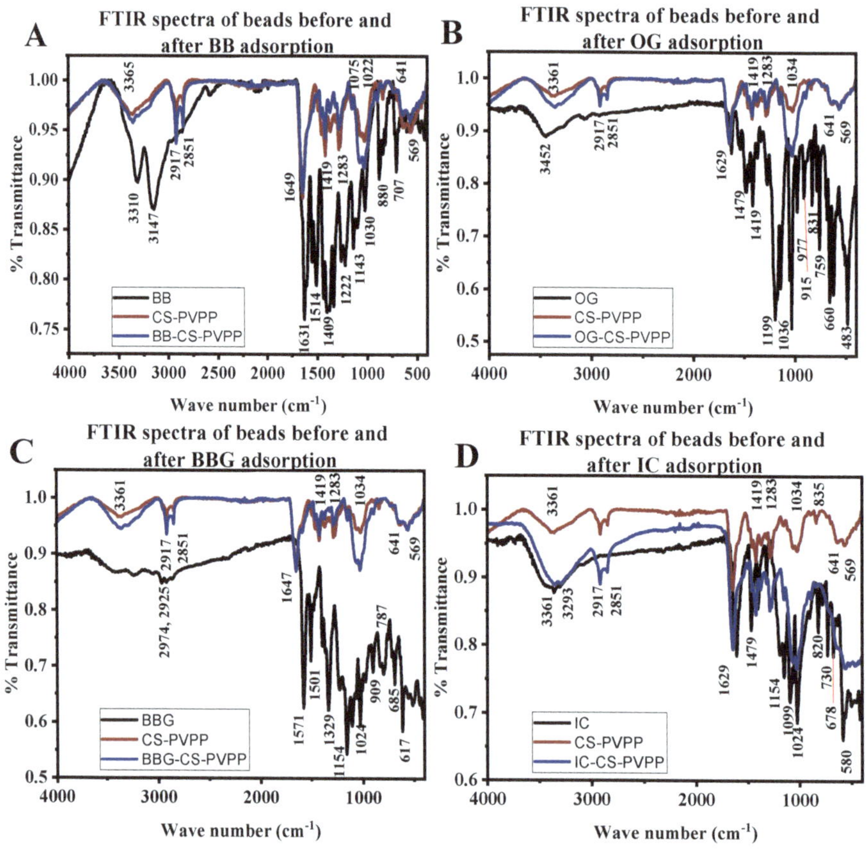

Figure 5. FTIR spectra of beads before and after (**A**) BB, (**B**) OG, (**C**) IC, and (**D**) BBG adsorption.

The polymer composite also displayed characteristic bands at 3365 cm^{-1} for O-H alcohol (intermolecular bonded), a doublet at 2917 and 2851 cm^{-1} for C-H stretch carbonyl, 1649 cm^{-1} for N-H bending (amine), 1419 cm^{-1} for O-H bending alcohol, 1283 cm^{-1} for C-N stretching aromatic amine, 1075 cm^{-1} for C-O stretching alkyl ether, 1022 cm^{-1} for

C-O stretching primary alcohol, and 569 cm^{-1} for C=O stretching alkene [58,62–64]. After dye adsorption, there is a shift in the band at 3310 to 3365, and all other signatures of the composite in the presence of the dyes, especially BB, OG, and BBG, significantly decreased. This could be attributed to hydrogen bonding and intermolecular forces between amino groups, carbonyl, hydroxyl, and sulfo groups on the dyes and adsorption sites [55,62,63].

3.2. *Effect of Contact Time on Adsorption of Dye*

Adsorption is a surface phenomenon where molecules (adsorbates) from a liquid or gas adhere to the surface of a solid (adsorbent) [65] facilitated by van der Waals forces, electrostatic interactions, hydrogen bonding, and hydrophobic interactions [66]. Its effectiveness depends on the physiochemical characteristics of the adsorbate and the adsorbent, such as morphology, chemical structure, surface charge, and functional groups [67]. Chitosan is rich in amino and hydroxyl groups [68], which are easily protonated in acidic conditions leading to increased electrostatic interactions with anionic molecules [69,70]. PVPP contains pyrrolidone rings with amide groups that can participate in hydrogen bonding and van der Waals interactions [71,72]. Due to the lack of charged groups, it relies more on hydrogen bonding and van der Waals forces, making it effective for a broader range of both cationic and anionic dyes but often less efficient than chitosan for strong electrostatic adsorption of anionic dyes [73,74].

It was observed that there was rapid adsorption of the dyes from the solution by the beads in the first 1 h (Figure 6), which could be due to readily available active sites and functional groups for electrostatic interactions [75]. The adsorption rate decreased in the next 3 h as more active sites were occupied and eventually reached equilibrium after 6 h where the rate of adsorption became equal to the rate of desorption.

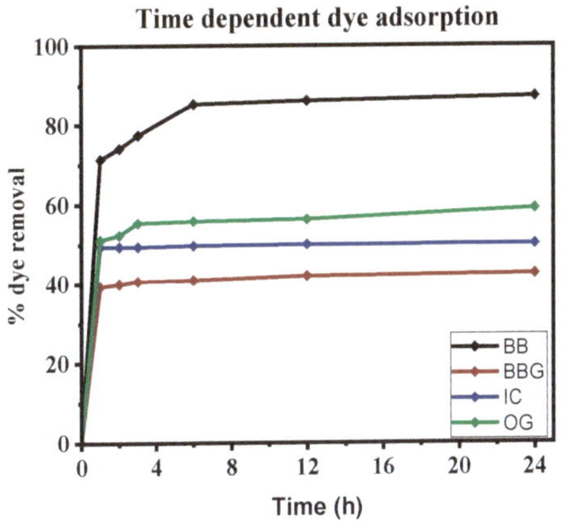

Figure 6. Time-dependent degradation of BB, IC, BBG, and OG over 24 h.

BB showed the highest removal (87%) from the solution compared to all the other dyes and this could be attributed to its cationic nature and large number of amine groups that facilitate electrostatic attraction and hydrogen bonding with functional groups on the bead surfaces, hence enhancing adsorption capacity [76]. Also, the amine groups on BB react with water to establish an equilibrium ammonium salt and hydroxide ion through protonation of the amine group leading to increased interaction with the negatively charged surface of the beads [77]. The adsorption of OG (58%), BBG (42%), and IC (49%) mainly occurred in the first hour, and no adsorption was observed in the subsequent hours. Their low adsorption was attributed to their anionic nature contributed by the negatively charged sulfonate and hydroxyl groups that create electrostatic repulsion between the dye molecules and bead surface [78]. The presence of amino groups on IC and BBG facilitates electrostatic interactions of the dyes with the functional groups on the beads' surface [79,80].

3.3. The Effect of pH

The effect of pH on the adsorption of different concentrations of dyes (2.5–100 mg/L) from the solution was determined at three different pH values of 7, 9.25, and 12.5 by adjusting the initial reaction pH using hydrochloric acid or sodium hydroxide solutions. The equilibrium concentration on the adsorbent Q_e was measured against the equilibrium concentration in the solution Ce for all pH levels.

It was observed that an increase in pH increased the adsorption capabilities for BB, decreased the adsorption of IC, and had no significant effect on the adsorption of BBG and OG (Figure 7). This could be attributed to the cationic nature of BB, with a low pKa, and the anionic nature of BBG, OG, and IC, due to their high pKa values. An increase in pH leads to deprotonation of the amine groups making the surface of the CS-PVPP more negatively charged, leading to increased electrostatic interaction with cationic BB [81,82]. On the other hand, an increase in pH leads to deprotonation of the amine groups, making the surface of the CS-PVPP more negatively charged, leading to increased electrostatic repulsive forces with the anionic BBG, OG, and IC dyes [81,82]. This was further confirmed by the change in zeta potential with a change in pH, as earlier reported in Figure 3, where the surface of the beads became more negatively charged as the pH increased.

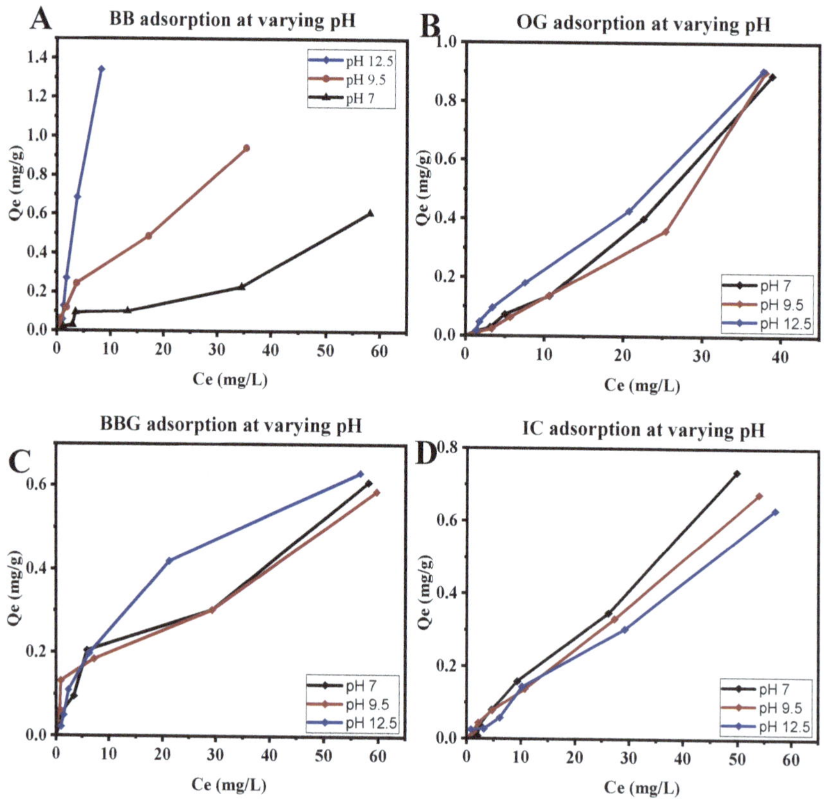

Figure 7. Adsorption of (**A**) Bismarck brown, (**B**) orange G, (**C**) brilliant blue G, and (**D**) indigo carmine onto the beads at varying pH values.

3.4. The Effect of Temperature

The effect of temperature on the adsorption of dyes onto PVPP–chitosan beads varies depending on the dye's chemical nature and interactions with the adsorbent. For BB, IC, and BBG, the equilibrium adsorption capacity (Qe) increases with rising temperature (Figure 8), indicating an endothermic adsorption process. This suggests that higher temperatures enhance the kinetic energy and diffusion rate of these dye molecules, allowing them to overcome activation barriers and interact more effectively with the adsorption sites on the beads [83,84]. Additionally, the adsorption process for these dyes might involve stronger chemical interactions at elevated temperatures, such as increased hydrogen bonding or other endothermic interactions, leading to higher Qe values [85,86]. On the contrary, an increase in temperature has no significant effect on the rate or amount of adsorption of OG from the solution. The stable adsorption behaviour of OG across different temperatures may be attributed to the process being exothermic which has been reported in other works [87–89].

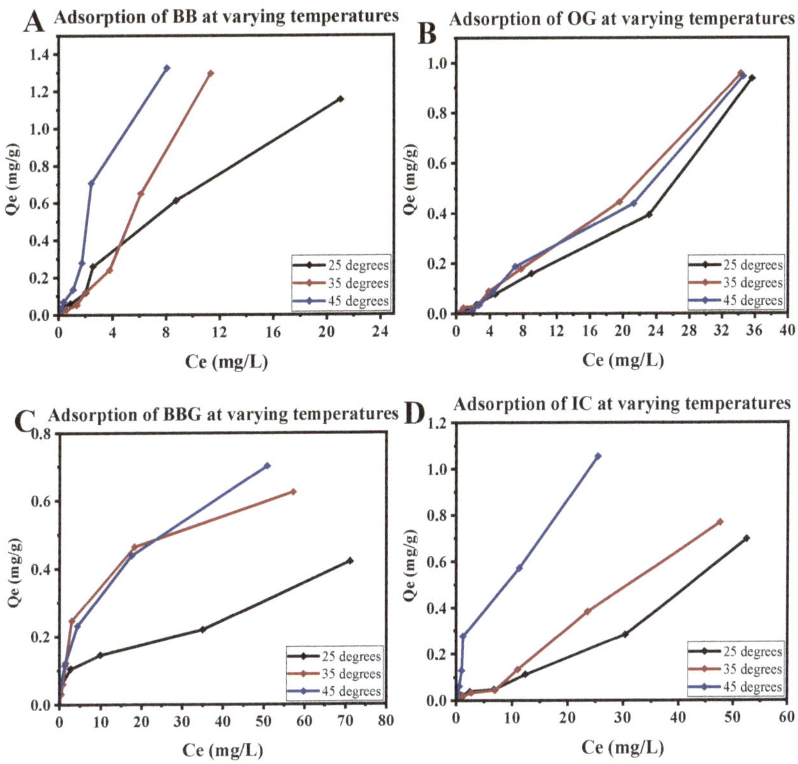

Figure 8. Adsorption of (**A**) Bismarck brown, (**B**) orange G, (**C**) brilliant blue G, and (**D**) indigo carmine at varying temperatures.

3.5. Adsorption Isotherms

Dye solutions of varying concentrations (from 2.5 mg/L to 100 mg/L) were contacted with 1 g of composite beads at room temperature for 24 h to obtain equilibrium. The amount of dye adsorbed at equilibrium, Q_e (mg/g), was plotted as a function of the equilibrium concentration, C_e (mg/L). To analyze the equilibrium data and adsorptive capacity of the beads, three isotherm models, namely, Langmuir, Freundlich and BET, were used in their non-linear forms in Equations (4), (5), and (6), respectively.

$$Q_e = \frac{Q_{m1} K_L C_e}{1 + K_L C_e} \tag{4}$$

$$Q_e = K_F C_e^{1/n} \tag{5}$$

$$Q_e = \frac{Q_m K_s C_e}{(1 - K_L C_e)(1 - K_L C_e + K_s C_e)} \quad (6)$$

where Q_{m1} (mg/g) is the monolayer adsorption capacity, K_L (L/mg) is the Langmuir affinity constant, K_F ((mg/g) (L/mg)$^{1/n}$) and n the adsorption potential and strength constants of the Freundlich isotherm model, and K_L and K_S the BET equilibrium constants for adsorption of the first and upper layers, respectively. The isotherm model parameters were obtained by non-linear regression and the fitting parameters are provided in Table 2.

Table 2. Fitting parameters of isotherm models.

Model		BB	OG	BBG	IC
Langmuir	Best-fit values				
	Q_m	3.106	5305 *	0.5927	4493 *
	K_L	0.02809	4.367×10^{-6}	0.02632	2.691×10^{-6}
	Goodness of Fit				
	R^2	0.9937	0.9472	0.8668	0.9637
Freundlich	Best-fit values				
	K_F	0.1026	0.002916	0.04808	0.002293
	n	1.254	0.6219	2.045	0.6946
	Goodness of Fit				
	R^2	0.9914	0.9844	0.9323	0.9920
BET	Best-fit values				
	Q_m	0.9650	0.2694	0.1470	0.2944
	K_S	0.1059	0.07362	1.010	0.03098
	K_L	0.01682	0.02070	0.009145	0.01241
	Goodness of Fit				
	R^2	0.9926	0.9995	0.9725	0.9980

* These values are unrealistically large and are artefacts of the model fit which do not reach saturation.

The Langmuir isotherm model assumes homogenous surface of adsorbates, identical energy adsorption sites, no interaction between adsorbed molecules, and a monolayer adsorption phenomenon where a molecule adsorbs and no other molecule may lay on top of it [90,91]. The BET model was then developed as an extension of the Langmuir accounting for adsorption on multiple layers with the adsorbed molecules also serving as adsorption sites. It also assumes perfectly flat homogeneous adsorbate surfaces and negligible interaction between the adsorbed molecules [91–93]. On the other hand, the Freundlich model takes into account the heterogeneity of surfaces with exponential distribution of adsorption sites and adsorption site energies [94,95]. On fitting the adsorption data on the models, it was observed that the BET model gave the best fit values for all investigated dyes (Figure A2), as demonstrated with the highest R^2 values implying that adsorption was due to multiple adsorptions sites. The Langmuir isotherm models was considered unstable in fitting the data since it did not tend to a constant Q_m value for some of the dye solutions (OG and IC).

3.6. Adsorption Kinetics and Mass Transfer Studies

In order to predict the rate of dye removal from the solution and the mechanism of adsorption process, Lagergren's pseudo first-order (PFO) [96] and Ho and McKay's pseudo second-order (PSO) [97] (Equations (7) and (8), respectively) were used to fit the kinetics data, as shown in Figure A3.

$$Q_t = Q_e\left(1 - e^{-k_1 t}\right) \quad (7)$$

$$Q_t = \frac{k_2 {Q_e}^2 t}{1 + k_2 Q_e t} \quad (8)$$

where k_1 (L/h) is the PFO rate constant and k_2 (g/mg/h) is the PSO rate constant. The kinetic parameters for the kinetic and mass transfer models are summarized in Table 3.

Table 3. Fitting parameters for kinetic and mass transfer models.

Model		BB	OG	BBG	IC
Pseudo first order	Best-fit values				
	Q_e	0.6170	0.4139	0.3038	0.3677
	k_1	1.732	2.040	2.973	4.826
	Goodness of Fit				
	R^2	0.9720	0.9733	0.9935	0.9993
Pseudo second order	Best-fit values				
	Q_e	0.6489	0.4286	0.3098	0.3692
	k_1	5.634	12.71	38.48	168.9
	Goodness of Fit				
	R^2	0.9867	0.9782	0.9958	0.9994
Boyd	Best-fit values				
	Plateau	0.6170	0.4139	0.3038	0.3677
	K	1.732	2.040	2.973	4.826
	Goodness of Fit				
	R^2	0.9720	0.9733	0.9935	0.9993
Multiple linear regression	Best-fit values				
	k_{WM1}	0.5241	0.3661	0.2891	0.3651
	k_{WM2}	0.07301	0.05025	0.01114	0.0002281
	k_{WM3}	0.01037	0.01104	0.003359	0.003164
	Goodness of Fit				
	R^2	0.9909	0.9805	0.9966	0.9997

The PFO and PSO models are based on the assumption that the difference between the actual and equilibrium surface concentration are the driving force for adsorption [98]. The PFO assumes that the rate of adsorption is directly proportional to the distance from equilibrium while the PSO model assumes that the concentration of the adsorbate remains constant during the process, the adsorption process is not limited by diffusion and its reaction is controlled with negligible desorption [99,100]. From the parameters presented in Table 3, both models gave high R^2 values, but the PFO model adequately described the kinetics of the experiments since it gave Q_e values that were consistent with the experimental results obtained.

Since the PFO and PSO models are purely descriptive and do not provide specific information about the mechanism of adsorption [99], the mechanism of adsorption of the dyes on the beads was studied using the intra-particle diffusion model (Equation (9)) [101] and Boyd's single-resistance model of film diffusion (Equation (10)) [102,103] for internal and external mass transfer, respectively.

$$Q_t = k_p t^{\frac{1}{2}} + C \tag{9}$$

$$-ln\left(1 - \frac{Q_t}{Q_e}\right) = k_{fd}t \tag{10}$$

where k_p (mg/(g·h$^{0.5}$)) is the intra-particle diffusion (IPD) constant and C is a constant (mg/g) that is proportional to the boundary layer thickness and k_{fd} is the adsorption rate constant.

The adsorption process consists of three major steps, namely, boundary layer diffusion, diffusion of adsorbate into the pores of adsorbent, and adsorption of the adsorbate onto the active sites of the internal pores of the adsorbent. Adsorption of adsorbate onto internal pores is assumed to be very rapid implying that the overall rate of adsorption is controlled by one or both of the first two steps. The Boyd model suggests that the main resistance to diffusion is in the boundary layer surrounding the adsorbent particle, while the IPD model suggests negligible external resistance to film diffusion and constant intraparticle diffusivity that does not change with time and position [104–106]. From the graphs obtained (Figure A3), both the Boyd and IPD plots are multilinear, implying that the reaction rate is dependent on both external and internal mass transfer. More so, extrapolation of the line segments produced positive intercepts implying that rapid adsorption occurred within a short time (the first hour) [107].

3.7. Adsorption Thermodynamics

Equations (11) and (12) were used to obtain the adsorption parameters for standard Gibbs free energy change (ΔG°), entropy change (ΔS°), and change in enthalpy (ΔH°).

$$\Delta G^{\circ} = -RTln\left(R_{eq}\right) \quad (11)$$

$$ln\left(K_{eq}\right) = \frac{\Delta S^{\circ}}{R} - \frac{\Delta H^{\circ}}{RT} \quad (12)$$

where R is the universal gas constant in (J/mol/K), T (K) is the absolute temperature, and K_L is a non-dimensional equilibrium constant derived from the Langmuir isotherm model (Table 2).

The modified dual-site Langmuir isotherm equilibrium constants were non-dimensionalised by employing Equation (13) as follows:

$$K_{eq} = K_L \times M_w \times 10^3 \times \frac{C_s^o}{\gamma_s^o} \quad (13)$$

where Mw is the molecular mass of the adsorbate, C_s^o is the standard adsorbate concentration, and γ_s^o is the standard adsorbate activity. C_s^o and γ_s^o are assumed to be unity. The Van't Hoff plot (Figure A1) was used to obtain the thermodynamic parameters that are also presented in Table 4.

Table 4. Thermodynamic parameters for dye adsorption.

Dye	ΔG° (kJ/mol) 25 °C	35 °C	45 °C	ΔH° (kJ/mol)	ΔS° (J/mol)	R^2
BB	−23.24	−27.53	−28.99	62.95	290.55	0.8611
OG	−13.53	−16.40	−16.23	27.37	138.74	0.5091
BBG	−25.11	−30.28	−29.10	35.42	206.35	0.2824
IC	−9.62	−10.85	−27.27	250.88	865.79	0.7756

For all the dye adsorption processes, the Gibbs free energy is negative, implying that the adsorption was thermodynamically favoured. The positive ΔH° values showed the reversible pathway for adsorption of the dyes is endothermic by nature, and thus adsorption is favoured at high temperatures [82]. There is a possibility of structural changes in the adsorbent and adsorbate due to increased randomness at the solid–liquid interface suggested by the positive ΔS° [108].

3.8. Reusability

The reusability of a material is essential for possible industrial-scale application and hence four subsequent adsorption experiments were carried out using the same beads that were collected after every degradation experiment while keeping the same operating conditions. To study the reusability of the beads, the beads were washed with deionized water and dried using absorbent paper sheets after each experiment cycle and used to consecutively adsorb freshly prepared dye solution of 50 mg/L. The initial extent of degradation was recorded as 100%.

As observed in Figure 9, there is no loss in the adsorption capacity of the beads up to the fourth cycle.

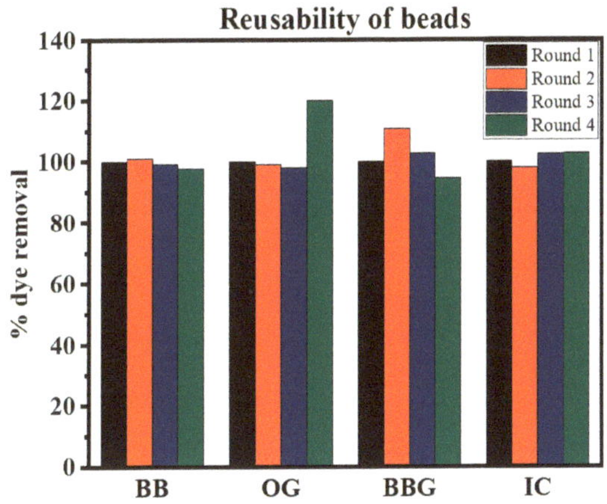

Figure 9. Reusability of the beads for dye removal for 24 h cycles.

3.9. Degradation of Dye Mixtures

In most dye degradation studies reported, single-test dye solution experiments have been reported that do not depict the actual conditions in real industrial wastewater. Industrial effluents from dye industries contain a mixture of dyes which makes the effluent treatment challenging. Investigations using stimulated dye wastewater are helpful in understanding the colour-removal process of the materials for application in actual industrial effluents [109,110]. This study explored the applicability of the synthesized beads in degrading dye mixtures. Using the dyes explored in single study experiments, the beads were used to degrade binary and quaternary dye mixtures for a total nominal dye concentration of 50 mg/L for 24 h while shaking at 150 rpm; their degradation over time is depicted in Figure 10.

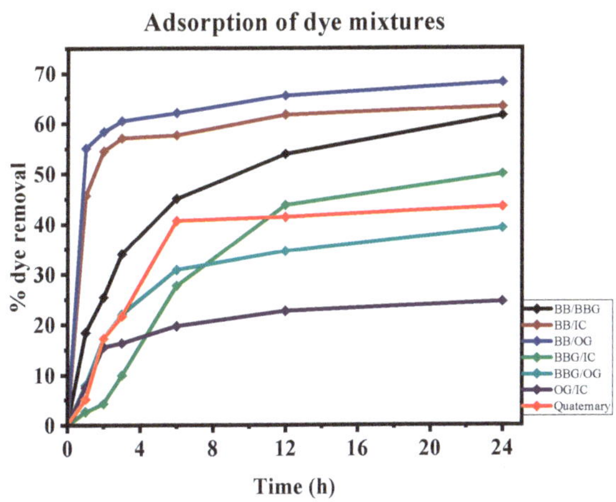

Figure 10. Time-dependent degradation of dye mixtures using the beads.

Adsorption of the dyes followed a rapid dye removal from solution in the first 6 h which could have been due to the availability of active adsorbent sites. It is worth noting that adsorption of the mixed dyes follows the adsorption nature of the individual dyes with mixtures containing BB showing higher percentages of removal than those without it. This suggests that BB potentially retards the degradation of other dyes and that only when it is largely adsorbed do the other dyes start competing for the remaining adsorption site [110,111].

4. Conclusions

This study demonstrates the efficacy of chitosan–polyvinylpolypyrrolidone (PVPP) beads in the adsorption of various dyes from aqueous solutions. The results indicate that these beads are highly effective in adsorbing dyes, with performance varying based on the type of dye used showing high affinity for cationic dyes due to the negative surface charge at high pH values. The investigation into the effect of pH revealed that, for cationic dyes such as Bismarck Brown, an increase in pH leads to a higher equilibrium concentration of dye on the adsorbent. Conversely, for anionic dyes such as Indigo Carmine, an increase in pH results in decreased dye adsorption.

Temperature also plays a significant role in dye adsorption. For dyes like Indigo Carmine, Brilliant Blue G, and Bismarck Brown, an increase in temperature enhances the kinetic energy and diffusion rate of dye molecules, thereby increasing the equilibrium concentration on the adsorbent. However, the adsorption capacity for Orange G remains relatively unaffected by temperature changes, suggesting that its adsorption onto PVPP–chitosan beads is independent of temperature variations.

The TGA analysis showed that the beads can withstand high temperatures up to 400 °C, which provides a platform for the use of the beads in the identified optimum conditions, that is, high pH and high temperatures.

Therefore, PVPP–chitosan beads show promising potential for dye removal from aqueous solutions, with their performance influenced by factors such as dye type, pH, temperature, and dye mixture composition. Due to the porosity of the beads and the reasonable surface area, the beads can be explored for continuous adsorption processes for pollutant removal. These insights can guide the optimization of dye removal processes in various industrial applications.

Author Contributions: Conceptualization, H.D.K., H.G.B. and U.F.; methodology, H.D.K., W.M. and M.A.; software, H.G.B.; validation, H.D.K., H.G.B. and U.F.; formal analysis, H.D.K., W.M., M.A., H.G.B. and U.F.; investigation, H.D.K., W.M. and M.A.; resources, H.G.B.; data curation, H.D.K., W.M., M.A. and H.G.B.; writing—original draft preparation, W.M., M.A., N.H.H. and H.D.K.; writing—review and editing, H.D.K., H.G.B., N.H.H. and U.F.; supervision, H.G.B. and U.F.; project administration, H.G.B. and U.F.; funding acquisition, H.G.B., N.H.H. and H.D.K. All authors have read and agreed to the published version of the manuscript.

Funding: This study was funded by the National Research Foundation (NRF) of South Africa [MND210426597525 and CSRP2204204025], Margaret McNamara Education grants, and the Schlumberger Foundation Faculty for the Future program. This work was further supported by the Austrian Federal Ministry of Education, Science and Research (BMBWF) through Austria's Agency for Education and Internationalization (OeAD) [Grant numbers: Africa UNINET P056 and P058 as well as APPEAR Project 341]. APPEAR is a program of the Austrian Development Organization.

Institutional Review Board Statement: Not applicable

Data Availability Statement: The raw data supporting the conclusions of this article will be made available by the authors upon request.

Conflicts of Interest: The authors declare no conflicts of interest. The funders had no role in the design of the study; in the collection, analyses, or interpretation of data; in the writing of the manuscript; or in the decision to publish the results.

Appendix A

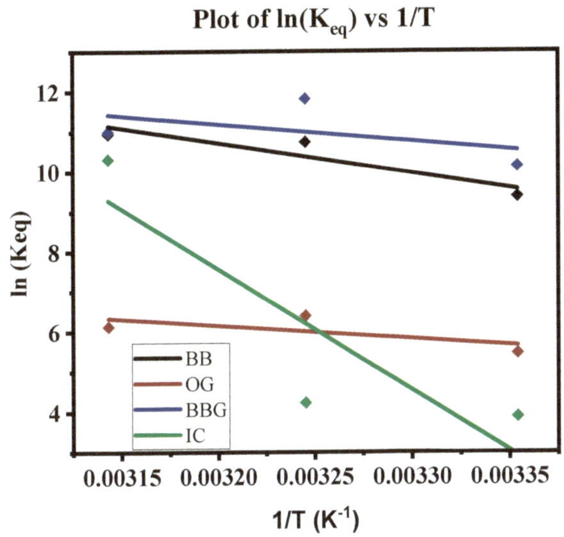

Figure A1. Van't Hoff plot for the adsorption of dyes onto the beads.

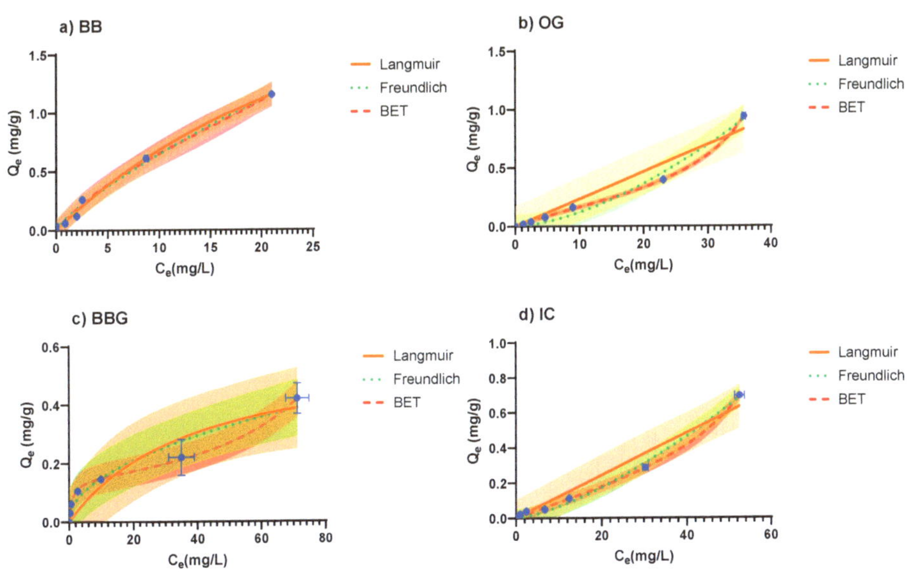

Figure A2. Langmuir, Freundlich, and BET isotherm models for (**a**) Bismarck brown, (**b**) orange G, (**c**) Brilliant blue G, and (**d**) indigo carmine.

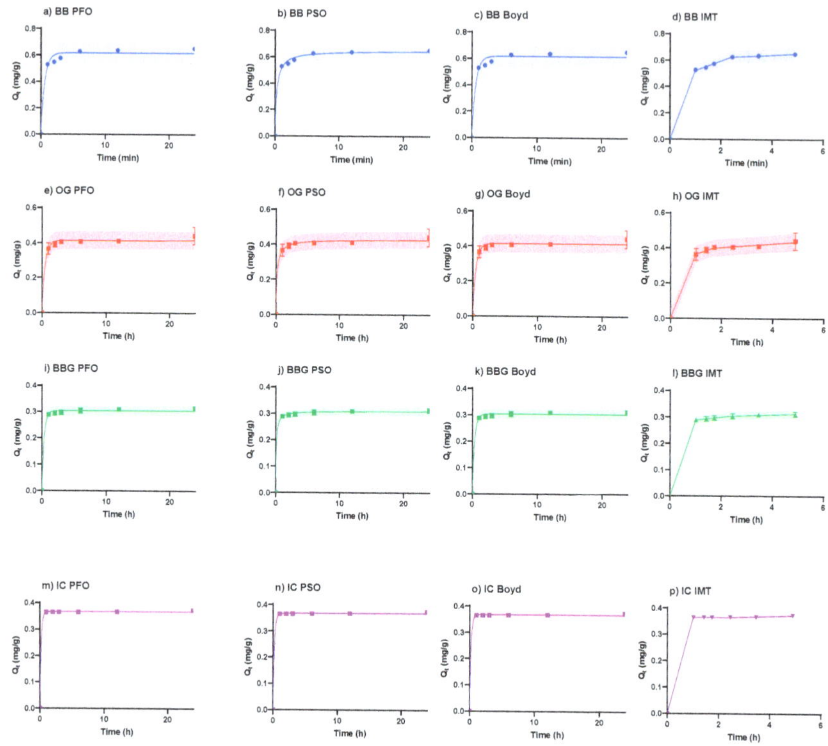

Figure A3. PFO, PSO, Boyd, and IPD graphs for dye adsorption.

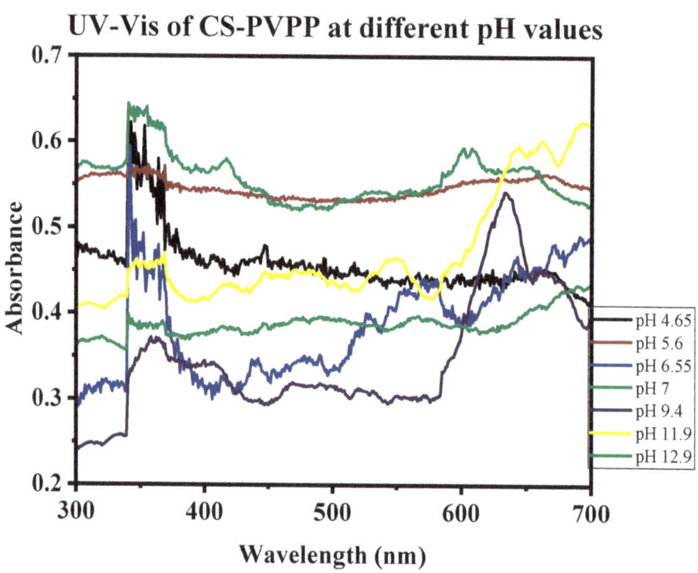

Figure A4. UV-Vis spectroscopy of CS-PVPP at different pH values.

References

1. Varjani, S.; Rakholiya, P.; Ng, H.Y.; You, S.; Teixeira, J.A. Microbial degradation of dyes: An overview. *Bioresour. Technol.* **2020**, *314*, 123728. [CrossRef] [PubMed]
2. Wang, X.; Jiang, J.; Gao, W. Reviewing textile wastewater produced by industries: Characteristics, environmental impacts, and treatment strategies. *Water Sci. Technol.* **2022**, *85*, 2076–2096. [CrossRef]
3. Zhai, L.; Bai, Z.; Zhu, Y.; Wang, B.; Luo, W. Fabrication of chitosan microspheres for efficient adsorption of methyl orange. *Chin. J. Chem. Eng.* **2018**, *26*, 657–666. [CrossRef]
4. Al-Tohamy, R.; Ali, S.S.; Li, F.; Okasha, K.M.; Mahmoud, Y.A.-G.; Elsamahy, T.; Jiao, H.; Fu, Y.; Sun, J. A critical review on the treatment of dye-containing wastewater: Ecotoxicological and health concerns of textile dyes and possible remediation approaches for environmental safety. *Ecotoxicol. Environ. Saf.* **2022**, *231*, 113160. [CrossRef]
5. Kishor, R.; Purchase, D.; Saratale, G.D.; Saratale, R.G.; Ferreira, L.F.R.; Bilal, M.; Chandra, R.; Bharagava, R.N. Ecotoxicological and health concerns of persistent coloring pollutants of textile industry wastewater and treatment approaches for environmental safety. *J. Environ. Chem. Eng.* **2021**, *9*, 105012. [CrossRef]
6. Tkaczyk, A.; Mitrowska, K.; Posyniak, A. Synthetic organic dyes as contaminants of the aquatic environment and their implications for ecosystems: A review. *Sci. Total Environ.* **2020**, *717*, 137222. [CrossRef]
7. Rai, A.; Chauhan, P.S.; Bhattacharya, S. Remediation of Industrial Effluents. In *Water Remediation*; Bhattacharya, S., Gupta, A.B., Gupta, A., Pandey, A., Eds.; Springer: Singapore, 2018; pp. 171–187. [CrossRef]
8. Islam, M.; Mostafa, M. Textile Dyeing Effluents and Environment Concerns—A Review. *J. Environ. Sci. Nat. Resour.* **2019**, *11*, 131–144. [CrossRef]
9. Karri, R.R.; Ravindran, G.; Dehghani, M.H. Chapter 1-Wastewater—Sources, Toxicity, and Their Consequences to Human Health. In *Soft Computing Techniques in Solid Waste and Wastewater Management*; Karri, R.R., Ravindran, G., Dehghani, M.H., Eds.; Elsevier: Amsterdam, The Netherlands, 2021; pp. 3–33. [CrossRef]
10. Khan, S.; Malik, A. Environmental and Health Effects of Textile Industry Wastewater. In *Environmental Deterioration and Human Health: Natural and Anthropogenic Determinants*; Malik, A., Grohmann, E., Akhtar, R., Eds.; Springer: Dordrecht, The Netherlands, 2014; pp. 55–71. [CrossRef]
11. Liang, C.-Z.; Sun, S.-P.; Li, F.-Y.; Ong, Y.-K.; Chung, T.-S. Treatment of highly concentrated wastewater containing multiple synthetic dyes by a combined process of coagulation/flocculation and nanofiltration. *J. Membr. Sci.* **2014**, *469*, 306–315. [CrossRef]
12. Hussain, S.; Kamran, M.; Khan, S.A.; Shaheen, K.; Shah, Z.; Suo, H.; Khan, Q.; Shah, A.B.; Rehman, W.U.; Al-Ghamdi, Y.O.; et al. Adsorption, kinetics and thermodynamics studies of methyl orange dye sequestration through chitosan composites films. *Int. J. Biol. Macromol.* **2021**, *168*, 383–394. [CrossRef]
13. Bastami, T.R.; Khaknahad, S.; Malekshahi, M. Sonochemical versus reverse-precipitation synthesis of CuxO/Fe2O3/MoC nano-hybrid: Removal of reactive dyes and evaluation of smartphone for colorimetric detection of organic dyes in water media. *Environ. Sci. Pollut. Res.* **2020**, *27*, 9364–9381. [CrossRef]
14. Joseph, J.; Radhakrishnan, R.C.; Johnson, J.K.; Joy, S.P.; Thomas, J. Ion-exchange mediated removal of cationic dye-stuffs from water using ammonium phosphomolybdate. *Mater. Chem. Phys.* **2020**, *242*, 122488. [CrossRef]
15. Al-Othman, A.; Darwish, N.N.; Qasim, M.; Tawalbeh, M.; Darwish, N.A.; Hilal, N. Nuclear desalination: A state-of-the-art review. *Desalination* **2019**, *457*, 39–61. [CrossRef]
16. Moradihamedani, P. Recent advances in dye removal from wastewater by membrane technology: A review. *Polym. Bull.* **2022**, *79*, 2603–2631. [CrossRef]
17. Selim, M.T.; Salem, S.S.; Mohamed, A.A.; El-Gamal, M.S.; Awad, M.F.; Fouda, A. Biological Treatment of Real Textile Effluent Using Aspergillus flavus and Fusarium oxysporium and Their Consortium along with the Evaluation of Their Phytotoxicity. *J. Fungi* **2021**, *7*, 193. [CrossRef]
18. Rafiq, A.; Ikram, M.; Ali, S.; Niaz, F.; Khan, M.; Khan, Q.; Maqbool, M. Photocatalytic degradation of dyes using semiconductor photocatalysts to clean industrial water pollution. *J. Ind. Eng. Chem.* **2021**, *97*, 111–128. [CrossRef]
19. Katheresan, V.; Kansedo, J.; Lau, S.Y. Efficiency of various recent wastewater dye removal methods: A review. *J. Environ. Chem. Eng.* **2018**, *6*, 4676–4697. [CrossRef]
20. Bal, G.; Thakur, A. Distinct approaches of removal of dyes from wastewater: A review. *Mater. Today Proc.* **2022**, *50*, 1575–1579. [CrossRef]
21. Adesanmi, B.; Hung, Y.-T.; Paul, H.; Huhnke, C. Comparison of dye wastewater treatment methods: A review. *GSC Adv. Res. Rev.* **2022**, *10*, 126–137. [CrossRef]
22. García-Montaño, J.; Torrades, F.; García-Hortal, J.A.; Domènech, X.; Peral, J. Combining photo-Fenton process with aerobic sequencing batch reactor for commercial hetero-bireactive dye removal. *Appl. Catal. B Environ.* **2006**, *67*, 86–92. [CrossRef]
23. Zeb, S.; Hussain, S.; Khan, H.A.; Ali, Z.; Khan, N.; Khan, K.I.; Ali, F.; khan, S.; del Pilar Taboada Sotomayor, M.; Gul, S. Electrochemical Oxidation of Acid Brown 98 using Ti/Ru0.3Ti0.7O2 Composite Anode. *Int. J. Electrochem. Sci.* **2018**, *13*, 9428–9440. [CrossRef]
24. Wang, J.; Yao, J.; Wang, L.; Xue, Q.; Hu, Z.; Pan, B. Multivariate optimization of the pulse electrochemical oxidation for treating recalcitrant dye wastewater. *Sep. Purif. Technol.* **2020**, *230*, 115851. [CrossRef]
25. Mok, C.F.; Ching, Y.C.; Muhamad, F.; Osman, N.A.A.; Hai, N.D.; Hassan, C.R.C. Adsorption of Dyes Using Polyvinyl alcohol PVA and PVA-Based Polymer Composite Adsorbents: A Review. *J. Polym. Environ.* **2020**, *28*, 775–793. [CrossRef]

26. Sadiq, A.C.; Olasupo, A.; Ngah, W.S.W.; Rahim, N.Y.; Suah, F.B.M. A decade development in the application of chitosan-based materials for dye adsorption: A short review. *Int. J. Biol. Macromol.* **2021**, *191*, 1151–1163. [CrossRef] [PubMed]
27. Thakur, S. An overview on alginate based bio-composite materials for wastewater remedial. *Mater. Today Proc.* **2021**, *37*, 3305–3309. [CrossRef]
28. Tolkou, A.K.; Tsoutsa, E.K.; Kyzas, G.Z.; Katsoyiannis, I.A. Sustainable use of low-cost adsorbents prepared from waste fruit peels for the removal of selected reactive and basic dyes found in wastewaters. *Environ. Sci. Pollut. Res.* **2024**, *31*, 14662–14689. [CrossRef]
29. Sonal, S.; Acharya, S.; Mishra, B.K. Mesoporous carbon structure impregnated with 2D engineered zirconium: A sustainable adsorbent for the removal of dyes from the aqueous solution. *J. Environ. Manag.* **2022**, *314*, 115009. [CrossRef]
30. Kavitha, V.U.; Kandasubramanian, B. Tannins for wastewater treatment. *SN Appl. Sci.* **2020**, *2*, 1081. [CrossRef]
31. Chenab, K.K.; Sohrabi, B.; Jafari, A.; Ramakrishna, S. Water treatment: Functional nanomaterials and applications from adsorption to photodegradation. *Mater. Today Chem.* **2020**, *16*, 100262. [CrossRef]
32. Chai, W.S.; Cheun, J.Y.; Kumar, P.S.; Mubashir, M.; Majeed, Z.; Banat, F.; Ho, S.-H.; Show, P.L. A review on conventional and novel materials towards heavy metal adsorption in wastewater treatment application. *J. Clean. Prod.* **2021**, *296*, 126589. [CrossRef]
33. Dutta, S.; Gupta, B.; Kumar, S.; Kumar, A. Recent advances on the removal of dyes from wastewater using various adsorbents: A critical review. *Mater. Adv.* **2021**, *2*, 4497–4531. [CrossRef]
34. Raninga, M.; Mudgal, A.; Patel, V.K.; Patel, J.; Sinha, M.K. Modification of activated carbon-based adsorbent for removal of industrial dyes and heavy metals: A review. *Mater. Today: Proc.* **2023**, *77*, 286–294. [CrossRef]
35. Sultana, M.; Rownok, M.H.; Sabrin, M.; Rahaman, M.H.; Alam, S.M.N. A review on experimental chemically modified activated carbon to enhance dye and heavy metals adsorption. *Clean. Eng. Technol.* **2022**, *6*, 100382. [CrossRef]
36. Adegoke, K.A.; Akinnawo, S.O.; Adebusuyi, T.A.; Ajala, O.A.; Adegoke, R.O.; Maxakato, N.W.; Bello, O.S. Modified biomass adsorbents for removal of organic pollutants: A review of batch and optimization studies. *Int. J. Environ. Sci. Technol.* **2023**, *20*, 11615–11644. [CrossRef]
37. Mishra, S.; Cheng, L.; Maiti, A. The utilization of agro-biomass/byproducts for effective bio-removal of dyes from dyeing wastewater: A comprehensive review. *J. Environ. Chem. Eng.* **2021**, *9*, 104901. [CrossRef]
38. Ewis, D.; Ba-Abbad, M.M.; Benamor, A.; El-Naas, M.H. Adsorption of organic water pollutants by clays and clay minerals composites: A comprehensive review. *Appl. Clay Sci.* **2022**, *229*, 106686. [CrossRef]
39. Zhang, T.; Wang, W.; Zhao, Y.; Bai, H.; Wen, T.; Kang, S.; Song, G.; Song, S.; Komarneni, S. Removal of heavy metals and dyes by clay-based adsorbents: From natural clays to 1D and 2D nano-composites. *Chem. Eng. J.* **2021**, *420*, 127574. [CrossRef]
40. Waheed, A.; Baig, N.; Ullah, N.; Falath, W. Removal of hazardous dyes, toxic metal ions and organic pollutants from wastewater by using porous hyper-cross-linked polymeric materials: A review of recent advances. *J. Environ. Manag.* **2021**, *287*, 112360. [CrossRef]
41. Stejskal, J. Interaction of conducting polymers, polyaniline and polypyrrole, with organic dyes: Polymer morphology control, dye adsorption and photocatalytic decomposition. *Chem. Pap.* **2020**, *74*, 1–54. [CrossRef]
42. Biswas, S.; Pal, A. Application of biopolymers as a new age sustainable material for surfactant adsorption: A brief review. *Carbohydr. Polym. Technol. Appl.* **2021**, *2*, 100145. [CrossRef]
43. Crini, G. Non-conventional low-cost adsorbents for dye removal: A review. *Bioresour. Technol.* **2006**, *97*, 1061–1085. [CrossRef]
44. Liu, X.; Zhao, X.; Liu, Y.; Zhang, T. Review on preparation and adsorption properties of chitosan and chitosan composites. *Polym. Bull.* **2022**, *79*, 2633–2665. [CrossRef]
45. Desbrières, J.; Guibal, E. Chitosan for wastewater treatment. *Polym. Int.* **2018**, *67*, 7–14. [CrossRef]
46. Elzahar, M.M.H.; Bassyouni, M. Removal of direct dyes from wastewater using chitosan and polyacrylamide blends. *Sci. Rep.* **2023**, *13*, 15750. [CrossRef] [PubMed]
47. Mohan, K.; Rajan, D.K.; Rajarajeswaran, J.; Divya, D.; Ganesan, A.R. Recent trends on chitosan based hybrid materials for wastewater treatment: A review. *Curr. Opin. Environ. Sci. Health* **2023**, *33*, 100473. [CrossRef]
48. Aramesh, N.; Bagheri, A.R.; Bilal, M. Chitosan-based hybrid materials for adsorptive removal of dyes and underlying interaction mechanisms. *Int. J. Biol. Macromol.* **2021**, *183*, 399–422. [CrossRef]
49. Das, L.; Das, P.; Bhowal, A.; Bhattachariee, C. Synthesis of hybrid hydrogel nano-polymer composite using Graphene oxide, Chitosan and PVA and its application in waste water treatment. *Environ. Technol. Innov.* **2020**, *18*, 100664. [CrossRef]
50. Zhao, X.; Wang, X.; Lou, T. Simultaneous adsorption for cationic and anionic dyes using chitosan/electrospun sodium alginate nanofiber composite sponges. *Carbohydr. Polym.* **2022**, *276*, 118728. [CrossRef]
51. Wu, S.; Shi, W.; Li, K.; Cai, J.; Xu, C.; Gao, L.; Lu, J.; Ding, F. Chitosan-based hollow nanofiber membranes with polyvinylpyrrolidone and polyvinyl alcohol for efficient removal and filtration of organic dyes and heavy metals. *Int. J. Biol. Macromol.* **2023**, *239*, 124264. [CrossRef]
52. Khorshidi, A.G.; Khalaji, A.D. Chitosan-polyvinylpyrrolidone composite as an efficient adsorbent for eosin Y dye from aqueous solution. *J. Polym. Environ.* **2024**, *32*, 4577–4588. [CrossRef]
53. Zhang, Y.; Wang, F.; Wang, Y. Recent developments of electrospun nanofibrous materials as novel adsorbents for water treatment. *Mater. Today Commun.* **2021**, *27*, 102272. [CrossRef]

54. Alwi, M.A.M.; Normaya, E.; Ismail, H.; Iqbal, A.; Piah, B.M.; Samah, M.A.A.; Ahmad, M.N. Two-Dimensional Infrared Correlation Spectroscopy, Conductor-like Screening Model for Real Solvents, and Density Functional Theory Study on the Adsorption Mechanism of Polyvinylpolypyrrolidone for Effective Phenol Removal in an Aqueous Medium. *ACS Omega* **2021**, *6*, 25179–25192. [CrossRef] [PubMed]
55. Ougiya, H.; Hioki, N.; Watanabe, K.; Morinaga, Y.; Yoshinaga, F.; Samejima, M. Relationship between the Physical Properties and Surface Area of Cellulose Derived from Adsorbates of Various Molecular Sizes. *Biosci. Biotechnol. Biochem.* **1998**, *62*, 1880–1884. [CrossRef] [PubMed]
56. Spence, K.L.; Venditti, R.A.; Rojas, O.J.; Habibi, Y.; Pawlak, J.J. The effect of chemical composition on microfibrillar cellulose films from wood pulps: Water interactions and physical properties for packaging applications. *Cellulose* **2010**, *17*, 835–848. [CrossRef]
57. da Silva, P.M.M.; Camparotto, N.G.; de Figueiredo Neves, T.; Lira, K.T.G.; Mastelaro, V.R.; Picone, C.S.F.; Prediger, P. Effective removal of basic dye onto sustainable chitosan beads: Batch and fixed-bed column adsorption, beads stability and mechanism. *Sustain. Chem. Pharm.* **2020**, *18*, 100348. [CrossRef]
58. Zhang, W.; Wang, P.; Deng, Y.; He, X.; Yang, X.; Chen, R.; Lei, Z. Preparation of superabsorbent polymer gel based on PVPP and its application in water-holding in sandy soil. *J. Environ. Chem. Eng.* **2021**, *9*, 106760. [CrossRef]
59. Qiao, C.; Ma, X.; Wang, X.; Liu, L. Structure and properties of chitosan films: Effect of the type of solvent acid. *LWT* **2021**, *135*, 109984. [CrossRef]
60. Gupta, S.; Vasanth, D.; Kumar, A. Physicochemical analysis of chitosan oligosaccharide revealed its usefulness in effective delivery of drugs. *J. Biomater. Sci. Polym. Ed.* **2024**, *9*, 1–19. [CrossRef]
61. Han, X.; Zheng, Z.; Yu, C.; Deng, Y.; Ye, Q.; Niu, F.; Chen, Q.; Pan, W.; Wang, Y. Preparation, characterization and antibacterial activity of new ionized chitosan. *Carbohydr. Polym.* **2022**, *290*, 119490. [CrossRef] [PubMed]
62. Sudatta, B.P.; Sugumar, V.; Varma, R.; Nigariga, P. Extraction, characterization and antimicrobial activity of chitosan from pen shell, Pinna bicolor. *Int. J. Biol. Macromol.* **2020**, *163*, 423–430. [CrossRef]
63. Zhang, W.; Cao, J.; Jiang, W. Analysis of film-forming properties of chitosan with different molecular weights and its adhesion properties with different postharvest fruit surfaces. *Food Chem.* **2022**, *395*, 133605. [CrossRef]
64. Jabli, M.; Hamdaoui, M.; Jabli, A.; Ghandour, Y.; Hassine, B.B. A comparative study on the performance of dye removal, from aqueous suspension, using 2-hydroxypropyl-β-cyclodextrin-CS, PVP-PVA-CS, PVA-CS, PVP-CS and plain CS microspheres. *J. Text. Inst.* **2014**, *105*, 661–675. [CrossRef]
65. Pourhakkak, P.; Taghizadeh, A.; Taghizadeh, M.; Ghaedi, M.; Haghdoust, S. Chapter 1-Fundamentals of adsorption technology. In *Interface Science and Technology*; Ghaedi, M., Ed.; Elsevier: Amsterdam, The Netherlands, 2021; pp. 1–70. [CrossRef]
66. Tang, H.; Zhao, Y.; Shan, S.; Yang, X.; Liu, D.; Cui, F.; Xing, B. Theoretical insight into the adsorption of aromatic compounds on graphene oxide. *Environ. Sci. Nano* **2018**, *5*, 2357–2367. [CrossRef]
67. Yang, X.; Wan, Y.; Zheng, Y.; He, F.; Yu, Z.; Huang, J.; Wang, H.; Ok, Y.S.; Jiang, Y.; Gao, B. Surface functional groups of carbon-based adsorbents and their roles in the removal of heavy metals from aqueous solutions: A critical review. *Chem. Eng. J.* **2019**, *366*, 608–621. [CrossRef] [PubMed]
68. Mokhtar, A.; Abdelkrim, S.; Djelad, A.; Sardi, A.; Boukoussa, B.; Sassi, M.; Bengueddach, A. Adsorption behavior of cationic and anionic dyes on magadiite-chitosan composite beads. *Carbohydr. Polym.* **2020**, *229*, 115399. [CrossRef]
69. Zhang, Y.; Zhao, M.; Cheng, Q.; Wang, C.; Li, H.; Han, X.; Fan, Z.; Su, G.; Pan, D.; Li, Z. Research progress of adsorption and removal of heavy metals by chitosan and its derivatives: A review. *Chemosphere* **2021**, *279*, 130927. [CrossRef]
70. Wang, X.; Liu, Y.; Zheng, J. Removal of AsIII and AsV from water by chitosan and chitosan derivatives: A review. *Environ. Sci. Pollut. Res.* **2016**, *23*, 13789–13801. [CrossRef]
71. Timaeva, O.; Pashkin, I.; Mulakov, S.; Kuzmicheva, G.; Konarev, P.; Terekhova, R.; Sadovskaya, N.; Czakkel, O.; Prevost, S. Synthesis and physico-chemical properties of polyN-vinyl pyrrolidone-based hydrogels with titania nanoparticles. *J. Mater. Sci.* **2020**, *55*, 3005–3021. [CrossRef]
72. Du, R.; Cao, H.; Wang, G.; Dou, K.; Tsidaeva, N.; Wang, W. PVP modified rGO/CoFe2O4 magnetic adsorbents with a unique sandwich structure and superior adsorption performance for anionic and cationic dyes. *Sep. Purif. Technol.* **2022**, *286*, 120484. [CrossRef]
73. Sirajudheen, P.; Poovathumkuzhi, N.C.; Vigneshwaran, S.; Chelaveettil, B.M.; Meenakshi, S. Applications of chitin and chitosan based biomaterials for the adsorptive removal of textile dyes from water—A comprehensive review. *Carbohydr. Polym.* **2021**, *273*, 118604. [CrossRef]
74. Qamar, S.A.; Ashiq, M.; Jahangeer, M.; Riasat, A.; Bilal, M. Chitosan-based hybrid materials as adsorbents for textile dyes–A review. *Case Stud. Chem. Environ. Eng.* **2020**, *2*, 100021. [CrossRef]
75. Ahmadi, Y.; Kim, K.-H. Hyperbranched polymers as superior adsorbent for the treatment of dyes in water. *Adv. Colloid Interface Sci.* **2022**, *302*, 102633. [CrossRef] [PubMed]
76. Folch-Cano, C.; Olea-Azar, C.; Speisky, H. Structural and thermodynamic factors on the adsorption process of phenolic compounds onto polyvinylpolypyrrolidone. *Colloids Surf. A Physicochem. Eng. Asp.* **2013**, *418*, 105–111. [CrossRef]
77. Solanki, S.; Sinha, S.; Seth, C.S.; Tyagi, S.; Goyal, A.; Singh, R. Enhanced adsorption of Bismark Brown R dye by chitosan conjugated magnetic pectin loaded filter mud: A comprehensive study on modeling and mechanisms. *Int. J. Biol. Macromol.* **2024**, *270*, 131987. [CrossRef]

78. Dai, M. The Effect of Zeta Potential of Activated Carbon on the Adsorption of Dyes from Aqueous Solution: I. The Adsorption of Cationic Dyes: Methyl Green and Methyl Violet. *J. Colloid Interface Sci.* **1994**, *164*, 223–228. [CrossRef]
79. Sivashankar, R.; Sathya, A.B.; Vasantharaj, K.; Sivasubramanian, V. Magnetic composite an environmental super adsorbent for dye sequestration—A review. *Environ. Nanotechnol. Monit. Manag.* **2014**, *1–2*, 36–49. [CrossRef]
80. Huang, C.; Liao, H.; Ma, X.; Xiao, M.; Liu, X.; Gong, S.; Shu, X.; Zhou, X. Adsorption performance of chitosan Schiff base towards anionic dyes: Electrostatic interaction effects. *Chem. Phys. Lett.* **2021**, *780*, 138958. [CrossRef]
81. Gajera, R.; Patel, R.V.; Yadav, A.; Labhasetwar, P.K. Adsorption of cationic and anionic dyes on photocatalytic flyash/TiO2 modified chitosan biopolymer composite. *J. Water Process Eng.* **2022**, *49*, 102993. [CrossRef]
82. Lohrentz, L.; Bhaumik, M.; Brink, H.G. High-capacity adsorption of hexavalent chromium by a polyaniline-Ni0 nanocomposite adsorbent: Expanding the Langmuir-Hinshelwood kinetic model. *J. Mol. Liq.* **2023**, *389*, 122931. [CrossRef]
83. Adeyemo, A.A.; Adeoye, I.O.; Bello, O.S. Adsorption of dyes using different types of clay: A review. *Appl. Water Sci.* **2017**, *7*, 543–568. [CrossRef]
84. Mangla, D.; Sharma, A.; Ikram, S. Synthesis of ecological chitosan/PVP magnetic composite: Remediation of amoxicillin trihydrate from its aqueous solution, isotherm modelling, thermodynamic, and kinetic studies. *React. Funct. Polym.* **2022**, *175*, 105261. [CrossRef]
85. Zhou, Y.; Lu, J.; Zhou, Y.; Liu, Y. Recent advances for dyes removal using novel adsorbents: A review. *Environ. Pollut.* **2019**, *252*, 352–365. [CrossRef] [PubMed]
86. Yagub, M.T.; Sen, T.K.; Afroze, S.; Ang, H.M. Dye and its removal from aqueous solution by adsorption: A review. *Adv. Colloid Interface Sci.* **2014**, *209*, 172–184. [CrossRef] [PubMed]
87. Ouachtak, H.; El Guerdaoui, A.; Haounati, R.; Akhouairi, S.; El Haouti, R.; Hafid, N.; Addi, A.A.; Šljukić, B.; Santos, D.M.F.; Taha, M.L. Highly efficient and fast batch adsorption of orange G dye from polluted water using superb organo-montmorillonite: Experimental study and molecular dynamics investigation. *J. Mol. Liq.* **2021**, *335*, 116560. [CrossRef]
88. Mondal, M.K.; Singh, S.; Umareddy, M.; Dasgupta, B. Removal of Orange G from aqueous solution by hematite: Isotherm and mass transfer studies. *Korean J. Chem. Eng.* **2010**, *27*, 1811–1815. [CrossRef]
89. Banerjee, S.; Dubey, S.; Gautam, R.K.; Chattopadhyaya, M.C.; Sharma, Y.C. Adsorption characteristics of alumina nanoparticles for the removal of hazardous dye, Orange G from aqueous solutions. *Arab. J. Chem.* **2019**, *12*, 5339–5354. [CrossRef]
90. Sircar, S. Comments on practical use of Langmuir gas adsorption isotherm model. *Adsorption* **2017**, *23*, 121–130. [CrossRef]
91. Alafnan, S.; Awotunde, A.; Glatz, G.; Adjei, S.; Alrumaih, I.; Gowida, A. Langmuir adsorption isotherm in unconventional resources: Applicability and limitations. *J. Pet. Sci. Eng.* **2021**, *207*, 109172. [CrossRef]
92. Ebadi, A.; Mohammadzadeh, J.S.S.; Khudiev, A. What is the correct form of BET isotherm for modeling liquid phase adsorption? *Adsorption* **2009**, *15*, 65–73. [CrossRef]
93. van Erp, T.S.; Martens, J.A. A standardization for BET fitting of adsorption isotherms. *Microporous Mesoporous Mater.* **2011**, *145*, 188–193. [CrossRef]
94. Al-Ghouti, M.A.; Da'ana, D.A. Guidelines for the use and interpretation of adsorption isotherm models: A review. *J. Hazard. Mater.* **2020**, *393*, 122383. [CrossRef]
95. Skopp, J. Derivation of the Freundlich Adsorption Isotherm from Kinetics. *J. Chem. Educ.* **2009**, *86*, 1341. [CrossRef]
96. Lagergren, S.K. About the Theory of So-called Adsorption of Soluble Substances. *Sven. Vetenskapsakad. Handingarl.* **1898**, *24*, 1–39.
97. Ho, Y.S.; McKay, G. Pseudo-second order model for sorption processes. *Process Biochem.* **1999**, *34*, 451–465. [CrossRef]
98. Rudzinski, W.; Plazinski, W. Studies of the Kinetics of Solute Adsorption at Solid/Solution Interfaces: On the Possibility of Distinguishing between the Diffusional and the Surface Reaction Kinetic Models by Studying the Pseudo-First-order Kinetics. *J. Phys. Chem. C* **2007**, *111*, 15100–15110. [CrossRef]
99. Salvestrini, S. Analysis of the Langmuir rate equation in its differential and integrated form for adsorption processes and a comparison with the pseudo first and pseudo second order models. *Reac Kinet. Mech. Cat.* **2018**, *123*, 455–472. [CrossRef]
100. Xiao, Y.; Azaiez, J.; Hill, J.M. Erroneous Application of Pseudo-Second-Order Adsorption Kinetics Model: Ignored Assumptions and Spurious Correlations. *Ind. Eng. Chem. Res.* **2018**, *57*, 2705–2709. [CrossRef]
101. Weber, W.J.; Morris, J.C. Kinetics of Adsorption on Carbon from Solution. *J. Sanit. Eng. Div.* **1963**, *89*, 31–59. [CrossRef]
102. Boyd, G.E.; Adamson, A.W.; Myers, L.S. The Exchange Adsorption of Ions from Aqueous Solutions by Organic Zeolites. II. Kinetics [1]. *J. Am. Chem. Soc.* **1947**, *69*, 2836–2848. [CrossRef]
103. Reichenberg, D. Properties of Ion-Exchange Resins in Relation to their Structure. III. Kinetics of Exchange. *J. Am. Chem. Soc.* **1953**, *75*, 589–597. [CrossRef]
104. Malash, G.F.; El-Khaiary, M.I. Piecewise linear regression: A statistical method for the analysis of experimental adsorption data by the intraparticle-diffusion models. *Chem. Eng. J.* **2010**, *163*, 256–263. [CrossRef]
105. Hu, Q.; Ma, S.; He, Z.; Liu, H.; Pei, X. A revisit on intraparticle diffusion models with analytical solutions: Underlying assumption, application scope and solving method. *J. Water Process Eng.* **2024**, *60*, 105241. [CrossRef]
106. Wang, J.; Guo, X. Rethinking of the intraparticle diffusion adsorption kinetics model: Interpretation, solving methods and applications. *Chemosphere* **2022**, *309*, 136732. [CrossRef] [PubMed]
107. Wu, F.-C.; Tseng, R.-L.; Juang, R.-S. Initial behavior of intraparticle diffusion model used in the description of adsorption kinetics. *Chem. Eng. J.* **2009**, *153*, 1–8. [CrossRef]

108. Saha, P.; Chowdhury, S. Insight Into Adsorption Thermodynamics. In *Thermodynamics*; Tadashi, M., Ed.; InTechOpen Limited: London, UK, 2011. [CrossRef]
109. Cristóvão, R.O.; Tavares, A.P.M.; Ferreira, L.A.; Loureiro, J.M.; Boaventura, R.A.R.; Macedo, E.A. Modeling the discoloration of a mixture of reactive textile dyes by commercial laccase. *Bioresour. Technol.* **2009**, *100*, 1094–1099. [CrossRef] [PubMed]
110. Routoula, E.; Patwardhan, S.V. Degradation of Anthraquinone Dyes from Effluents: A Review Focusing on Enzymatic Dye Degradation with Industrial Potential. *Environ. Sci. Technol.* **2020**, *54*, 647–664. [CrossRef]
111. Tilli, S.; Ciullini, I.; Scozzafava, A.; Briganti, F. Differential decolorization of textile dyes in mixtures and the joint effect of laccase and cellobiose dehydrogenase activities present in extracellular extracts from *Funalia trogii*. *Enzym. Microb. Technol.* **2011**, *49*, 465–471. [CrossRef]

polymers

MDPI

Article

Adsorbent-Embedded Polymeric Membranes for Efficient Dye-Water Treatment

Junaid Saleem [1,*,†], Zubair Khalid Baig Moghal [2,†], Snigdhendubala Pradhan [1], Ahsan Hafeez [2], Mohammad Shoaib [2], Johaina Alahmad [2] and Gordon McKay [1]

1 Division of Sustainable Development, College of Science and Engineering, Hamad Bin Khalifa University, Qatar Foundation, Doha 5825, Qatar
2 Center for Advanced Materials, Qatar University, Doha 2713, Qatar
* Correspondence: jsaleem@hbku.edu.qa
† These authors contributed equally to this work.

Abstract: Traditional bulk adsorbents, employed for the removal of dyes and metal ions, often face the drawback of requiring an additional filtration system to separate the filtrate from the adsorbent. In this study, we address this limitation by embedding the adsorbent into the polymer matrix through a process involving dissolution–dispersion, spin-casting, and heat-stretching. Selective dissolution and dispersion facilitate the integration of the adsorbent into the polymer matrix. Meanwhile, spin-casting ensures the formation of a uniform and thin film structure, whereas heat-induced stretching produces a porous matrix with a reduced water contact angle. The adsorbent selectively captures dye molecules, while the porous structure contributes to water permeability. We utilized inexpensive and readily available materials, such as waste polyethylene and calcium carbonate, to fabricate membranes for the removal of methylene blue dye. The effects of various parameters, such as polymer-adsorbent ratio, initial dye concentration, and annealing temperature, were investigated. Equilibrium data were fitted to Langmuir, Freundlich, Temkin, and Dubinin–Radushkevich isotherms. The equilibrium data were best represented by the Langmuir isotherm, with maximum adsorption capacity of 35 mg/g and 43 mg/g at 25 °C and 45 °C, respectively. The membranes can be regenerated and recycled with a 97% dye removal efficiency. The study aims to present a template for adsorbent-embedded polymeric membranes for dye removal, in which adsorbent can be tailored to enhance adsorption capacity and efficiency.

Keywords: mineral-filled membranes; polymeric films; plastic waste; polyethylene and mixed-matrix membranes

1. Introduction

Mineral-filled or heterogeneous polymeric films have become integral in the realm of separation applications [1–3]. These minerals are used as adsorbents in gas separation techniques [4–7] and as fillers in the hygiene market [8]. Specifically, calcium carbonate ($CaCO_3$) emerges as the predominant filler in polymeric films within the hygiene industry [9]. The role of filler is limited to heat-stretching, in which it ensures uniform pores formation at the interface between the polymer and itself, imparting a microporous structure [10–14]. This unique design allows for the diffusion of water vapor while simultaneously acting as a barrier against liquid water [15]. The microporous configuration facilitates better air circulation and minimizes trapped moisture, playing a pivotal role in preventing skin rashes and irritations. However, the filler does not participate in the separation phenomena—adsorption or absorption.

In addition to its role as a filler, $CaCO_3$ is widely used as an adsorbent for the removal of dyes and metal ions from wastewater [3]. Other commonly employed adsorbents are activated carbon, zeolites, and molecular sieves. The effectiveness of these adsorbents relies on two things: (a) high internal surface area and (b) the presence of active sites in the form of functional groups [16,17]. Although they are effective in the sorption of pollutants, their bulk form necessitates an additional filtration system to separate the filtrate from the adsorbent. This limitation could be overcome by employing a method that involves embedding adsorbent, such as $CaCO_3$, into the polymer matrix.

This study focuses on the development of adsorbent-embedded polymeric membranes tailored for the efficient treatment of water contaminated with dyes. Our approach draws inspiration from the conventional utilization of $CaCO_3$, which serves as a filler in polymeric films within the hygiene industry and an adsorbent for the removal of dyes and metal ions from wastewater. Unlike traditional filler applications, where $CaCO_3$ is typically used in smaller quantities—ranging from 1–25% [4,15,18], often around 10%—our methodology distinguishes itself by employing a significantly larger amount, approximately 300% (three times the mass of the polymer). This increase aims to maximize the adsorption of dyes within the polymer matrix and sets our approach apart from conventional practices in filler applications.

The novelty in our approach lies in crafting a porous adsorbent–polymer composite membrane. This involves a sequential process of dissolution–dispersion, spin-casting, and heat-induced stretching. Most reported literature preferred extrusion–stretching routes in cases where semi-crystalline polymers—such as polyolefins—are involved [19–22]. Among polyolefins, polyethylene (PE) is the most chosen one due to its favorable attributes, which include low cost, processability, scalability, and versatility [12]. The integration of the adsorbent into the polymer matrix is facilitated by selective dissolution and dispersion. Simultaneously, spin-casting ensures the formation of a uniform and thin film structure [23,24], while heat-induced stretching results in a porous matrix. Notably, without the formation of a porous structure, this composite becomes unsuitable for dye-water separation, as the absence of pores renders the films impermeable to the passage of dye-water. The adsorbent selectively captures dye molecules, while the porous structure contributes to water permeability.

We opted for inexpensive and readily available materials, such as waste polyethylene and $CaCO_3$, in the fabrication of membranes designed for the removal of methylene blue (MB) dye. MB is an organic dye with a cationic nature and is prevalent as a contaminant in industrial wastewater originating from various sectors, including plastics, textiles, cosmetics, and food manufacturing [1]. This holistic process demonstrates the efficiency and practicality of our membrane development methodology.

The purpose of this work was to evaluate the adsorption potential of adsorbent-embedded polymeric membranes using inexpensive and readily available materials—waste polyethylene and $CaCO_3$—for MB dye. The equilibrium data of the adsorption isotherm were then studied to understand the mechanism of dye molecules onto the prepared composite membrane. This study presents a template for adsorbent-embedded polymeric membranes for dye removal, in which adsorbent can be tailored to enhance adsorption capacity and efficiency. These membranes hold significant promise for applications in industries such as textiles, dyeing, and wastewater treatment, where the removal of dyes from water is essential for both environmental sustainability and regulatory compliance.

2. Experimental Section

2.1. Materials

p-cymene, sourced from Njnq Bio-Tech Ltd. (Beijing, China), was used as a solvent for dissolving polyethylene without additional purification. Locally collected waste high-density polyethylene (HDPE) bottles were employed. Ultrahigh molecular weight polyethylene (UHMWPE) with a molecular weight ranging from 3–6 million g/mol was obtained from Sigma (St. Louis, MO, USA). $CaCO_3$, procured from Sigma, served as an adsorbent without additional purification. Methylene Blue (MB) was supplied by Sigma. For membrane fabrication, a customized glass plate with a surface area of 25 cm^2 was used as the solid substrate. Thin membrane films were annealed on a Heidolph hotplate (Schwabach, Germany).

2.2. Characterization Techniques

Tensile strength was measured using a friction/peel tester from Lloyd Instruments Ltd. (London, UK). Scanning electron microscope (SEM) images was captured with FEI Quanta650FEG (Hillsboro, OR, USA). Differential scanning calorimetry (DSC) determined the thermal behavior of hydrophobic thin films. X-ray diffraction (XRD) measurements were carried out using PANalytical Empyrean multipurpose XRD by Malvern Panalytical, Malvern, UK. Fourier-transform infrared spectroscopy (FTIR) was performed using the PerkinElmer Frontier instrument (Waltham, MA, USA). Thickness was measured with a micrometer and cross-referred with Deflesko FS3 PosiTector 6000 (Ogdensburg, NY, USA) using an iron metal base.

2.3. Membrane Preparation

A composite solution denoted as PE:$CaCO_3$ (1:1), was prepared by dissolving 500 mg of UHMWPE and 500 mg of HDPE in 100 mL of p-cymene, followed by the addition of 1 g of dried $CaCO_3$. The solution was heated to 130 °C for 20 min to obtain a completely dispersed composite solution. Simultaneously, a cleaned glass substrate was heated to 120 °C, and the hot composite solution was spin-coated onto the substrate. The spin coating process involved four steps: 300 rpm for 10 s, 700 rpm for 30 s, 1000 rpm for 60 s, and finally, 3000 rpm for 120 s. This spinning allows the formation of a thin membrane and also expels the solvent from the formed thin membrane. The glass substrate with the polymer composite was then placed on a hot plate at 130 °C for 2–5 min. The resulting polymer-composite membrane was peeled off and subjected to uniaxial stretching to obtain the desired thin film membranes. Similar procedures were followed to prepare PE:$CaCO_3$ (1:2) and PE:$CaCO_3$ (1:3) composite solutions, adjusting the amount of added $CaCO_3$ accordingly.

2.4. Batch Equilibrium Studies

Methylene blue (MB) was used as an adsorbate. All solutions were prepared using distilled water. The choice of MB was based on its recognized strong affinity for adsorption onto solid surfaces [25]. The polymer-composite membranes, namely PE:$CaCO_3$ (1:1), PE:$CaCO_3$ (1:2), and PE:$CaCO_3$ (1:3), were employed for the removal of MB from water at varying concentrations (1, 2, 5, 10, 20, and 100 mg/L). Each experiment was repeated under identical conditions. The concentrations of MB in the solutions before and after adsorption were determined using a double-beam UV–vis spectrophotometer. The amount of adsorption at equilibrium, q_e (mg/g), was calculated by

$$q_e = \frac{(C_0 - C_e)V}{W} \tag{1}$$

where C_0 and C_e (mg/L) are the liquid-phase concentrations of dye at initial and equilibrium, respectively. V is the volume of the solution (L), and W is the mass (g) of dry adsorbent used. Equilibrium data were fitted to Langmuir, Freundlich, Temkin, and Dubinin–Radushkevich adsorption isotherms.

2.5. Membrane Regeneration

For regeneration, the membrane was immersed in methanol and stirred for 10–15 min or until the stains of dye were removed. This process dissolved MB in methanol, allowing the membrane to be regenerated and recycled.

3. Results and Discussions

The primary objective of incorporating the adsorbent into the polymer matrix is to augment adsorption capacity; thus, the stretching step exposes the adsorbent to the dye water. Moreover, in order to ensure optimal integration within the polymer matrix, as opposed to mere surface adherence, it is crucial to note that the selected adsorbents for this system should exhibit a density equal to or greater than that of the polymer–solvent matrix.

3.1. Morphology and Surface Properties

SEM was employed to analyze the porous structure of the adsorbent-embedded polymeric membrane. The presented images in Figure 1 depict varying PE:$CaCO_3$ ratios—(a–b) 1:1, (c–d) 1:2, and (e–f) 1:3. In all three instances, micropores were formed through heat-stretching at the interface between the polymer and the adsorbent. The third case (1:3) reveals the highest amount of adsorbent distributed uniformly throughout the polymer matrix. The formation of pores facilitates the entry of dye water into the matrix, enabling efficient dye adsorption. Importantly, the heat-stretching procedure allows for in-depth adsorption of dye molecules, moving beyond mere surface adsorption. In essence, this unique design promotes adsorption in all three dimensions.

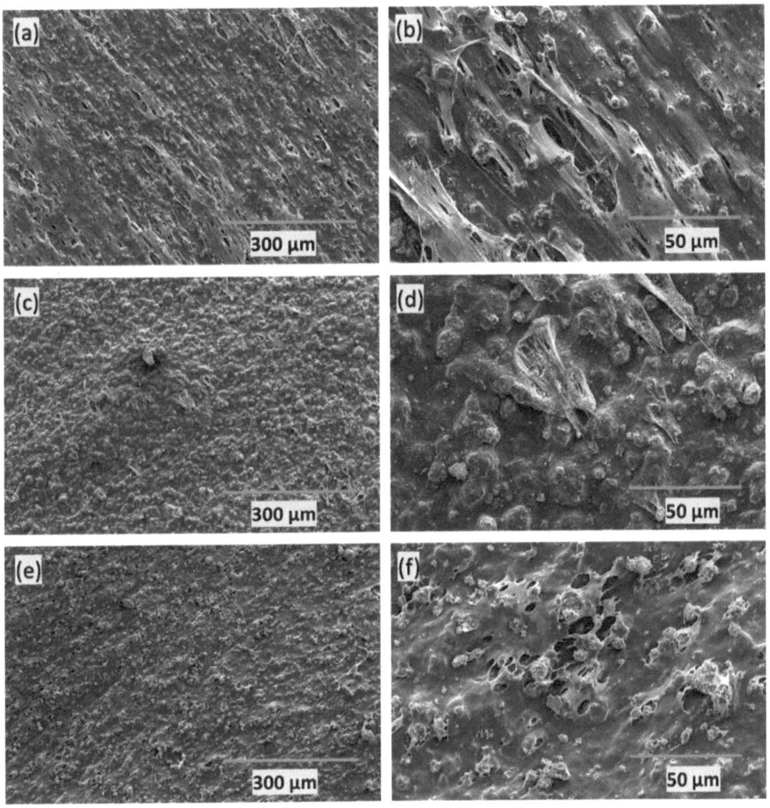

Figure 1. SEM images of membranes with different PE:$CaCO_3$ ratios (**a**,**b**) 1:1, (**c**,**d**) 1:2, and (**e**,**f**) 1:3.

Figure 2 illustrates the contact angle of the composite membrane. In order to improve dye-water penetration, the membrane should not be superhydrophobic (contact angle ~150°). Our spin-casting process yielded an extremely rough membrane surface with a water contact angle close to 148°, as reported previously [24]. However, the heat-stretching step facilitated in lowering of the water contact angle due to the heating of the polymer, which created oxygen moieties, as reported in our previous study [26].

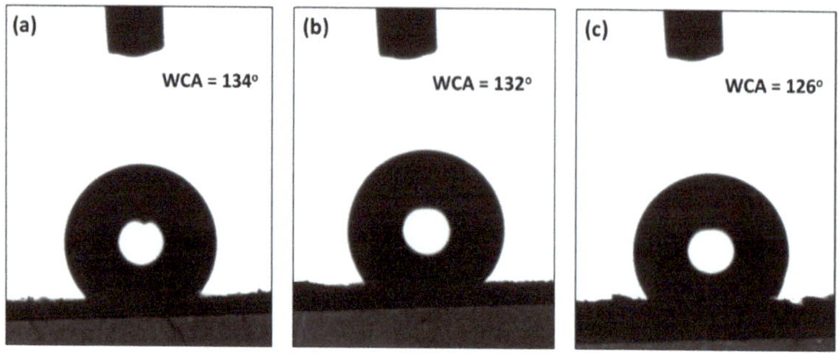

Figure 2. Contact angle of water on PE and $CaCO_3$ (**a**) 1:1, (**b**) 1:2, and (**c**) 1:3 composite membranes.

3.2. Chemical Composition and Thermal Behaviour

To validate the integration of the adsorbent into the polymer matrix without altering the PE structure, we conducted a characterization of the membrane using XRD. Figure 3 presents the spectra of PE both with and without the adsorbent. The XRD analysis provides further confirmation of the adsorbent's presence within the membrane structure.

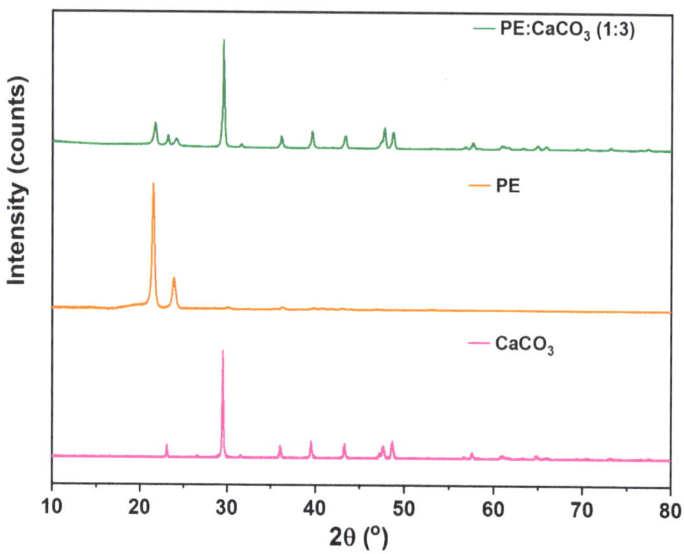

Figure 3. XRD patterns for PE and $CaCO_3$ (1:3) composite membrane and pure PE.

The significance of this examination lies in addressing the potential release of the adsorbent during synthesis, particularly given its considerably higher density compared to the polymer. The polymer (PE) exhibits characteristic peaks at 22° and 24°, serving as reference points. All other identifiable peaks in the XRD spectrum are attributed to the adsorbent. Notably, the intensity of the PE characteristic peaks appeared smaller due to the relatively higher amount of the adsorbent.

The incorporation of the adsorbent into the polymer matrix is further verified through FTIR. This confirmation is achieved by comparing the spectra of the composite membrane with those of pure PE, as illustrated in Figure 4. The adsorbent's peaks are distinctly observed at around 1440 (asymmetric CO stretching), 873–898 cm^{-1} (out-of-plane deformation of carbonate), and 712 (OCO bending in-plane deformation vibrations of $CaCO_3$), respectively, confirming the presence of $CaCO_3$ and its successful integration with the polymer [18]. At 1500 cm^{-1}, the PE peak merges with the peak of $CaCO_3$.

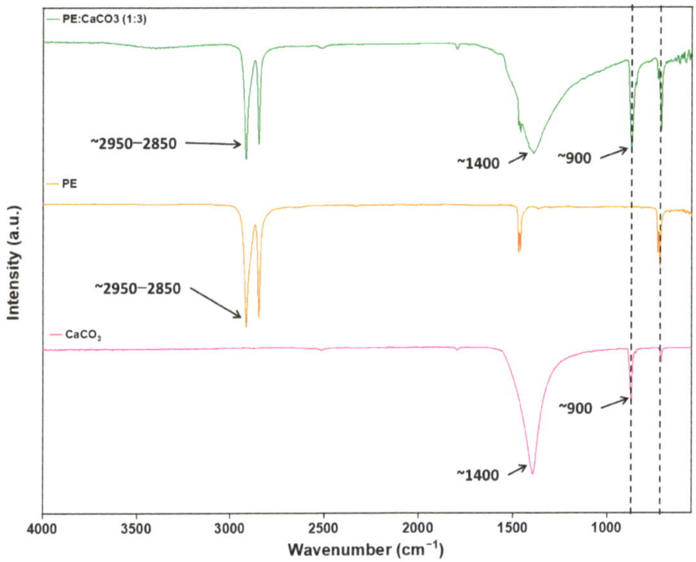

Figure 4. FTIR spectra of PE and $CaCO_3$ (1:3) composite membrane, pure PE and pure $CaCO_3$.

Notably, the absence of new peaks in the spectrum suggests that there are no chemical changes occurring within the membrane. Instead, the interaction between the polymer and the adsorbent is predominantly intermolecular dispersion forces. This is supported by the observation that both PE and $CaCO_3$ retain their characteristic peaks, signifying the preservation of their individual chemical identities within the composite structure.

The thermal behavior of the membrane was examined by DSC and is presented in Figure 5. As observed, the melting point of the PE did not change due to the incorporation of the adsorbent, which further confirms that no new formulation or compound formed, and PE retains its characteristics.

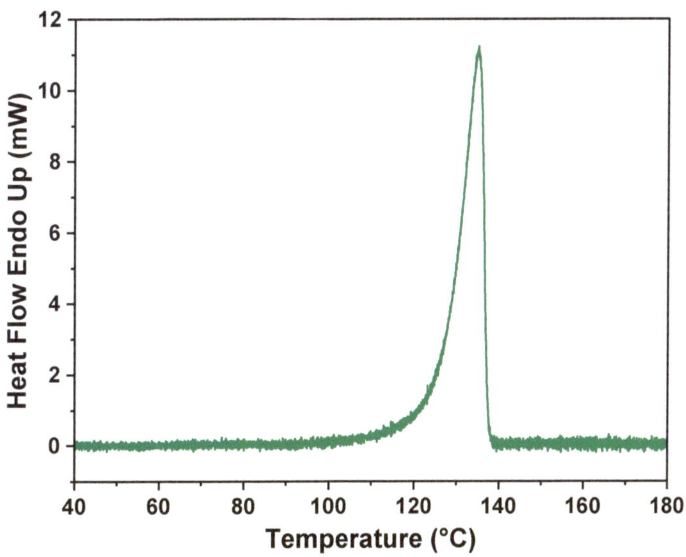

Figure 5. DSC spectra of PE and $CaCO_3$ (1:3) composite membrane.

3.3. Mechanical Properties

Figure 6 presents the impact of the adsorbent on the tensile strength of the composite membrane. Four cases were studied with an increasing amount of adsorbent in the polymer matrix—1:0, 1:1, 1:2, and 1:3. As anticipated, the tensile strength decreases as the number of adsorbent increases in the polymer matrix, attributed to the increase in breakpoints within the structure.

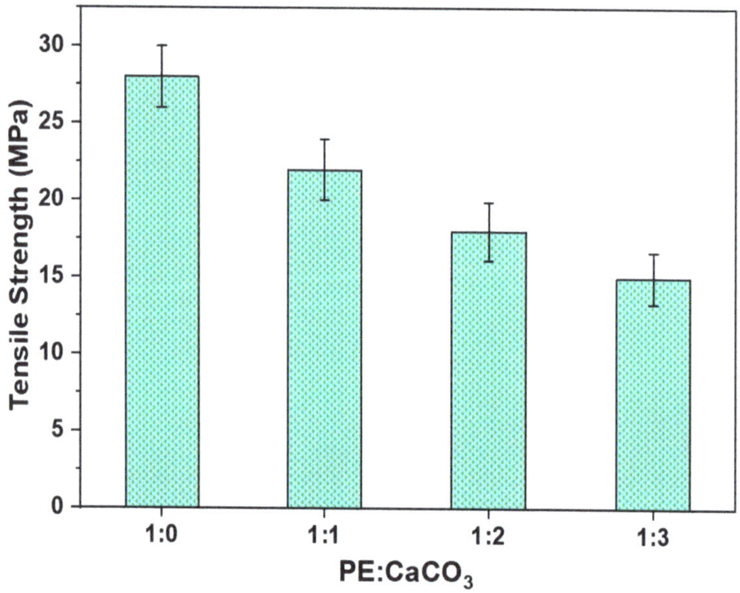

Figure 6. Tensile strength of the membranes with different PE:$CaCO_3$ ratios.

The membrane exhibited the highest tensile strength (28 MPa) when the adsorbent was not added to its structure, whereas it showed the lowest value (14 MPa) when the polymer–adsorbent ratio was 1:3. The tensile strength of 14 MPa was sufficient for the application of the composite membrane in dye-water treatment. This is particularly crucial for the membrane's regeneration and recyclability, as evident through recyclability studies that will be discussed further.

3.4. Adsorption Performance of Embedded Membrane

Figure 7a,b depicts the performance test of the adsorbent-embedded membrane, demonstrating a rise in adsorption capacity with an increasing amount of adsorbent and equilibrium concentration of the adsorbate. The maximum experimental adsorption values were 35 mg/g at 25 °C and 43 mg/g at 45 °C. In Figure 7b, the adsorption equilibrium is illustrated against various equilibrium MB concentrations against various temperatures. It was observed that for low MB equilibrium concentrations, the impact of temperature on adsorption equilibrium was not significant. However, at higher equilibrium concentrations (>60 mg/L), the adsorption equilibrium increased with a rise in temperature. This observed temperature-dependent adsorption trend aligns with the notion that elevated temperatures expedite the diffusion rates of dye molecules into the porous structure of the membrane. The heightened adsorption capacity may be attributed to chemical interactions between adsorbates and adsorbent, the creation of new adsorption sites, or an increased rate of intraparticle diffusion of MB molecules into the adsorbent pores at higher temperatures [25]. This improved diffusion allows for a more effective utilization of available adsorption sites, leading to increased adsorption. Furthermore, the temperature increase potentially enhances the chemical interactions between the surface functional groups of the membrane—

facilitated through $CaCO_3$—and the dye molecules. This, in turn, results in stronger adsorption bonds and increased adsorption efficiency.

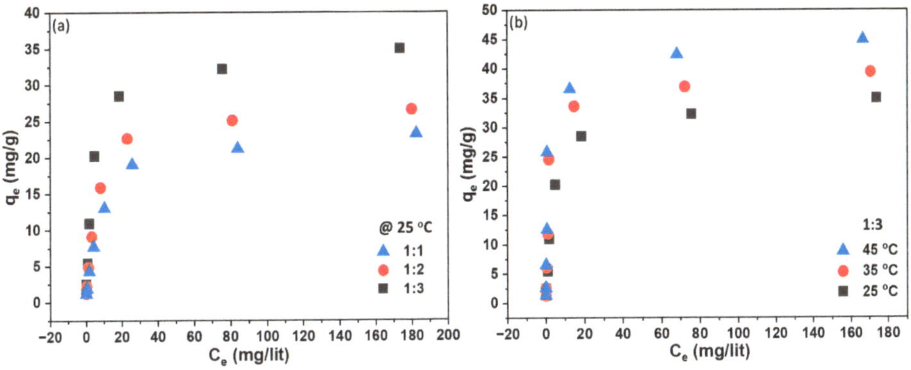

Figure 7. Plot of adsorption isotherm at equilibrium capacity: (**a**) with different membranes at 25 °C; (**b**) 1:3 membrane at different temperatures.

3.5. Isotherm Study

Adsorption isotherms are determined under equilibrium conditions. The experimental values were compared with four adsorption isotherms, namely Langmuir, Freundlich, Temkin, and Dubinin–Radushkevich. The equilibrium data were best represented by the Langmuir isotherm, with maximum theoretical monolayer adsorption capacity of 34.8 mg/g and 42.4 mg/g at 25 °C and 45 °C, respectively. The results of the isotherm study are presented in Figures 8–10, whereas the model equations, parameters, and their values are tabulated in Tables 1–3.

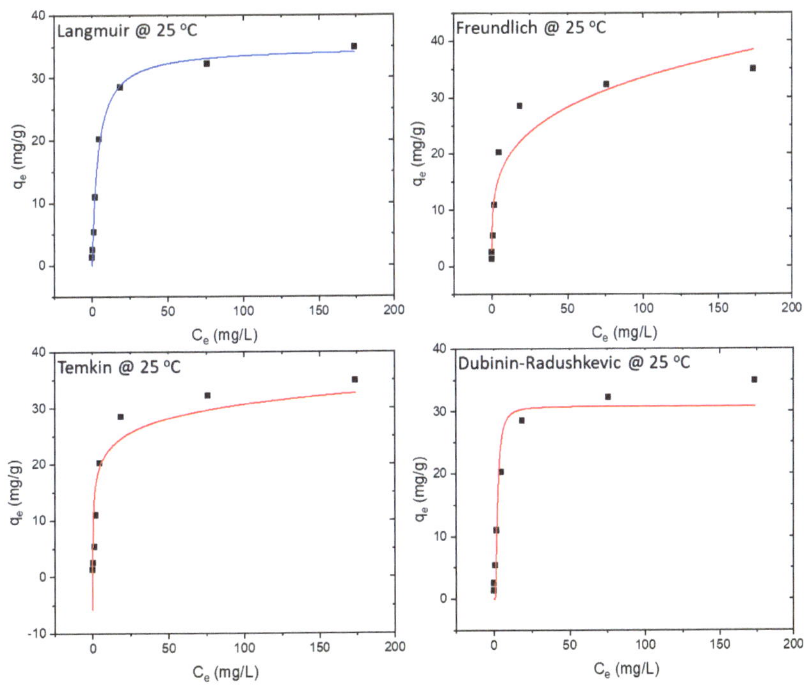

Figure 8. Isotherm models at 25 °C.

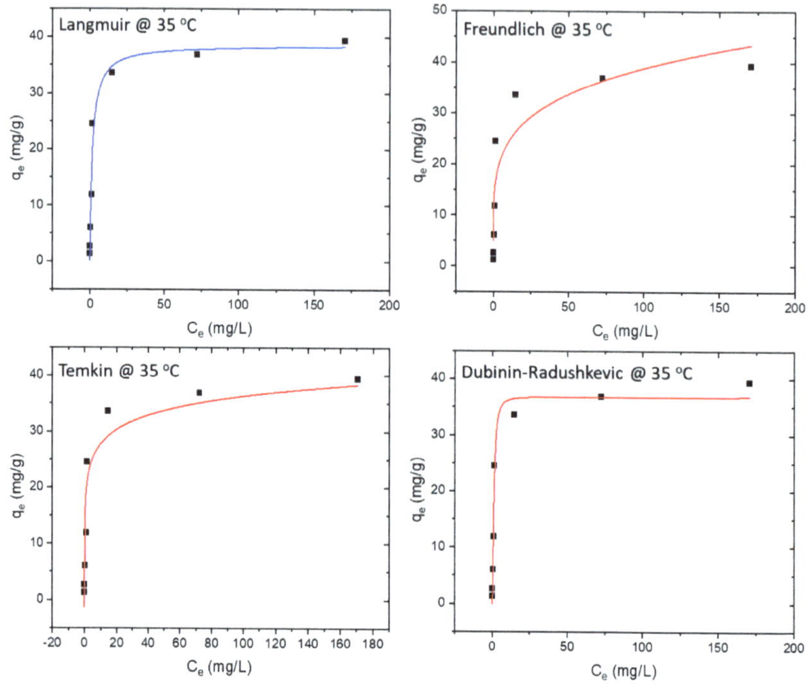

Figure 9. Isotherm models at 35 °C.

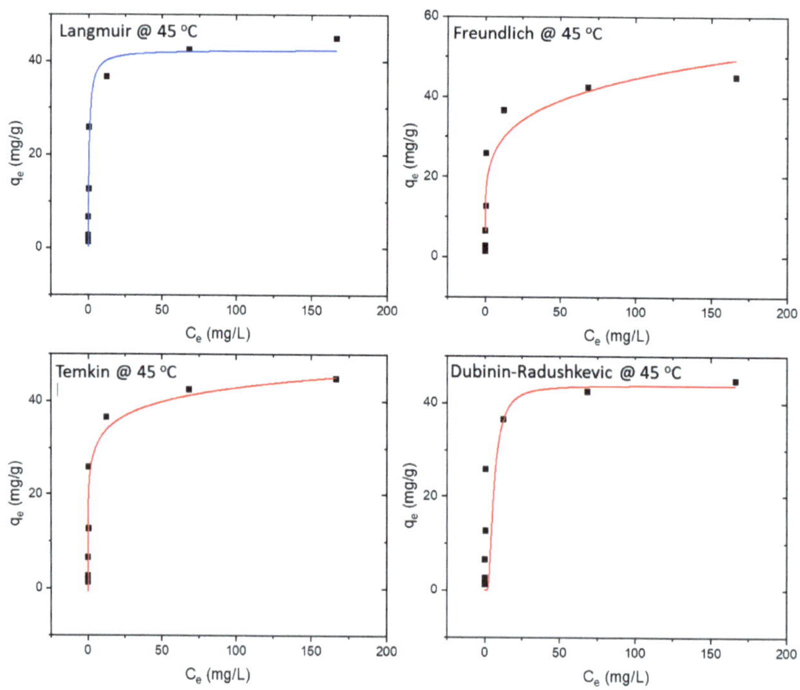

Figure 10. Isotherm models at 45 °C.

Table 1. Isotherm model equations with their respective parameters and values at 25 °C.

Model	Independent	Dependent	Equation	Parameters	Values
Langmuir Non-linear	C_e	q_e	$\frac{(K_L)(q_m)(Ce)}{1+((K_L)(Ce))}$	q_m	34.84136
				K_L	0.2561
				R^2	0.99329
Freundlich Non-linear	C_e	q_e	$(K_F)(Ce^{1/n})$	n	4.06845
				K_F	10.81772
				R^2	0.91254
Temkin	C_e	q_e	$\left(\frac{RT}{b_T}\right)\ln(A_T Ce)$	b_T	683.10816
				A_T	47.25523
				R^2	0.86343
Dubinin–Radushkevich	C_e	q_e	$(q_{DR})\exp(-B_{DR}\cdot\varepsilon^2)$	q_{DR}	30.82449
			$\varepsilon = RT\ln[1+\frac{1}{C_e}]$	B_{DR}	9.17907
				ε	738.05128
				R^2	0.94599

Table 2. Isotherm models and parameters at 35 °C.

Model	Parameters	Values
Langmuir Non-linear	q_m	38.63215
	K_L	0.60493
	R^2	0.96651
Freundlich Non-linear	n	4.86823
	K_F	15.09673
	R^2	0.88229
Temkin	b_T	685.05074
	A_T	163.04295
	R^2	0.87783
Dubinin–Radushkevich	q_{DR}	36.80169
	B_{DR}	3.28022
	ε	1234.62124
	R^2	0.96695

Table 3. Isotherm models and parameters at 45 °C.

Model	Parameters	Values
Langmuir Non-linear	q_m	42.41439
	K_L	1.38943
	R^2	0.95667
Freundlich Non-linear	n	5.17984
	K_F	18.33436
	R^2	0.90241
Temkin	b_T	610.96986
	A_T	198.08295
	R^2	0.93852
Dubinin–Radushkevich	q_{DR}	43.82223
	B_{DR}	4.46211
	ε	334.74559
	R^2	0.61681

The Langmuir isotherm assumes monolayer adsorption onto a surface with a finite number of uniform adsorption sites, involving no transmigration of adsorbate within the surface plane [27]. The Langmuir equation, as detailed in Table 1, features C_e as the equilibrium concentration of the adsorbate (mg/L) and q_e as the amount of adsorbate adsorbed per unit mass of adsorbent (mg/g). The R^2 value of 0.993 indicates that the adsorption data for MB onto the membrane at all three temperatures—25 °C, 35 °C, and 45 °C adhered best to the Langmuir isotherm model.

The Freundlich isotherm, on the other hand, assumes heterogeneous surface energies, wherein the energy term in the Langmuir equation varies with surface coverage [27]. Represented by the equation in Table 1, the Freundlich isotherm involves C_e as the equilibrium concentration of the adsorbate (mg/L) and q_e as the amount of adsorbate adsorbed per unit mass of adsorbent (mg/g). The parameter 'n' provides an indication of the favorability of the adsorption process, with a value below one indicating a normal Langmuir isotherm and a value above one suggestive of cooperative adsorption.

The Temkin model (TM) assumes that adsorption heat as a function of all molecules' temperature in the layer declines linearly rather than logarithmically due to the surface coverage increase [28,29]. The Temkin isotherm, detailed in Table 1, reflects these considerations.

For isotherms displaying a high degree of rectangularity, another widely used equation is the Dubinin–Radushkevich isotherm [27], also presented in Table 1.

Figure 11 illustrates the regeneration cycles of the composite membrane. As demonstrated in the tensile strength section, the membrane exhibits sufficient strength to maintain its structure throughout the regeneration and recycling processes. In the regeneration experiment, it was observed that $CaCO_3$ was susceptible to sonication, and the process could even break its particles, allowing them to dislodge from the polymer matrix. Therefore, it is recommended to employ normal stirring to effectively remove dye from the membrane. In recycling experiments, we observed a 97% dye removal efficiency even after the fifth cycle.

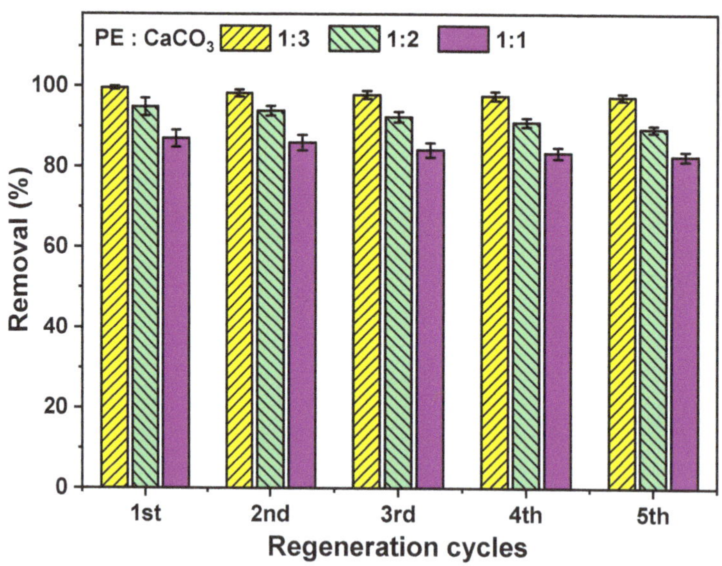

Figure 11. Regeneration cycles of PE:$CaCO_3$ membranes for dye removal.

4. Conclusions

Our study successfully addresses the limitations associated with traditional bulk adsorbents by introducing an innovative approach in the form of adsorbent-embedded polymeric membranes. The incorporation of $CaCO_3$ into a polyethylene matrix through a meticulous process involving dissolution–dispersion, spin-casting, and heat-induced stretching results in a porous composite membrane with significant advantages in dye-water treatment, especially in both selective dye molecule capture and water permeability. This porous configuration allows for three-dimensional adsorption, enhancing the overall efficiency of dye removal.

The adsorbent-embedded polymeric membranes exhibit promising efficiency in the removal of MB dye from water. The Langmuir isotherm fitting of equilibrium data reveals a maximum monolayer adsorption capacity of 34.8 mg/g at 25 °C and 42.4 mg/g at 45 °C. Notably, the membranes can be regenerated and recycled with a 97% dye removal efficiency, demonstrating their sustainability and practicality.

Moreover, our choice of inexpensive and readily available materials, such as waste polyethylene and $CaCO_3$, adds to the cost-effectiveness and feasibility of the proposed membrane fabrication methodology. The membranes demonstrate significant promise for applications in industries such as textiles, dyeing, and wastewater treatment, addressing the critical need for efficient dye removal for environmental sustainability and regulatory compliance.

The adsorbent-embedded polymeric membranes presented in this study provide a template for an efficient approach to dye-water treatment, combining the advantages of adsorbent materials with the versatility of polymeric matrices. This research contributes to the ongoing efforts to develop sustainable and effective solutions for water treatment challenges in various industrial sectors.

Author Contributions: Conceptualization, J.S. and Z.K.B.M.; Methodology, Z.K.B.M., S.P., A.H., M.S., J.A. and G.M.; Validation, J.S.; Formal analysis, Z.K.B.M.; Investigation, J.S., S.P., A.H., M.S., J.A. and G.M.; Writing—original draft, J.S.; Supervision, J.S.; Project administration, J.S. All authors have read and agreed to the published version of the manuscript.

Funding: This publication was made possible by NPRP grant number NPRP12S-0325-190443 from the Qatar National Research Fund (a member of the Qatar Foundation). Open access funding is provided by the Qatar National Library.

Institutional Review Board Statement: Not applicable.

Data Availability Statement: The original contributions presented in the study are included in the article, further inquiries can be directed to the corresponding author.

Acknowledgments: The authors would also like to acknowledge Core Labs, Qatar Environment and Energy Research Institute, Hamad Bin Khalifa University, Qatar Foundation for providing assistance in SEM.

Conflicts of Interest: The authors declare no conflicts of interest.

References

1. Liu, R.; Zhang, J.; Fu, H.; Yin, L.; Song, Y.; He, G. A comparative study of methylene blue adsorption and removal mechanisms by calcium carbonate from different sources. *Bioresour. Technol.* **2023**, *387*, 129603. [CrossRef] [PubMed]
2. Zhao, M.; Chen, Z.; Lv, X.; Zhou, K.; Zhang, J.; Tian, X.; Ren, X.; Mei, X. Preparation of core–shell structured $CaCO_3$ microspheres as rapid and recyclable adsorbent for anionic dyes. *R. Soc. Open Sci.* **2017**, *4*, 170697. [CrossRef] [PubMed]
3. Brinza, L.; Maftei, A.E.; Tascu, S.; Brinza, F.; Neamtu, M. Advanced removal of Reactive Yellow 84 azo dye using functionalised amorphous calcium carbonates as adsorbent. *Sci. Rep.* **2022**, *12*, 3112. [CrossRef]
4. Hradil, J.; Krystl, V.; Hrabánek, P.; Bernauer, B.; Kočiřík, M. Heterogeneous membranes based on polymeric adsorbents for separation of small molecules. *React. Funct. Polym.* **2004**, *61*, 303–313. [CrossRef]
5. Nagy, E. *Basic Equations of Mass Transport through a Membrane Layer*; Elsevier: Amsterdam, The Netherlands, 2019. [CrossRef]
6. Punpee, S.; Tongpadungrod, P.; Suttikul, T.; Phalakornkule, C. Characteristics of CO_2 adsorption anddesorption on activated carbon in comparisonwith zeolite 13X and carbon molecular sieveand applications in biogas upgrading usingvacuum pressure swing adsorption. *J. Chem. Technol. Biotechnol.* **2023**, *98*, 2677–2690. [CrossRef]
7. Chen, S.; Wu, W.; Niu, Z.; Kong, D.; Li, W.; Tang, Z.; Zhang, D. High adsorption selectivity of activated carbon and carbon molecular sieve boosting CO_2/N_2 and CH_4/N_2 separation. *Chin. J. Chem. Eng.* **2024**, *67*, 282–297. [CrossRef]
8. Shafiur Rahman, G.M.; Aftab, H.; Shariful Islam, M.; Bin Mukhlish, M.Z.; Ali, F. Enhanced physico-mechanical properties of polyester resin film using $CaCO_3$ filler. *Fibers Polym.* **2016**, *17*, 59–65. [CrossRef]
9. Hale, W.R.; McGuire, J.; Sand, I.D.; Dohrer, K.K. Heat setting of stretched and microvoided PE/$CaCO_3$ films. *J. Appl. Polym. Sci.* **2001**, *82*, 2454–2471. [CrossRef]
10. Zeng, Y.; Yang, X.; Yu, H.; Zhang, X.; Ma, F. Comparative Studies on Thermochemical Characterization of Corn Stover Pretreated by White-Rot and Brown-Rot Fungi. *J. Agric. Food Chem.* **2011**, *59*, 9965–9971. [CrossRef]
11. Young-Keun, L.; Jang-Weon, R.; Jung-Moon, S.; Byoung-Cheon, J. Microporous High Density Polyethylene Film and Preparing Method Thereof. US 2007/0218271 A1, 20 November 2012.

12. Takeoka, S.; Saito, A.; Zhang, H.; Takamizawa, N. Ultra-Thin Polymer Film and Porous Ultra-Thin Polymer Film. US 10,858,490 B2, 8 December 2020.
13. Ho-Seok, C. Preparation Method for Free Standing Polymer Film with Through-Pore Structured Micropores. KR101714621B1, 9 March 2017.
14. Kono, K.; Mori, S.; Miyasaka, K.; Tabuchi, J. Polyethylene Microporous Membrane of Ultra High Molecular Weight. U.S. Patent 4,588,633, 13 May 1986.
15. Özen, İ.; Şİmşek, S. Changing Breathability of Polyethylene Composite Films with Different Porous Structures Depending on Stretching Processes. *Adv. Polym. Technol.* **2017**, *36*, 203–210. [CrossRef]
16. Parthasarathy, P.; Sajjad, S.; Saleem, J.; Alherbawi, M.; McKay, G. A Review of the Removal of Dyestuffs from Effluents onto Biochar. *Separations* **2022**, *9*, 139. [CrossRef]
17. Hu, Z.; Srinivasan, M.P.; Ni, Y. Preparation of Mesoporous High-Surface-Area Activated Carbon. *Adv. Mater.* **2000**, *12*, 62–65. [CrossRef]
18. Şirin, K.; Doğan, F.; Balcan, M.; Kaya, I. Effect of $CaCO_3$ filler component on solid state decomposition kinetic of PP/LDPE/$CaCO_3$ composites. *J. Macromol. Sci. Part A Pure Appl. Chem.* **2009**, *46*, 949–958. [CrossRef]
19. Herkenberg, W. Thin Flexible Sheet Sorption Material for the Removal of Oil from Oil Spills. WO Pat. 1,991,008,347, 1991. pp. 1–7. Available online: http://patentscope.wipo.int/search/en/WO1991008347 (accessed on 22 August 2013).
20. Saleem, J.; Bazargan, A.; Barford, J.; McKay, G. Application of strong porous polymer sheets for superior oil spill recovery. *Chem. Eng. Technol.* **2015**, *38*, 482–488. [CrossRef]
21. Saleem, J.; Bazargan, A.; Barford, J.; McKay, G. Super-fast oil uptake using porous ultra-high molecular weight polyethylene sheets. *Polym. Adv. Technol.* **2014**, *25*, 1181–1185. [CrossRef]
22. Funaoka, H.; Takita, K.; Kaimai, N.; Kobayashi, S.; Kono, K. Method of Producing of Microporous Polyolefin Membrane. US8075818B2, 13 December 2011.
23. Saleem, J.; Moghal, Z.K.B.; Luque, R.; McKay, G. Towards green membranes: Repurposing waste polypropylene with a single plant-based solvent via tandem spin-casting and annealing. *Chem. Eng. J.* **2024**, *481*, 148560. [CrossRef]
24. Saleem, J.; Moghal, Z.K.B.; Sun, L.; McKay, G. Valorization of mixed plastics waste for the synthesis of flexible superhydrophobic films. *Adv. Compos. Hybrid Mater.* **2024**, *7*, 11. [CrossRef]
25. Tan, I.A.W.; Hameed, B.H.; Ahmad, A.L. Equilibrium and kinetic studies on basic dye adsorption by oil palm fibre activated carbon. *Chem. Eng. J.* **2007**, *127*, 111–119. [CrossRef]
26. Saleem, J.; Moghal, Z.K.B.; Luyt, A.S.; Shakoor, R.A.; McKay, G. Free-Standing Porous and Nonporous Polyethylene Thin Films Using Spin Coating: An Alternate to the Extrusion–Stretching Process. *ACS Appl. Polym. Mater.* **2023**, *5*, 2177–2184. [CrossRef]
27. Rengaraj, S.; Kim, Y.; Joo, C.K.; Choi, K.; Yi, J. Batch adsorptive removal of copper ions in aqueous solutions by ion exchange resins: 1200H and IRN97H. *Korean J. Chem. Eng.* **2004**, *21*, 187. [CrossRef]
28. Márquez, C.O.; García, V.J.; Guaypatin, J.R.; Fernández-Martínez, F.; Ríos, A.C. Cationic and Anionic Dye Adsorption on a Natural Clayey Composite. *Appl. Sci.* **2021**, *11*, 5127. [CrossRef]
29. Hosseini, M.; Mertens, S.F.; Ghorbani, M.; Arshadi, M.R. Asymmetrical Schiff bases as inhibitors of mild steel corrosion in sulphuric acid media. *Mater. Chem. Phys.* **2003**, *78*, 800. [CrossRef]

polymers

MDPI

Article

NH_2-MIL-125-Derived N-Doped TiO_2@C Visible Light Catalyst for Wastewater Treatment

Wenbin Wang [1,2,†], Wei Qiang [1,†], Chuntao Chen [1] and Dongping Sun [1,*]

[1] Institute of Chemicobiology and Functional Materials, School of Chemistry and Chemical Engineering, Nanjing University of Science and Technology, Nanjing 210094, China; 18302596016@163.com (W.W.); zhnlg@sina.cn (W.Q.); chchunt@njust.edu.cn (C.C.)

[2] Guizhou Panjiang Civil Explosion Co., Ltd., Guiyang 551404, China

* Correspondence: hysdp@njust.edu.cn

† These authors contributed equally to this work.

Abstract: The utilization of titanium dioxide (TiO_2) as a photocatalyst for the treatment of wastewater has attracted significant attention in the environmental field. Herein, we prepared an NH_2-MIL-125-derived N-doped TiO_2@C Visible Light Catalyst through an in situ calcination method. The nitrogen element in the organic connector was released through calcination, simultaneously doping into the sample, thereby enhancing its spectral response to cover the visible region. The as-prepared N-doped TiO_2@C catalyst exhibited a preserved cage structure even after calcination, thereby alleviating the optical shielding effect and further augmenting its photocatalytic performance by increasing the reaction sites between the catalyst and pollutants. The calcination time of the N-doped TiO_2@C-450 °C catalyst was optimized to achieve a balance between the TiO_2 content and nitrogen doping level, ensuring efficient degradation rates for basic fuchsin (99.7%), Rhodamine B (89.9%) and tetracycline hydrochloride (93%) within 90 min. Thus, this study presents a feasible strategy for the efficient degradation of pollutants under visible light.

Keywords: titanium dioxide; NH_2-MIL-125; N-doped TiO_2@C; visible light photocatalysis; wastewater pollutants

1. Introduction

The 21st century is all about global economic development, but there are some problems that we cannot ignore. One of the common concerns is how to maintain or even improve the quality of water. According to the reports of the United Nations Educational, Scientific and Cultural Organization in the World 2021 Water Resources Development Report, as our society continues to grow, we use more and more fresh water in industry, agriculture and urban life. Since 1980, its usage has been increasing by approximately 1% every year, which exacerbates the issue of water pollution [1]. Around 80% of wastewater is being discharged into the environment without any treatment. This leads to water pollution that causes approximately 2 million deaths worldwide annually and leaves many others suffering from chronic diseases [2]. Therefore, we must consider how we can upgrade our existing methods of treating water to meet the growing demand for wastewater disposal.

The composition of wastewater usually includes organic dye pollution and antibiotic pollution, which are typical troublemakers [3]. Organic dye pollution is mainly caused by industries such as textiles, cosmetics, food and paper. It has a strong color, high toxicity to living things, exhibits excellent resistance to oxidation and is tough to break down naturally. All these factors make it a big threat to our ecosystem [4,5]. Basic fuchsin and Rhodamine B are representative organic dyes. The molecular formula of basic fuchsin is $C_{20}H_{20}ClN_3$. Long-term exposure to basic fuchsin can cause skin cells to become cancerous because its carcinogenic triphenylmethane structure makes it a cationic dye, which is more toxic than anionic dyes. Rhodamine B, with a molecular formula of $C_{28}H_{31}ClN_2O_3$, is an alkaline industrial dye that poses direct harm to human health. It belongs to the third class of carcinogens in the list published by the International Agency for Research on Cancer of the World Health Organization (WHO). Antibiotic contamination usually originates from livestock and pharmaceutical industries [6]. It is difficult for animals and humans to fully absorb or transform tetracycline hydrochloride, resulting in over 50% of it being excreted through feces and urine into the environment. As a result, tetracycline hydrochloride can be detected in surface water, groundwater, sediment, soil and even drinking water. This is the main reason for the increase in bacterial resistance among humans. Growing drug resistance leads to higher healthcare costs and makes diseases more challenging to treat. Therefore, the degradation of basic fuchsin, Rhodamine B and tetracycline hydrochloride was studied in this paper's experiments.

Currently, various technologies are being widely studied to address the challenge of water pollution, including biological methods, adsorption methods, membrane separation and photocatalysis. The biological method decomposes pollutants through the adsorption capacity and metabolic processes of bacteria, fungi, yeast, actinomycetes, algae and other microorganisms. This degradation process is simple and stable but requires strict environmental conditions. If researchers operate improperly or without cautionary measures in place, it can easily lead to secondary pollution; therefore, implementing this method on a large scale is not practical [7]. The adsorption method involves the use of physicochemical interactions between the adsorbent material and the pollutant, resulting in the adsorption of pollutants onto the material. Hani et al. reported the processing of MOFs via the three-dimensional (3D) printing of cellulose MOFs as adsorbents and catalysts for water treatment, and the materials offered the complete (>99%) removal of organic dyes within 10 min toward anionic dyes, e.g., methyl blue (MeB). The reaction conditions are mild and fast, and the adsorbent materials can be reused. However, changes in reaction conditions make it easier for them to be released again [8]. Membrane separation is a novel technology that utilizes selective permeation of membranes to separate pollutants from wastewater, including reverse osmosis, microfiltration, ultrafiltration, nanofiltration and other specific methods. It offers simple and pollution-free operation. However, when the membrane becomes blocked by pollutants, restoring it to its original state using existing cleaning technologies can be challenging. As a result, the treatment efficiency of the membrane continuously decreases, and reusing it becomes difficult. In practical applications, membrane separation is often combined with other techniques to enhance utilization [9]. Therefore, photocatalysis, which uses solar excitation to produce strong oxidizing substances, is an effective and promising means for purifying phenolic water. The photocatalytic performance of TiO_2 is affected by the large band gap (3.2 eV of anatase), which can only absorb ultraviolet light, accounting for about 5% of the solar spectrum [10–12]. In addition, the rapid recombination of photogenerated electron–hole pairs in the catalytic process also reduces the photoutilization rate of TiO_2, resulting in a decrease in the degradation efficiency of organic pollutants [7].

Anatase TiO_2 belongs to the tetragonal system. The octahedral unit TiO_6^{2-} has four edges. The band gap of anatase TiO_2 is 3.2 eV, which is higher than that of rutile TiO_2 (3.0 eV). Although this means that anatase TiO_2 has a narrower range of light absorption, the difference in the conduction band and valence band positions gives the electron–hole pair of anatase TiO_2 a greater positive or negative potential, thus possessing a stronger oxidation capacity. After performing theoretical calculations, Georg et al. [13] demonstrated that the surface of anatase TiO_2 demonstrates a high adsorption capacity for H_2O, O_2 and OH^-, leading to the increased production of active free radicals during catalytic reactions and enhanced photocatalytic activity. During the crystallization process, anatase TiO_2 tends to form small-sized particles with a large specific surface area [14], which is also beneficial for the photocatalytic reaction [15].

Metal–organic frameworks (MOFs) are materials that consist of a self-assembled metal–organic skeleton formed by clusters of metal ions and organic ligands [16]. NH_2-MIL-125 is a type of MOF that has a large specific surface area and pore volume, and it can facilitate transmission and diffusion and expose as many active sites as possible [17]. The porous structure of NH_2-MIL-125 allows for a shorter distance for charge carrier transfer, effectively separating photogenerated electron pairs [18]. Furthermore, the large, interconnected 3D open cavities in NH_2-MIL-125 facilitate easy penetration of light, effectively reducing light shielding [19]. The catalyst prepared using NH_2-MIL-125 as the template combines TiO_2 with a carbon matrix, which not only maintains the morphological advantage of the MOF but also limits the aggregation of TiO_2 nanocrystals and improves the electrical conductivity of the TiO_2@C composite [20–24].

In this work, NH_2-MIL-125 was used as a template to synthesize N-doped TiO_2@C composite catalyst with high photocatalytic efficiency via calcination. TiO_2 grew in situ with titanium elements in NH_2-MIL-125 as the core, and the carbon skeleton of NH_2-MIL-125 as a template still existed after calcination. The structure of N/TiO_2@C catalytic materials was characterized using XRD, SEM, TEM, XPS and other testing methods. The natural and photocatalytic properties of N-doped TiO2@C composites were evaluated.

2. Materials and Methods

2.1. Materials

2-amino-terephthalic acid (99%) was bought from Beijing Bailingwei Technology Co., Ltd., Beijing, China. Dimethylformamide (99.9%) was bought from Aladdin Chemical Co., Ltd., Shanghai, China. Methanol (99.9%) was bought from Jinan Century Tongda Co., Ltd., Jinan, Chian. Isopropyl titanate (95%) was obtained from Syntechem Co., Ltd., Guangzhou, China.

2.2. Preparation of NH_2-MIL-125

NH_2-MIL-125 was prepared through a hydrothermal reaction. An amount of 1 g of 2-amino-terephthalic acid was added to a mixture of 18 mL of dimethylformamide and 2 mL of methanol. The mixture was stirred and subjected to ultrasonication for 5 min, followed by the addition of 1.2 mL of isopropyl titanate. After stirring and ultrasonication for 5 min, it was transferred to a Teflon autoclave with a volume capacity of 50 mL for heat treatment at 150 °C for 15 h. The resulting yellow powder was collected, centrifuged, washed three times each with dimethylformamide/ethanol and subsequently dried under vacuum conditions at 50 °C for 20 h.

2.3. Preparation of N-Doped TiO_2@C Nanomaterials

N-doped TiO_2@C was prepared via the one-step pyrolysis method. A porcelain boat containing 500 mg of NH_2-MIL-125 powder was positioned at the center of a high-temperature tubular furnace for air-based heat treatment. The tubular furnace operated with a heating rate set to 2 °C·min^{-1} within a target temperature range spanning from 200 °C to 550 °C while maintaining the desired temperature for two hours. Following the completion of the reaction, we retrieved the resulting powder.

2.4. Characterization

Chemical groups within samples were investigated using a Nicolet IS 20 FTIR spectrometer (Thermo Fisher, Waltham, MA, USA) in the range of 4000–500 cm^{-1}. The X-ray diffraction (XRD) of samples was conducted with an X-ray diffractometer (D8 Advance Bruker, Berlin, Germany). The morphological structure of the cryogels was observed with a JSM-IT500HR scanning electron microscope (SEM) and a transmission electron microscope (TEM) (JEM-2100, JEOL, Kyoto, Japan). The Brunauer–Emmett–Teller (BET) surface area and pore size were tested via nitrogen adsorption and desorption at 77 K with an ASAP automated micromeritics system. The elemental composition of samples was measured via X-ray photoelectron spectroscopy (PHI QUANTERA II, Ulvac-Phi, Kanagawa, Japan). Thermogravimetric analysis (TGA, METTLER TOLEDO TGA/SDTA851) was carried out at a constant heating rate of 10 °C min^{-1} from 50 °C to 700 °C under a nitrogen atmosphere. The optical performance of the samples was conducted on a UV-3600 spectrophotometer (Shimadzu, Kyoto, Japan). The electrochemical tests, including the electrochemical impedance (EIS), transient photocurrent and Mott–Schottky (M–S), were pictured with an electrochemical workstation (RST5200F, Restile, Gujarat, India). The response of hydroxyl radicals was tested with an electron paramagnetic resonance (EPR) spectrometer (EMXPLUS, Bruker). The contents of C, N and H in the resultant cellulose samples were determined with a Eurovector EA 3000 elemental analyzer in CHN mode.

2.5. Photocatalytic Performance Measurements

Visible light irradiation (more than 400 nm) degraded the basic fuchsin, Rhodamine B and tetracycline hydrochloride, and the photocatalytic properties of various catalysts were studied. All photocatalytic experiments were repeated three times. A 300 W xenon lamp with a filter was used to simulate AM 1.5 G lighting (100 mW/cm^2) as a light source. The circulating water flow in the reactor ensures that the photocatalytic reaction takes place at room temperature. During the degradation process, 20 mg of each photocatalyst was separately dispersed into 30 mL of basic fuchsin, Rhodamine B and tetracycline hydrochloride at a concentration of 15 mg/L. The mixture was subsequently stirred in darkness for 60 min to achieve adsorption–desorption equilibrium. During irradiation, the samples were regularly extracted and filtered using a 0.45 μm PTFE needle filter. The resulting solution was measured for absorbance at wavelengths of 546 nm, 554 nm and 370 nm utilizing an ultraviolet–visible spectrophotometer. After degradation, the catalyst underwent recovery through filtration before being washed with deionized water and ethanol. Subsequently, it was dried under vacuum conditions at a temperature of 60 °C for twenty hours. Finally, the treated catalyst could undergo multiple tests to evaluate its stability.

3. Results

3.1. Synthesis and Characterization of N-Doped TiO_2@C

The overall fabrication of the N-doped TiO_2@C composite is schematically illustrated in Figure 1a. In order to explore the crystal structure of the sample prepared in the experiment, X-ray diffraction (XRD) tests were first conducted. NH_2-MIL-125 samples show typical diffraction peaks at 6.67°, 9.62°, 11.48°, 14.90°, 16.50°, 17.89° and 19.42°. The outcomes indicated that the synthesis of NH_2-MIL-125 was successful and that the produced MOF exhibited acceptable crystallinity. N-doped TiO_2@C-450 °C was obtained via the calcination of NH_2-MIL-125, and the characteristic peak of NH_2-MIL-125 totally vanished during the process. The XRD pattern of N-doped TiO_2@C-450 °C was compared with the standard diffraction card of acanite TiO_2 (JCPDSNo.21−1272). It was found that peaks of 25.06°, 38.10°, 47.89°, 53.70° and 61.46° corresponded to the (101), (004), (200), (105) and (204) planes of anatase, respectively. The results show that the pyrolysis of NH_2-MIL-125 to produce anatase TiO_2 was successful. In addition, the XRD characterization of N-doped TiO_2@C was significantly different from different calcination temperatures (Figure 1b,c). The calcination temperature of N-doped TiO_2@C-250 °C was too low, which caused insufficient pyrolysis. The characteristic peaks of NH_2-MIL-125 can still be seen at 6.63°, 9.65° and 11.64°, but the peak intensity was reduced significantly. This may have been due to the partial disintegration and recombination of the NH_2-MIL-125 structure. When the temperature of the calcination reached 350 °C, only a carbon sheath peaked at 13.19°, and no TiO_2-related peak pattern was visible, indicating that the NH_2-MIL-125 structure was totally broken down. TiO_2 had not yet formed crystals, and the carbon element that was liberated during the calcination was doped into the compound. The distinctive peak of anatase TiO_2 was clearly seen when the temperature hit 450 °C. Following that, the intensity of the anatase TiO_2 characteristic peak no longer varied considerably with the rise in calcination temperature, but the crystallinity of TiO_2 was more excellent.

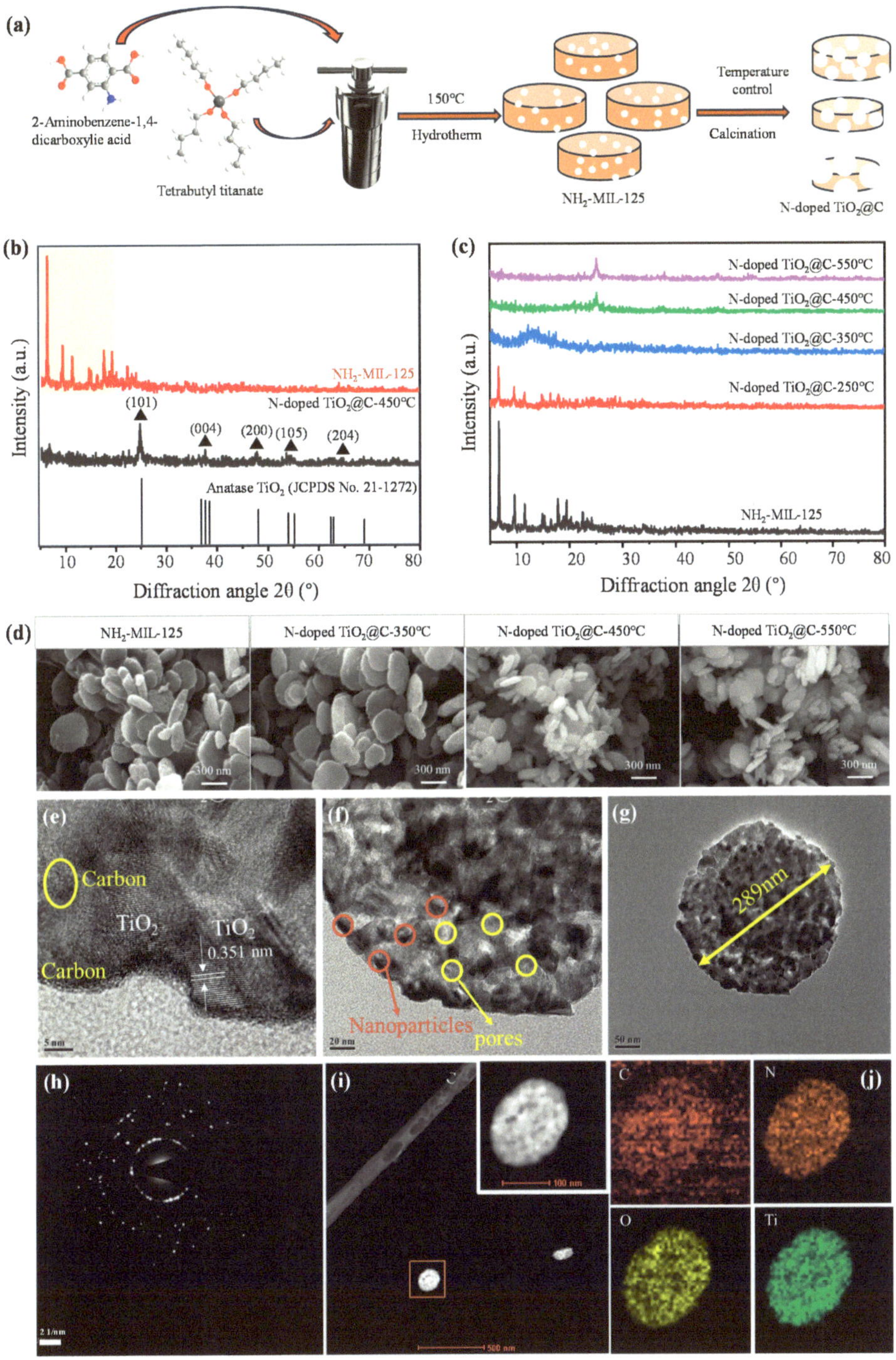

Figure 1. Schematic diagram and characterization of N-doped TiO_2@C composite. (**a**) Fabrication of N-doped TiO_2@C composite. (**b**,**c**) NH_2-MIL-125 and N-doped TiO_2@C composites with calcination

temperatures of 250 °C, 350 °C, 450 °C and 550 °C. (**d**) SEM images of NH_2-MIL-125 and N-doped TiO_2@C composites with calcination temperatures of 250 °C, 350 °C, 450 °C and 550 °C. (**e–g**) TEM images of N-doped TiO_2@C-450 °C. (**h**) Electron diffraction image of N-doped TiO_2@C-450 °C. (**i**,**j**) Element mapping of N-doped TiO_2@C-450 °C.

From the SEM images, the prepared NH_2-MIL-125 showed a circular flake shape with an average diameter of about 422 nm (Figure 1d). In N-doped TiO_2@C-450 °C, the thickness of TiO_2-like circular tablets was about 50 nm, and the diameter was about 140 nm. The sample had a thickness of around 60 nm and a diameter of about 370 nm for N-doped TiO_2@C-350 °C. And for the N-doped TiO_2@C-250 °C sample, its thickness was around 70 nm, and its diameter was about 450 nm. It is evident that, as the calcination temperature rose, the average particle size of the sample particles shrank to some extent while maintaining its round pill shape. This might have been because, as the calcination temperature rose, the network structure of NH_2-MIL-125 gradually lost its capacity to crystallize, and the organic molecules in the skeleton broke down.

The shape and crystal structure of N-doped TiO_2@C-450 °C composites were obtained via TEM. The observation of amorphous carbon shows that a large amount of TiO_2 was distributed in the carbon matrix and finally formed the TiO_2@C composite. The sample obtained after calcination still maintained a porous structure, as seen by the presence of TiO_2 nanoparticles and pores (Figure 1e,f). This porous construction can increase photocatalytic efficiency by exposing as many active sites as feasible while also facilitating the transmission and diffusion of electrons during catalysis. In the adsorption experiment, more investigation into the porous structure was conducted on N/TiO_2@C-450 °C, on which a lattice spacing of 0.351 nm was found, compatible with the lattice spacing of the crystal planes of anatase TiO_2 (101) (Figure 1g). From TEM observations, the diameter of the N-doped TiO_2@C-450 °C was approximately 289 nm (Figure 1g). On N-doped TiO_2@C-450 °C, a lattice spacing of 0.351 nm was found, which was compatible with the lattice spacing of the crystal planes of anatase TiO_2 (101) (Figure 1h). Titanium, oxygen, carbon and nitrogen were all further confirmed by the element mapping (Figure 1i,j). It is evident that, on N-doped TiO_2@C-450 °C, the distribution of titanium, oxygen, carbon and nitrogen was quite uniform. It can be argued that, during calcination, the nitrogen element contained in the organic ligand 2-amino-terephtharic acid was released and then doped into the sample. The titanium element in NH_2-MIL-125 was grown in situ to form TiO_2, which was uniformly dispersed in the porous carbon skeleton, and ultimately, the nitrogen-doped TiO_2@C was successfully formed.

The elemental composition and valence information of N-doped TiO_2@C were subsequently examined using X-ray photoelectron spectroscopy (XPS), which was utilized to test the material composition (Figure 2a,b). When the temperature was too high, all nitrogen was released and was no longer doped into the TiO_2@C composite, and the peak value of carbon and nitrogen decreased with the increase in calcination temperature. When the temperature reached 550 °C, the peak of nitrogen was close to zero. However, the proper calcination temperature was crucial, as the XRD pattern analysis showed that anatase TiO_2 was difficult to produce or had poor crystallinity at too low temperatures. A calcination temperature of 450 °C could be used as the equilibrium temperature between nitrogen doping and TiO_2 crystallinity. Three peaks at 532.0 eV, 530.9 eV and 528.9 eV, which are associated with the C-O bond, C=O bond and Ti-O bond, respectively, can be seen in the XPS spectra of O 1s (Figure 2c). The surface hydroxyl oxygen, which is essential for the photodegradation of contaminants, was responsible for the development of the C-O bond. To fight contaminants, the hydroxyl radical OH• could be created when it interacted with photogenic holes. N 1s (Figure 2d) displayed a broad peak between 398.4 and 399.4 eV. It is consistent with other results and typical Ti-N structures in nitrogen-doped TiO_2 [25]. The C=C peaks (Figure 2e) at 288.2 eV, 286.4 eV, 285.6 eV and 284.2 eV correspond to C=O, C-O, C-N and C=C, respectively. For pure anatase TiO_2, Ti-O peaks were located at

458.5 eV and 464.5 eV. The Ti 2p peaks for N-doped TiO2@C-450 °C were 457.5 eV and 463.5 eV, which is a 6.0 eV energy split. The peaks of Ti 2p shown in Figure 2f centered at about 463.2 eV and 457.5 eV belonged to Ti 2p3/2 and Ti 2p1/2, respectively, which confirm the Ti^{4+} species in the form of TiO_2 nanoparticles. The Ti 2p peak for N-doped TiO_2@C-450 °C was slightly shifted toward a lower binding energy when compared to pure anatase TiO_2, which was brought on by nitrogen doping altering the local chemical environment of titanium ions. The successful doping of nitrogen into TiO_2@C, as shown by the test findings from XPS, has positive implications for reducing the band gap width of TiO_2 and increasing its visible light activity. We also tested the relative amount of C and N in the sample with an elemental analyzer. The content of nitrogen decreased with the increase in heat treatment temperature. A ratio as high as 0.12% N/C was obtained for the heat treatment of 450 °C.

The nitrogen adsorption–desorption isotherm samples for N-doped TiO2@C-450 °C (Figure 2g) exhibited the conventional type IV isotherm and type H3 hysteresis loop. The pore size distribution was calculated using the nonlocal density function model (NLDFT) based on data from N_2 adsorption (Figure 2h). N-doped TiO_2@C-450 °C has a specific surface area of 63 m^2/g, which is significantly less than that of NH_2-MIL-125 but still significantly larger than that of other TiO_2 catalysts. Therefore, basic fuchsin, rhodamine B and tetracycline hydrochloride can be degraded at more active sites thanks to the large specific surface area of N-doped TiO_2@C-450 °C.

BET data need to be interpreted in combination with thermogravimetric curves (Figure 2g,h, Table 1). The specific surface area of the samples decreased, and the average pore size increased with increasing roasting temperatures, but for different reasons. The sample changed at about 200 °C as a result of the elimination of any remaining solvent. When the temperature reached 350 °C, the porous structure collapsed, and the crystal skeleton began to disintegrate, greatly reducing the specific surface area. The decomposition of the crystal skeleton and the collapse of the porous structure made the specific surface area decrease sharply. When the calcination temperature was further increased to 450 °C, the BET data changed due to the recrystallization of TiO_2 and the formation of pyrolytic carbon substrates. However, the skeleton shrank significantly once more at 550 °C, and the specific surface area dropped to 13 m^2/g, making it difficult for contaminants and the catalyst to make contact. Therefore, by adjusting the calcination temperature, the N-doped TiO_2@C catalyst with the best photocatalytic effect was created at 450 °C.

Table 1. Textural characteristics of NH_2-MIL-125, N-doped TiO_2@C-200 °C, N-doped TiO_2@C-350 °C, N-doped TiO_2@C-450 °C and N-doped TiO_2@C-550 °C.

Sample	S_{BET} (m^2g^{-1})	Vtotal (cm^3g^{-1})	Average Pore Diameter (nm)	N/C *
NH_2-MIL-125	539	0.087	5.32	0.146
N-doped TiO_2@C-200 °C	763	0.168	6.54	0.0035
N-doped TiO_2@C-350 °C	89	0.157	9.30	0.0022
N-doped TiO_2@C-450 °C	63	0.141	10.89	0.0012
N-doped TiO_2@C-550 °C	13	0.046	12.26	0.0007

* The N/C value means atomic mass ratio, which was taken with a Eurovector EA 3000 elemental analyzer in CHN mode.

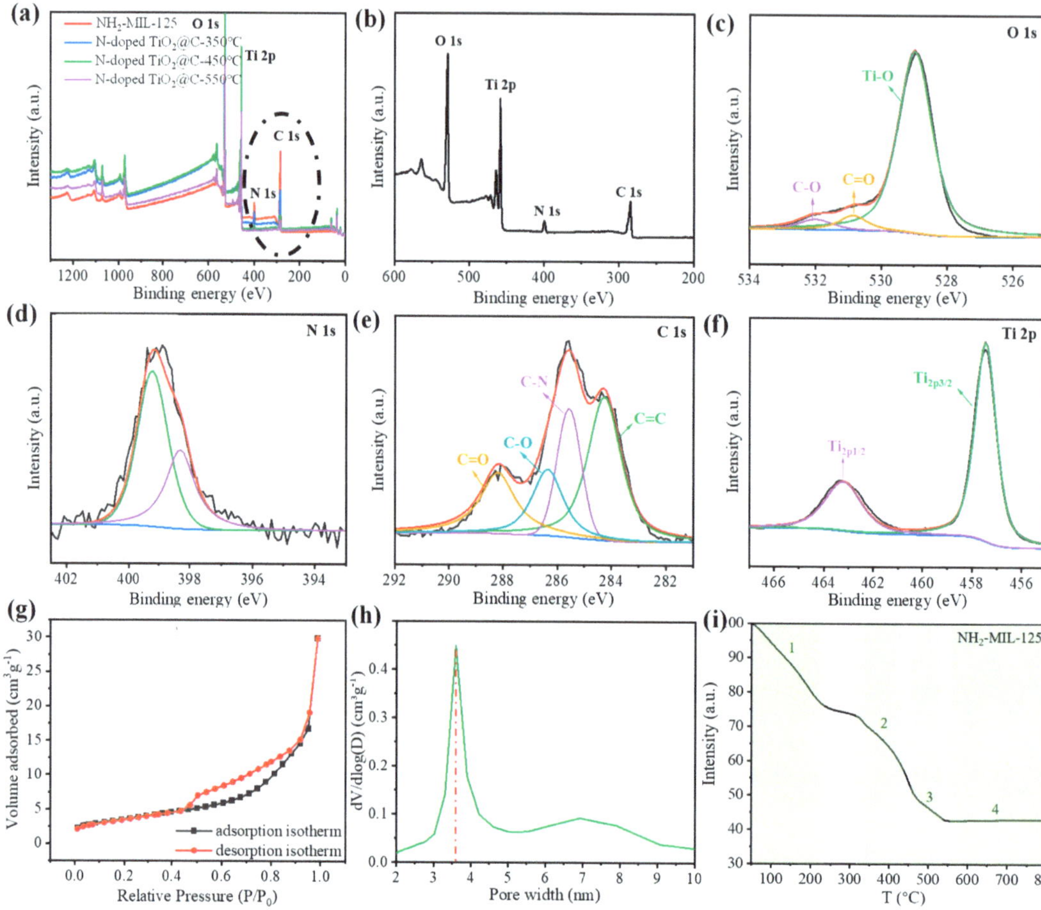

Figure 2. Chemical characterization of the N-doped TiO_2@C composites. (**a**,**b**) Survey XPS spectra of NH_2-MIL-125, N-doped TiO_2@C-250 °C, N-doped TiO_2@C-350 °C, N-doped TiO_2@C-450 °C and N-doped TiO_2@C-550 °C. High-resolution XPS spectra of (**c**) O 1s, (**d**) N 1s, (**e**) C 1s and (**f**) Ti 2p of N-doped TiO_2@C-450 °C. (**g**) N_2 adsorption–desorption isotherms of N-doped TiO_2@C-450 °C. (**h**) Pore size distributions of N-doped TiO_2@C-450 °C. (**i**) Thermogravimetric curve of NH_2-MIL-125.

3.2. Photocatalytic Degradation of N-Doped TiO_2@C

The light absorption capacity of N-doped TiO_2@C composites was investigated via a UV–Vis absorption test (Figure 3a). NH_2-MIL-125 was yellow, as already described. The sample's color continued to deepen as the calcination temperature rose due to the breakdown of the crystal skeleton and the collapse of the porous structure. The sample's hue changed to brown at 350 °C, when it exhibited the best ability to absorb visible light. As the calcination temperature increased after 350 °C, TiO_2 crystals started to form, and the sample's color lightened. The sample was nearly white at 550 °C, and visible light could no longer be absorbed. We know that P25, the white commercial-grade TiO_2, has poor absorption capabilities for visible light. The optimal catalyst for the best visible light absorption performance at 400–800 nm should be N-doped TiO_2@C-350 °C. However, the photocatalytic activity of N-doped TiO_2@C-350 °C was subpar due to the absence of TiO_2 crystals. The most acceptable visible light catalyst, according to a thorough comparison between the sample's photoabsorption capacity and TiO_2 content, was N-doped TiO_2@C-450 °C.

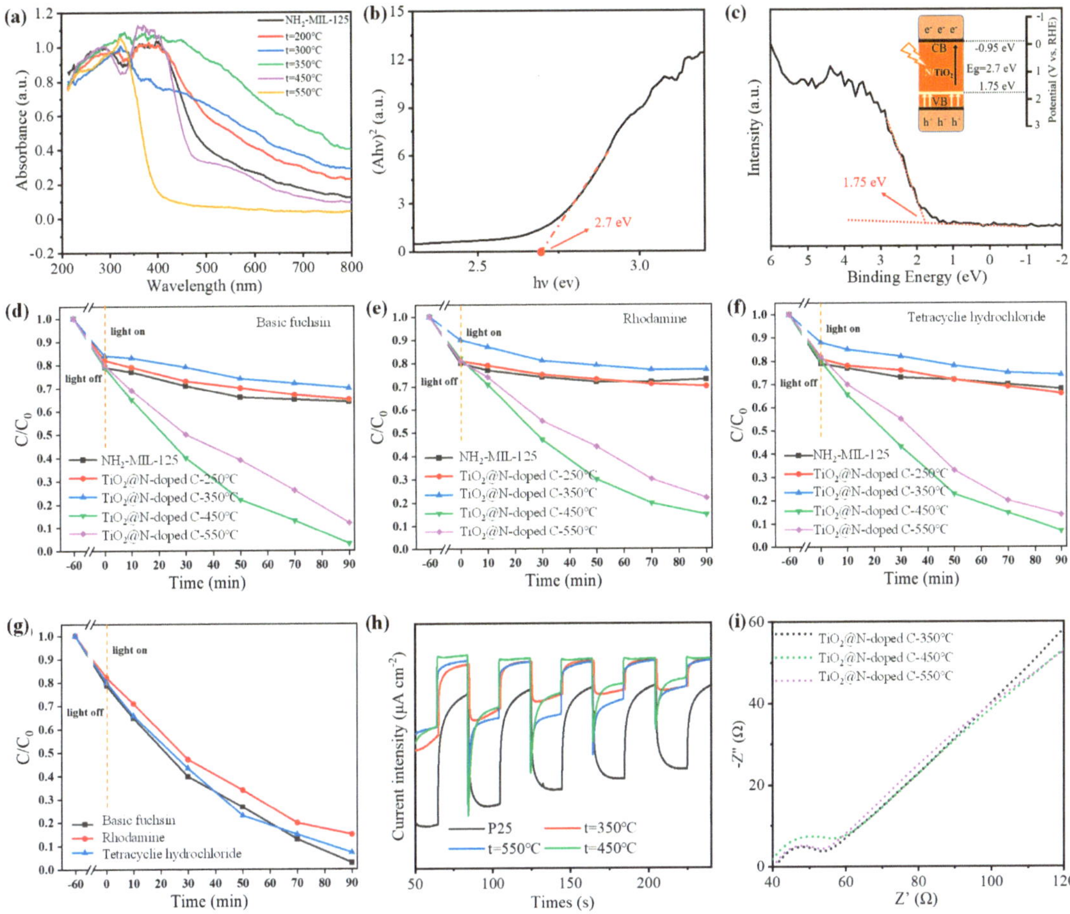

Figure 3. (**a**) DRS spectra for N-doped TiO_2@C with calcination temperatures of 200 °C, 300 °C, 350 °C, 450 °C and 550 °C and for NH_2-MIL-125. (**b**) Band gap and (**c**) valence band spectra of N-doped TiO_2@C-450 °C (with an energy level diagram of N-doped TiO_2@C-450 °C). Photocatalytic degradation efficiencies of (**d**) basic fuchsins, (**e**) Rhodamine B and (**f**) tetracycline hydrochloride over NH_2-MIL-125, N-doped TiO_2@C-200 °C, N-doped TiO_2@C-350 °C, N-doped TiO_2@C-450 °C and N-doped TiO_2@C-550 °C under xenon lamp irradiation. (**g**) Degradation plots of N-doped TiO_2@C-450 °C. (**h**) Transient photocurrent responses for N-doped TiO_2@C with calcination temperatures of 350 °C, 450 °C and 550 °C and for NH2-MIL-125. (**i**) EIS Nyquist plots of N-doped TiO_2@C with calcination temperatures of 350 °C, 450 °C and 550 °C.

Compared with P25, the as-prepared N-doped TiO_2@C-450 °C sample had a wider absorption range, which was because nitrogen in the organic connector was released and doped into the sample during calcination, which reduced the wide band gap of TiO_2 and caused the light absorption range to shift to the visible region. The prohibited band spectra (Figure 3b) and valence band spectra (Figure 3c) of N-doped TiO_2@C-450 °C can be used to compute the energies of the valence band, conduction band and band gap, which were −0.95 eV, 1.75 eV and 2.7 eV, respectively. The N-doped TiO_2@C0-450 °C band gap created in this work was narrower than that of anatase TiO_2, which had a band gap energy of 3.22 eV, making it more suitable for light absorption and photocatalytic processes.

Under xenon lamp illumination, the photocatalytic activity of N-doped TiO_2@C composites was assessed using dealkalized fuchsin, Rhodamine B and tetracycline hydrochloride. These three contaminants all degraded in a very similar way. Only NH_2-MIL-125 was capable of decomposing basic fuchsin (Figure 3d), lowering the pollutant concentration to 65%. This was because NH_2-MIL-125 can carry out a certain amount of physical adsorption due to its large specific surface area and micropore structure. The capability for deterioration varied significantly among N-doped TiO_2@C samples. At 350 °C, N-doped TiO_2@C exhibited the lowest degradation ability compared to other catalysts. N-doped TiO_2@C-350 °C showed the worst degradation ability, and 70% of basic fuchsin remained in the solution after degradation, even less than that of NH_2-MIL-125's adsorption effect. This was mainly due to two reasons. First, TiO_2 does not form at this temperature and prevents photocatalytic reactions; second, at 350 °C, NH_2-MIL-125 undergoes significant pyrolysis that rapidly changes its skeletal structure and reduces specific surface area sharply. Moreover, there was an overall increase in pore size, resulting in mesoporous pores and causing a decline in its adsorption effectiveness. Degradation of 99.7% of basic fuchsin with N-doped TiO2@C-450 °C was the highest among all samples within 90 min. However, anatase TiO_2 obtained via the high-temperature heat treatment at 550 °C had the highest content and the best crystal form. However, the above characterization results indicate that the high-temperature heat treatment was not conducive to nitrogen doping and reduced the utilization efficiency of visible light. Consequently, the degradation efficiency of basic fuchsin was not as good as that achieved with N-doped TiO_2@C-450 °C.

The degradation of Rhodamine B and tetracycline hydrochloride exhibited similar trends in the above samples (Figure 3e,f). NH_2-MIL-125 achieved a 73% reduction in the content of Rhodamine B through physical adsorption. Among the tested catalysts, N-doped TiO_2@C-350 °C demonstrated the lowest degradation efficiency with only 23% of pollutants being degraded, whereas N-doped TiO_2@C-450 °C exhibited the highest degradation effect by degrading 89.9% of Rhodamine B within 90 min. Physical adsorption by NH_2-MIL-125 resulted in a decrease in the content of tetracycline hydrochloride to 68%. Similarly, N-doped TiO_2@C-350 °C showed inferior performance with only a 26% degradation rate for pollutants, whereas N-doped TiO_2@C-450 °C displayed superior degradation capability by achieving a remarkable removal rate of 93% for tetracycline hydrochloride within the same time frame (Figure 3g). In Table 2, we summarize the recent works on the degradation performance of TiO_2 composites against tetracycline. In comparison, our materials show outstanding catalytic properties [26].

Table 2. Summary of works on the degradation performance of TiO_2 composites against tetracycline.

Order	Catalyst	Light Source	Concentrations of Catalysts (g/L)	Concentrations of Pollutants (mg/L)	Degradation Efficiency	Time (h)	Reference
1	$g\text{-}C_3N_4/TiO_2$	300 Xe (UV–Vis)	0.4	20	90.1%	1.0	[26]
2	TiO_2/Fe-MOF (15%)	300 Xe (UV–Vis, λ = 370 nm)	1	96	97%	4	[27]
3	TiO_2, $H_2Ti_3O_7$	Xe lamp	0.02	20	89%, 94%	1	[28]
4	Biofilm-UCPs-TiO_2	lamp (20 W) of 1800 Lux	1	40	82.1%	24	[29]
5	Black-TiO_2	SPD-16 UV–vis detector at 357 nm	0.5	10	66.2%	4.5	[30]
6	AgBreTiO_2-Pal (50%)	200–800 nm by UV-Vis DRS	0.5	10	89.6%	1.5	[31]
7	Defect-rich hydrogenated $g\text{-}C_3N_4/TiO_2$	300 Xe (λ > 400 nm)	0.6	30	60%	1.5	[32]
8	N-TiO_2/Ov carbon nitride doped with oxygen	300 Xe (λ > 420 nm)	0.4	30	79.9%	1.0	[33]
9	Oxygen vacancies modified TiO_2/O-terminated Ti_3C_2 composites	Vis (300 W)	0.4	20	88.5%	1.5	[34]
10	Ti_3C_2@TiO_2	125 W Xe (λ > 400 nm)	1	20	90%	1.5	[35]
11	N-doped TiO_2@C	300 W Xe (λ > 420 nm)	0.67	30	93%	1.5	This work

Under xenon lamp irradiation, we used a photocurrent test to investigate the phenomenon of photogenerated carrier transport in samples. The photocurrent intensity of all N-doped TiO_2@C samples was relatively high, indicating a slow photogenerated carrier recombination rate (Figure 3h). This was mainly because TiO_2 was combined with the carbon matrix, and the porous structure of the carbon skeleton shortened the carrier transfer distance so that the photogenerated electron pairs could be effectively separated. In addition, the large specific surface area of the carbon skeleton exposed more active sites, making carrier transmission and diffusion more convenient. In short, the combination of carbon skeletons and TiO_2 enhanced the photocurrent intensity of the composite compared with commercial-grade TiO_2 P25. It is well known that the higher the photocurrent density, the better the ability to separate photogenerated electrons and holes. Therefore, compared with P25, N-doped TiO_2@C improves carrier separation efficiency and inhibits photogenerated carrier recombination. To confirm this, we further studied the resistance of interfacial charge conversion via electrochemical impedance spectroscopy. As shown in Figure 3i, the radii of all N-doped TiO_2@C composites were small, indicating that the N-doped TiO_2@C catalyst had higher interfacial charge separation efficiency and better electrical conductivity, which is similar to the trend in the photocurrent response test results. The charge transfer resistance (Rct) of N-doped TiO_2@C in an equivalent circuit was smaller than that of pure TiO_2 [36], which further confirms the favored interfacial charge transfer.

In summary, the optimal balance between anatase TiO_2 formation and nitrogen doping was achieved when the calcination temperature reached 450 °C, and N-doped TiO_2@C-450 °C exhibited the highest photocatalytic capacity.

Tetracycline hydrochloride was used as an example to investigate the stability and reusability of the N-doped TiO_2@C-450 °C catalyst. The same sample was employed for five consecutive degradation cycles under identical conditions (Figure 4a). No significant changes were observed after five cycles, although the degradation capacity decreased from 93% to 81%. This decrease in degradation capacity was primarily attributed to the blockage of mesoporous channels in N-doped TiO_2@C-450 °C by pollutants over prolonged reaction times, making it challenging to effectively clean and negatively impacting the recyclability of the catalyst. Additionally, the inevitable loss of the photocatalyst during repeated use also resulted in a decrease in degradation efficiency. Moreover, a comparison of SEM images, XRD spectra and XPS spectra (Figure 4b–d) before and after the reaction revealed that there was no significant change in the structure, morphology characteristics and chemical composition of N-doped TiO_2@C-450 °C before and after testing. N-doped TiO_2@C-450 °C exhibited excellent photocatalytic ability, stability and reusability; it has great potential as a visible photocatalyst for various types of pollutants.

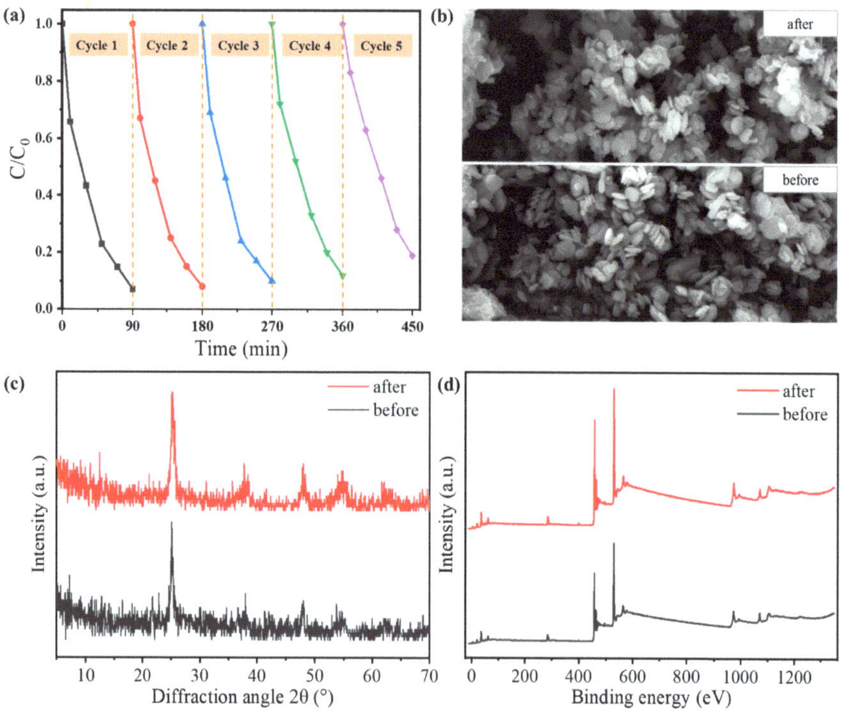

Figure 4. (**a**) Repeated experiments of tetracycline photodegradation over N-doped TiO_2@C-450 °C. (**b**) SEM images, (**c**) XRD patterns and (**d**) XPS spectra of N-doped TiO_2@C-450 °C before and after the photocatalytic reaction.

3.3. Photocatalytic Mechanism

The photocatalytic mechanism of N-doped TiO_2@C-450 °C nanomaterials during visible light degradation was investigated by conducting multiple trapping experiments using tetracycline hydrochloride as an example. Three additional trapping reagents were added to the reaction solution, among which ammonium oxalate (AO) was used to trap e^-, benzoquinone (BQ) was used to trap $O_2\bullet^-$ and isopropyl alcohol (IPA) was used to trap OH• (Figure 5a). The addition of each trapping agent affected the degradation rate of tetracycline hydrochloride, among which isopropyl alcohol had the most prominent inhibitory effect among all trapping agents. The degradation efficiency of N-doped TiO_2@C-450 °C was reduced significantly to 71% after the addition of isopropyl alcohol, indicating that OH• plays a leading role in the degradation reaction of tetracycline hydrochloride under visible light irradiation. In addition, the addition of ammonium oxalate and benzoquinone (BQ) also had a certain effect on the degradation rate, suggesting that e^- and $O_2\bullet^-$ were also involved in the process but did not play a major role. In conclusion, OH• was produced by the N-doped TiO_2@C-450 °C catalyst during the photocatalysis process, which enhanced the oxidation capacity of the active substance and improved its ability to degrade tetracycline hydrochloride greatly.

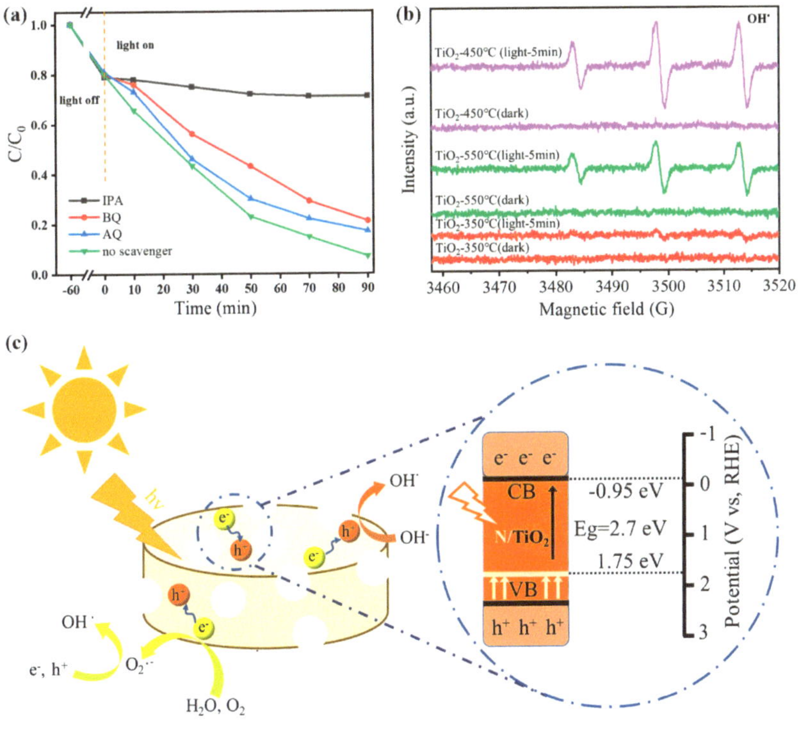

Figure 5. (**a**) Photocatalytic ability of N-doped TiO_2@C-450 °C for the degradation of tetracycline with or without adding IPA, BQ and AO under visible light. (**b**) EPR spectra of OH• on N-doped TiO_2@C-350 °C, N-doped TiO_2@C-450 °C and N-doped TiO2@C-550 °C catalysts. (**c**) Probable photocatalytic mechanism of tetracycline degradation by N-doped TiO_2@C-450 °C under visible light irradiation.

A DMPO trapping agent was added to the photocatalytic reaction system to detect active free radicals via EPR. Therefore, electron paramagnetic resonance spectroscopy was used to detect the signals of OH• radicals generated by N-doped TiO_2@C-350 °C, N-doped TiO_2@C-450 °C and N-doped TiO_2@C-550 °C under visible light irradiation (Figure 5b). Obviously, regardless of dark or light conditions, N-doped TiO_2@C-350 °C OH showed almost no occurrence of free radicals, and this was because the TiO_2 of N-doped TiO_2@C-350 °C had not formed. For N-doped TiO_2@C-450 °C and N-doped TiO_2@C-550 °C, almost no OH• free radicals increased. This indicates that the photogenerated electrons on anatase TiO_2 were transferred rapidly through the carbon skeleton, resulting in the production of a large number of OH• free radicals. N-doped TiO_2@C-450 °C exhibited the highest number of OH• free radicals due to nitrogen doping, which enhanced the sample's utilization rate of visible light. This phenomenon further confirms the excellent photocatalytic effect of nitrogen-doped TiO_2 supported by a carbon skeleton, which can serve as photosensitizer to absorb visible light and help to promote charge carriers' separation through cooperation with bulk/surface defects of TiO_2 [37,38].

Based on the test results and analysis mentioned above, a possible photocatalytic reaction mechanism of N-doped TiO_2@C-450 °C can be proposed (Figure 5c). Under visible light irradiation, nitrogen-doped TiO_2 nanoparticles can efficiently produce charge carriers. At this stage, the unique porous structure of N-doped TiO_2@C-450 °C provides additional pathways for the movement of charge carriers, facilitating the easy transfer of electrons to the carbon framework and effectively separating photogenerated electrons and holes. During the degradation process, $O_2\bullet^-$ free radicals are generated through reactions between electrons enriched on the carbon framework with H_2O and O_2. It can be found that the REDOX potential of N-doped TiO_2@C CB positions is more negative than that of $O_2/\bullet O_2^-$, and the REDOX potential of VB positions is more positive than that of $OH^-/\bullet OH$, thus contributing to the generation of reactive oxygen species under photoexcitation [39,40].

4. Conclusions

In summary, we successfully synthesized a N-doped TiO_2@C composite material through NH_2-MIL-125 template-assisted calcination. Supported by the NH_2-MIL-125 carbon framework, the catalyst exhibited a significantly enhanced specific surface area of 63 m^2/g, thereby facilitating an increased number of active sites for both the catalyst and reactant. The mesoporous structure (3.5 nm) of NH_2-MIL-125 facilitated enhanced light penetration, thereby reducing the occurrence of light shielding. Element mapping confirmed the presence of nitrogen, indicating that nitrogen was released from the organic ligand 2-amino-terephthalic acid and doped into the sample during calcination. Nitrogen doping in anatase shortened its wide band gap of 3.22 eV to 2.7 eV, resulting in a shift in the optical absorption range toward the visible region. As a result, the N-doped TiO_2@C catalyst exhibited extended optical absorption capabilities beyond just ultraviolet wavelengths. By carefully controlling the calcination temperature, we successfully synthesized various N-doped TiO_2@C composites. When calcined at 450 °C, the N-doped TiO_2@C catalyst achieved a balance between anatase content and nitrogen doping. The resulting N-doped TiO_2@C-450 °C composite exhibited outstanding performance in photocatalytic degradation. Under visible light for 90 min, basic fuchsin, Rhodamine B and tetracycline hydrochloride were degraded by 99.7%, 89.9% and 93% respectively. Moreover, the N-doped TiO_2@C-450 °C composite demonstrated excellent stability, as evidenced by negligible changes in SEM, XRD and XPS characteristics after five cycles of use; only a slight decrease in tetracycline hydrochloride degradation efficiency from 93% to 81% was observed. Last, we proposed the mechanism of photocatalytic degradation. This study presents a viable strategy for efficient pollution degradation under visible light.

Author Contributions: Conceptualization, C.C.; Methodology, W.W. and W.Q.; Validation, W.W. and W.Q.; Formal analysis, W.Q.; Investigation, W.W.; Resources, W.Q.; Data curation, W.W.; Writing—original draft preparation, W.W.; Writing—review and editing, C.C. and D.S.; Supervision, C.C. and D.S.; Funding acquisition, D.S. All authors have read and agreed to the published version of the manuscript.

Funding: This research was funded by the National Natural Science Foundation of China, grant numbers 51803092 and 51873087).

Institutional Review Board Statement: Not applicable.

Data Availability Statement: The data presented in this study are available on request from the corresponding author.

Conflicts of Interest: Author Wenbin Wang was employed by Guizhou Panjiang Civil Explosion Co., Ltd. The remaining authors declare that the research was conducted in the absence of any commercial or financial relationships that could be construed as potential conflicts of interest.

References

1. Hannah, D.M.; Abbott, B.W.; Khamis, K.; Kelleher, C.; Lynch, I.; Krause, S.; Ward, A.S. Illuminating the 'invisible water crisis' to address global water pollution challenges. *Hydrol. Process.* **2022**, *36*, e14525. [CrossRef]
2. Sehar, S.; Rasool, T.; Syed, H.M.; Mir, M.A.; Naz, I.; Rehman, A.; Shah, M.S.; Akhter, M.S.; Mahmood, Q.; Younis, A. Recent advances in biodecolorization and biodegradation of environmental threatening textile finishing dyes. *3 Biotech* **2022**, *12*, 186. [CrossRef] [PubMed]
3. Park, J.; Lee, T.H.; Kim, C.; Lee, S.A.; Choi, M.J.; Kim, H.; Yang, J.W.; Lim, J.; Jang, H.W. Hydrothermally obtained type-II heterojunction nanostructures of In_2S_3/TiO_2 for remarkably enhanced photoelectrochemical water splitting. *Appl. Catal. B Environ.* **2021**, *295*, 120276. [CrossRef]
4. Lu, Y.; Zang, Y.; Zhang, H.; Zhang, Y.; Wang, G.; Zhao, H. Meaningful comparison of photocatalytic properties of {001} and {101} faceted anatase TiO_2 nanocrystals. *Sci. Bull.* **2016**, *61*, 1003–1012. [CrossRef]
5. Tian, J.; Zhao, Z.; Kumar, A.; Boughton, R.I.; Liu, H. Recent progress in design, synthesis, and applications of one-dimensional TiO_2 nanostructured surface heterostructures: A review. *Chem. Soc. Rev.* **2014**, *43*, 6920–6937. [CrossRef]
6. Wang, J.; Wang, G.; Wang, X.; Wu, Y.; Su, Y.; Tang, H. 3D/2D direct Z-scheme heterojunctions of hierarchical TiO_2 microflowers/g-C_3N_4 nanosheets with enhanced charge carrier separation for photocatalytic H_2 evolution. *Carbon* **2019**, *149*, 618–626. [CrossRef]
7. Lin, L.; Yang, H.; Xu, X. Effects of water pollution on human health and disease heterogeneity: A review. *Surf. Interfaces* **2022**, *10*, 1088. [CrossRef]
8. Abdelhamid, H.N.; Sultan, S.; Mathew, A.P. Binder-free Three-dimensional (3D) printing of Cellulose-ZIF8 (CelloZIF-8) for water treatment and carbon dioxide (CO_2) adsorption. *Chem. Eng. J.* **2023**, *468*, 143567. [CrossRef]
9. Wang, Z.; Li, Y.; Cheng, Q.; Wang, X.; Wang, J.; Zhang, G. Sb-based photocatalysts for degradation of organic pollutants: A review. *J. Clean. Prod.* **2022**, *367*, 133060. [CrossRef]
10. Tobaldi, D.M.; Lajaunie, L.; Rozman, N.; Caetano, A.P.F.; Seabra, M.P.; Škapin, A.S.; Arenal, R.; Labrincha, J.A. Impact of the absolute rutile fraction on TiO_2 visible-light absorption and visible-light-promoted photocatalytic activity. *J. Photochem. Photobiol. A Chem.* **2019**, *382*, 111940. [CrossRef]
11. Sun, H.; Lu, W.; Zhao, J. Structure and reactivity of aanatase $TiO_2(001)$-(1×4) surface. *J. Phys. Chem. C* **2018**, *122*, 14528–14536. [CrossRef]
12. Wei, C.; Lin, W.Y.; Zainal, Z.; Williams, N.E.; Zhu, K.; Kruzic, A.P.; Smith, R.L.; Rajeshwar, K. Bactericidal Activity of TiO_2 Photocatalyst in Aqueous Media: Toward a Solar-Assisted Water Disinfection System. *Environ. Sci. Technol.* **1994**, *28*, 934–938. [CrossRef] [PubMed]
13. Playford, H.Y. Variations in the local structure of nano-sized anatase TiO_2. *J. Solid State Chem.* **2020**, *288*, 121414. [CrossRef]
14. Kong, X.; Li, J.; Yang, C.; Tang, Q.; Wang, D. Fabrication of Fe_2O_3/g-C_3N_4@N-TiO_2 photocatalyst nanotube arrays that promote bisphenol A photodegradation under simulated sunlight irradiation. *Sep. Purif. Technol.* **2020**, *248*, 116924. [CrossRef]
15. Chen, L.; Zhou, X.; Jin, B.; Luo, J.; Xu, X.; Zhang, L.; Hong, Y. Heterojunctions in g-C_3N_4/B-TiO_2 nanosheets with exposed {001} plane and enhanced visible-light photocatalytic activities. *Int. J. Hydrogen Energy* **2016**, *41*, 7292–7300. [CrossRef]
16. Akbari, M.; Ghasemzadeh, M.A.; Fadaeian, M. Synthesis and application of ZIF-8 MOF incorporated in a TiO_2@Chitosan nanocomposite as a strong nanocarrier for the drug delivery of acyclovir. *ChemistrySelect* **2020**, *5*, 14564–14571. [CrossRef]
17. Yang, X.; Wang, Y.; Hu, Y.; Zhao, H.; Sun, Y.; Hua, K.; Chen, G. Interior supported hierarchical TiO_2@Co_3O_4 derived from MOF-on-MOF architecture with enhanced electrochemical properties for lithium storage. *ChemElectroChem* **2019**, *6*, 3657–3666. [CrossRef]
18. Chen, K.; Guo, H.-N.; Li, W.-Q.; Wang, Y.J. MOF-derived core-shell CoP@NC@TiO_2 composite as a high-performance anode material for Li-ion batteries. *Chem. Asian J.* **2021**, *16*, 322–328. [CrossRef]
19. Wang, Y.; Yu, W.; Geng, F.; Zhao, L.; Wang, Y. TiO_2@MOF photocatalyst for the synergetic oxidation of microcystin-LR and reduction of Cr(VI) in a media. *Catalysts* **2021**, *11*, 1186. [CrossRef]
20. He, X.; Fang, H.; Gosztola, D.J.; Jiang, Z.; Jena, P.; Wang, W.N. Mechanistic insight into photocatalytic pathways of MIL-100(Fe)/TiO_2 composites. *ACS Appl. Mater. Interfaces* **2019**, *11*, 12516–12524. [CrossRef]
21. Lee, D.T.; Zhao, J.; Oldham, C.J.; Peterson, G.W.; Parsons, G.N. UiO-66-NH_2 Metal–Organic Framework (MOF) nucleation on TiO_2, ZnO, and Al_2O_3 atomic layer deposition-treated polymer fibers: Role of metal oxide on MOF growth and catalytic hydrolysis of chemical warfare agent simulants. *ACS Appl. Mater. Interfaces* **2017**, *9*, 44847–44855. [CrossRef] [PubMed]
22. He, K.; Wen, Q.; Wang, C.; Wang, B.; Yu, S.; Hao, C.; Chen, K. Porous TiO_2 nanoparticles derived from titanium Metal– Organic Framework and its improved electrorheological performance. *Ind. Eng. Chem. Res.* **2018**, *57*, 6888–6896. [CrossRef]
23. Sheng, H.; Chen, D.; Li, N.; Xu, Q.; Li, H.; He, J.; Lu, J. Urchin-inspired TiO_2@MIL-101 double-shell hollow particles: Adsorption and highly efficient photocatalytic degradation of hydrogen sulfide. *Chem. Mater.* **2017**, *29*, 5612–5616. [CrossRef]
24. Zhou, S.; Guo, J.; Dai, Z.; Liu, C.; Zhao, J.; Gao, Z.; Song, Y.-Y. Engineering homochiral MOFs in TiO_2 nanotubes as enantioselective photoelectrochemical electrode for chiral recognition. *Anal. Chem.* **2021**, *93*, 12067–12074. [CrossRef] [PubMed]
25. Fu, N.; Ren, X.-C. Synthesis of double-shell hollow TiO_2@ZIF-8 nanoparticles with enhanced photocatalytic activities. *Front. Chem.* **2020**, *8*, 578847. [CrossRef] [PubMed]
26. Zhao, J.; Wang, B.; Zhao, Y.; Hou, M.; Xin, C.; Li, Q.; Yu, X. High-performance visible-light photocatalysis induced by dye-sensitized Ti^{3+}-TiO_2 microspheres. *J. Phys. Chem. Solids* **2023**, *179*, 111374. [CrossRef]

27. Bi, R.; Liu, J.; Zhou, C.; Shen, Y.; Liu, Z.; Wang, Z. In situ synthesis of g-C_3N_4/TiO_2 heterojunction by a concentrated absorption process for efficient photocatalytic degradation of tetracycline hydrochloride. *Environ. Sci. Pollut. Res.* **2023**, *30*, 55044–55056. [CrossRef] [PubMed]
28. Feizpoor, S.; Habibi-Yangjeh, A.; Luque, R. Preparation of TiO_2/Fe-MOF n–n heterojunction photocatalysts for visible-light degradation of tetracycline hydrochloride. *Chemosphere* **2023**, *336*, 139101. [CrossRef]
29. Li, Y.; Liu, H.; Hou, C.; Xie, Y.; Wang, L.; Zhang, M. Transformation of titanium-based photocatalyst and its degradation of tetracycline hydrochloride. *J. Alloys Compd.* **2024**, *970*, 172644. [CrossRef]
30. Wang, Y.; Fan, H.; Wong, P.K.; Wu, Y.; Rittmann, B. Biodegradation of tetracycline using hybrid material (UCPs-TiO_2) coupled with biofilms under visible light. *Bioresour. Technol.* **2021**, *323*, 124638. [CrossRef]
31. Wu, S.; Li, X.; Tian, Y.; Lin, Y.; Hu, Y.H. Excellent photocatalytic degradation of tetracycline over black anatase-TiO_2 under visible light. *Chem. Eng. J.* **2021**, *406*, 126747. [CrossRef]
32. Shi, Y.; Yan, Z.; Xu, Y.; Tian, T.; Zhang, J.; Pang, J.; Peng, X.; Zhang, Q.; Shao, M.; Tan, W.; et al. Visible-light-driven AgBreTiO_2-Palygorskite photocatalyst with excellent photocatalytic activity for tetracycline hydrochloride. *J. Clean. Prod.* **2020**, *277*, 124021. [CrossRef]
33. Ning, P.; Chen, H.; Pan, J.; Liang, J.; Qin, L.; Chen, D.; Huang, Y. Surface defect-rich g-C_3N_4/TiO_2 Z-scheme heterojunction for efficient photocatalytic antibiotic removal: Rational regulation of free radicals and photocatalytic mechanism. *Catal. Sci. Technol.* **2020**, *10*, 8295–8304. [CrossRef]
34. Wang, Y.; Lei, R.B.; Pw, A.; Zs, A.; Lz, A. Photocatalytic activity of N-TiO_2/O-doped N vacancy g-C_3N_4 and the intermediates toxicity evaluation under tetracycline hydrochloride and Cr(VI) coexistence environment. *Appl. Catal. B Environ.* **2020**, *262*, 118308. [CrossRef]
35. Zhou, S.; Jiang, L.; Wang, H.; Yang, J.; Yuan, X.; Wang, H.; Liang, J.; Li, X.; Li, H.; Bu, Y. Oxygen Vacancies Modified TiO_2/O-Terminated Ti_3C_2 Composites: Unravelling the Dual Effects between Oxygen Vacancy and High-Work-Function Titanium Carbide. *Adv. Funct. Mater.* **2023**, *33*, 2307702. [CrossRef]
36. Biswal, L.; Mishra, B.P.; Das, S.; Acharya, L.; Nayak, S.; Parida, K. Nanoarchitecture of a Ti_3C_2@TiO_2 Hybrid for Photocatalytic Antibiotic Degradation and Hydrogen Evolution: Stability, Kinetics, and Mechanistic Insights. *Inorg. Chem.* **2023**, *62*, 7584–7597. [CrossRef] [PubMed]
37. Qu, X.; Lin, J.; Qiang, W.; Chen, C.; Sun, D. Self-doped defect-mediated TiO_2 with disordered surface for high-efficiency photodegradation of various pollutants. *Chemosphere* **2022**, *308*, 136239. [CrossRef] [PubMed]
38. Liu, J.; Han, L.; An, N.; Xing, L.; Ma, H.; Cheng, L.; Yang, J.; Zhang, Q. Enhanced visible-light photocatalytic activity of carbonate-doped anatase TiO_2 based on the electron-withdrawing bidentate carboxylate linkage. *Appl. Catal. B Environ.* **2017**, *202*, 642–652. [CrossRef]
39. Imparato, C.; Iervolino, G.; Fantauzzi, M.; Koral, C.; Macyk, W.; Kobielusz, M.; D'Errico, G.; Rea, I.; Di Girolamo, R.; De Stefano, L.; et al. Photocatalytic hydrogen evolution by co-catalyst-free TiO_2/C bulk heterostructures synthesized under mild conditions. *RSC Adv.* **2020**, *10*, 12519–12534. [CrossRef]
40. Pukdeejorhor, L.; Wannapaiboon, S.; Berger, J.; Rodewald, K.; Thongratkaew, S.; Impeng, S.; Warnan, J.; Bureekaew, S.; Fischer, R.A. Defect engineering in MIL-125-(Ti)-NH_2 for enhanced photocatalytic H_2 generation. *J. Mater. Chem. A* **2023**, *11*, 9143–9151. [CrossRef]

polymers

Article

Assessment of Molecularly Imprinted Polymers as Selective Solid-Phase Extraction Sorbents for the Detection of Cloxacillin in Drinking and River Water

Rosa Mª Garcinuño *, Eduardo José Collado, Gema Paniagua, Juan Carlos Bravo and Pilar Fernández Hernando

Department of Analytical Science, Faculty of Science, Universidad Nacional de Educación a Distancia, Las Rozas, 28232 Madrid, Spain; ejcollado@ccia.uned.es (E.J.C.); gpaniagua@ccia.uned.es (G.P.); juancarlos.bravo@ccia.uned.es (J.C.B.); pfhernando@ccia.uned.es (P.F.H.)
* Correspondence: rmgarcinuno@ccia.uned.es; Tel.: +34-91-398-73-66

Abstract: This paper describes a new methodology for carrying out quantitative extraction of cloxacillin from drinking and river water samples using a molecularly imprinted polymer (MIP) as a selective sorbent for solid-phase extraction (MISPE). Several polymers were synthesized via thermal polymerization using cloxacillin as a template, methacrylic acid (MAA) as a functional monomer, ethyleneglycoldimethacrylate (EGDMA) as a cross-linker and different solvents as porogens. Binding characteristics of the adequate molecularly imprinted and non-imprinted (NIP) polymers were evaluated via batch adsorption assays following the Langmuir and Freundlich isotherms and Scatchard assays. The parameters related to the extraction approach were studied to select the most appropriate polymer for cloxacillin determination. Using the optimized MIP as the SPE sorbent, a simple sample treatment methodology was combined with high-performance liquid chromatography (HPLC) to analyze cloxacillin residues in drinking and river water. Under the optimum experimental conditions, the MISPE methodology was validated using spiked samples. The linearity for cloxacillin was assessed within the limits of 0.05–1.5 μg L^{-1} and the recovery percentage was higher than 98% (RSD < 4%). The limits of detection and limits of quantification were 0.29 and 0.37 μg L^{-1} and 0.8 and 0.98 μg L^{-1} for drinking and river water, respectively. The selectivity of MIP against other ß-lactam antibiotics with similar structures (oxacillin, cefazoline, amoxicillin and penicillin V) was studied, obtaining a good recovery higher than 85% for all except cefazoline. The proposed MISPE-HPLC methodology was successfully applied for the detection of cloxacillin in drinking water from Canal de Isabel II (Madrid) and river water from the Manzanares River (Madrid).

Keywords: cloxacillin; molecularly imprinted polymer; solid-phase extraction; adsorption isotherms; water; HPLC-DAD

1. Introduction

ß-lactam antibiotics (BLAs) constitute a very important group of antibiotics belonging to the penicillin family, which have been broadly used as antimicrobial agents for almost a century. The increase in the consumption of pharmaceutical products in medicine and the veterinary field for the treatment and prevention of diseases, and to promote the growth of livestock, has contributed to a worrying growth in antibiotic-related problems in the aquatic environment [1]. Antibiotics can enter the aquatic system from different sources, including the widespread manufacturing of medical preparations, direct discharge of animal wastewater or seepage from manured farmland [2], and discharge of effluents from the treatment of communal wastewater containing antibiotics in sewage plants, which persist because of the high stability of some compounds or their metabolites. New studies have shown that antibiotics can be found in drinking water from groundwater [3,4]. The presence of antibiotic residues in foodstuffs, including drinking water, has harmful effects on consumers, including allergic reactions, and it affects the intestinal flora and the appearance and spread of antibiotic resistance [5–8]. So, monitoring residues of pharmaceuticals such as cloxacillin in environmental water samples is very important to avoid its hazard to human health [9].

At present, several screening tests (microbiological and immunological) are published in the literature to detect the presence of the maximum residue limits of ß-lactams in milk and other tissues [10]. However, these assays are highly time-consuming and expensive, have a high percentage of false positives and, in many cases, quantitative determination is not possible. Currently, analytical investigation of cloxacillin in biological fluids, environmental samples and food requires the use of high-performance liquid chromatography (HPLC) combined with ultraviolet (UV) [11], diode array detector (DAD) [12,13], fluorescence detection (FLD) [14], mass spectrometry (MS) [15,16] or tandem mass liquid chromatography coupled to tandem mass spectrometry (LC-MS/MS) [17–20], which is considered the preferred detection method for routine analysis in a complex matrix because of its great sensitivity and selectivity. Although these methods are rather efficient for the determination of cloxacillin in different matrices, some of them have disadvantages such as low sensitivity, time-consuming sample pretreatment, high-cost instrumentation and difficulties in sample preparation [16,17,19]. The analysis of antibiotics in different matrices usually requires clean-up and preconcentration of the sample. In that sense, solid-phase extraction (SPE) with different types of commercial sorbents and conditions is the most frequently used extraction method for water samples [3,21–25]. The use of molecular imprinting technology for sample preconcentration and efficient clean-up in food and environmental applications is constantly increasing [26–29]. Most of the present research in the molecular imprinting field is focused on SPE, which offers advantages such as the predetermined selectivity for the analyte(s) of interest, the relatively low amount of solvent and the stability of imprinted polymers, which affect the recovery and contribute highly to the low price of the analysis. These advantages make MISPE technology highly suitable for application to the analysis of antibiotics instead of SPE using commercial cartridges, which, in addition to their high cost, do not allow us to obtain the required selectivity or desired recoveries in some cases, as shown in the study by Bondi et al. [30]. Some MIPs synthesized for cloxacillin were applied to the determination of this antibiotic in milk [31,32], shrimp [33] and other food samples [34]. However, to date, there are no works reported in the literature that address the application of cloxacillin MIPs for water or environmental analyses.

This research describes the assessment of several synthetized MIPs for use as a solid-phase extraction sorbent (MISPE) applied to the quantitative detection of cloxacillin from drinking and river water samples. The developed MISPE-HPLC-DAD methodology was optimized for the detection and quantification of cloxacillin using fortified drinking and river water samples.

2. Experiment

2.1. Chemical, Reagents and Samples

All reagents were of analytical grade. For the synthesis of MIP, commercial antibiotics standards (cloxacillin, amoxicillin, penicillin V, oxacillin and cefazoline), ethyleneglycoldimethacrylate (EGDMA), 2,2′-azobisisobutyronitrile (AIBN) and methacrylic acid (MAA) were purchased from Sigma-Aldrich (Madrid, Spain). HPLC-grade acetonitrile (ACN), methanol (MeOH) and toluene were supplied by LAB-SCAN (Gliwice, Poland). Deionized water (18.2 MΩ cm) was procured from a Milli-Q water system (Millipore Ibérica, Barcelona, Spain).

Drinking water was collected from Canal de Isabel II in Madrid, which is the Spanish public company that supplies water for human consumption in Madrid. River water was obtained from the Manzanares River (Madrid, Spain).

2.2. Chromatographic Conditions

HPLC experiments were performed using a high-performance liquid chromatography instrument 1200 Series from Agilent Technologies (Agilent Technologies, Waldbronn, Germany) equipped with a 1260 diode-array detector, a 1260 quaternary pump and manual injector with a 20 µL sample loop. An analytical column packed with 3 µm C18-Pentafluorophenyl (15 cm × 4.6 mm) from ACE (Advanced Chromatography Technologies, Aberdeen, Scotland) was employed to perform the separation of the analytes. The mobile phase used combines water (solvent A) and acetonitrile (solvent B). The gradient elution was: 5 to 22% B at 0.4–0.7 mL min^{-1} (6 min), 22 to 10% B at 0.7–0.25 mL min^{-1} (0.5 min), 10% B at 0.25 mL min^{-1} (3.5 min), 10 to 22% B at 0.25–0.9 mL min^{-1} (0.5 min), 22% B at 0.9 mL min^{-1} (5.5 min). To achieve maximum sensitivity for cloxacillin determination, the detection wavelength was studied to determine the best quantitative peak areas employing a wavelength of 210 nm. Quantification of the analytes was achieved via external calibration and measurements of the peak area.

2.3. Synthesis of Cloxacillin Molecularly Imprinted Polymers

Three cloxacillin-imprinted polymers were synthesized via thermal polymerization. In a glass test tube, we mixed 0.2 mmol of template (cloxacillin), 1.6 mmol of functional monomer (MAA) and 8 mmol of cross-linker (EGDMA). The chemicals were dissolved in a volume of 4.6 mL of different porogens (water, ACN or toluene), with a template:functional monomer:cross-linker molar composition of 1:8:40. After mixture homogenization, 36.3 mg of AIBN (2.0% w/w) as the free radical initiator was added to the solution. The mixture was purged under a nitrogen stream (5 min) and the glass tube was sealed. Then, the polymerization was carried out at 60 °C in a water bath for 4 h. For reference, non-imprinted polymers (NIPs) were synthesized using the above-described protocol, in which the template molecule was not added to the polymerization mixture. The obtained polymer was ground and wet-sieved with acetone using a metal sieve (355 µm). The template was eliminated from the polymeric matrix using microwave-assisted extraction as the microwave energy provides fast extraction of analytes from solid matrices. The efficiency of this technique is comparable to classical extraction techniques such as Soxhlet or sonication [35]. The procedure used for template removal was that from a previous work [36] (1.5 mL of MeOH: acetic acid (95/5, v/v) at 800 W, 100 °C, 200 psi with stirring for 20 min). To complete the extraction of the template, this procedure was repeated 3 times, until cloxacillin was not detected in the extraction solvent during HPLC analysis.

2.4. Batch Binding Studies

The binding properties of the optimum selected polymer were studied using batch adsorption experiments and Scatchard analysis. The general procedure was as follows: 20 mg of MIP or NIP was incubated for 24 h in the dark in a screw-capped flask with 2.5 mL of water containing cloxacillin at various concentrations (0–300 mg L^{-1}). After incubation, the vials were centrifuged at 1200 rpm for 10 min and the free cloxacillin remaining in the supernatant was analyzed using HPLC-DAD at 210 nm.

2.5. MISPE Procedure

An amount of 100 mg of polymer was placed in SPE cartridges (3 mL) (Scharlau, Barcelona, Spain) between 2 polyethylene frits with 0.22 µm micropores (Supelco Analytical, Bellefonte, PA, USA). The SPE column containing the cloxacillin MIP was preconditioned with 2 mL of MeOH and next with 1 mL of water before loading 1 mL of sample. The selective washing step was performed with 1 mL of toluene. Finally, the target analyte adsorbed on the polymer was eluted with 1.5 × 3 mL of MeOH. The fractions were collected and evaporated to dryness at room temperature under a nitrogen stream. The residues obtained were reconstituted in a volume of 0.5 mL of water prior to HPLC analysis. The same procedure was used to evaluate the NIP.

3. Results and Discussion

3.1. Synthesis of Molecularly Imprinted Polymers and Binding Site Evaluation

The MIPs were synthesized following a non-covalent approach using bulk polymerization, as described in Section 2.3. As the functional monomer, methacrylic acid was used due to its capacity to form both ionic and hydrogen bonds with several functional groups from the template. This procedure allowed us to generate specific imprints inside the polymer matrix network. The molar ratio of functional monomer/template is crucial for obtaining specific binding sites. When an excess of monomer is used, elevated non-specific binding sites could be generated in the polymeric matrix [37]. The bifunctional cross-linker monomer EGDMA is mostly used to give a certain mechanical stability in the polymer, instilling appropriate rigidity in the matrix in a molar ratio of 1:5 (monomer:cross-linker). AIBN at 2.0% (*w*/*w*) was used as the heat-induced initiator for the thermal polymerization at 60 °C. The solvent is a key aspect in the polymerization due to the competitive effect with the monomer and the template, independently of the interaction established between these last ones. In this work, three polymers in different media (water, acetonitrile and toluene) were synthetized.

The selection of the most adequate MIP to be used as a selective adsorbent in the SPE process for the determination of cloxacillin was carried out studying the extraction factors, calculated with Equation (1).

$$\text{Extraction } \% = \frac{C_i - C_f}{C_i} \times 100 \quad (1)$$

C_i and C_f refer to the concentrations of cloxacillin in the solution before and after extraction, respectively.

To obtain the extraction factor data, the following procedure was performed: 1 mL of cloxacillin solution in water at 1 mgL^{-1} was passed through the SPE cartridge. Previously, a conditioning step was carried out with 2 mL of MeOH and next with 1 mL of water. The fractions of cloxacillin collected were analyzed using HPLC-DAD. Data obtained showed that the extraction factor for the water MIP was 54%, for the toluene MIP was 75% and for the ACN-MIP was 94%; meanwhile, the NIP extraction factors in these solvents were 16, 14 and 21%, respectively, lower that those obtained for the MIPs. This study showed that ACN was the optimum porogen for the MIP synthesis due to the strong affinity of ACN-MIP for cloxacillin and the good specificity obtained compared with the non-imprinted polymer (ACN-NIP). According to obtained results, ACN-MIP was chosen as the optimum sorbent in the SPE procedure to carry out cloxacillin determination.

3.2. Saturation Binding Curves and Scatchard Plot Analysis

The binding affinity of ACN-MIP for cloxacillin was evaluated using binding isotherm experiments and Scatchard plot analysis. Batch adsorption experiments were carried out as follows: 2.5 mL of solutions of cloxacillin at a concentration ranging from 0 to 300 mg L^{-1} were incubated with 20 mg of ACN-MIP or ACN-NIP for a period of 24 h. Then, the concentration of the remaining cloxacillin in the solution was quantified via chromatographic detection at 210 nm. Figure 1A represents the binding of cloxacillin per g of polymer (B), for both imprinted and non-imprinted polymers, versus the initial concentration of cloxacillin in the medium (C_0). The cloxacillin equilibrium concentration bound to ACN-MIP increases with the initial cloxacillin concentration in solution. The amount of cloxacillin bound to ACN-MIP is higher than the amount of cloxacillin bound to ACN-NIP, which means higher affinity of ACN-MIP to cloxacillin due to the specific sites generated in the synthesis.

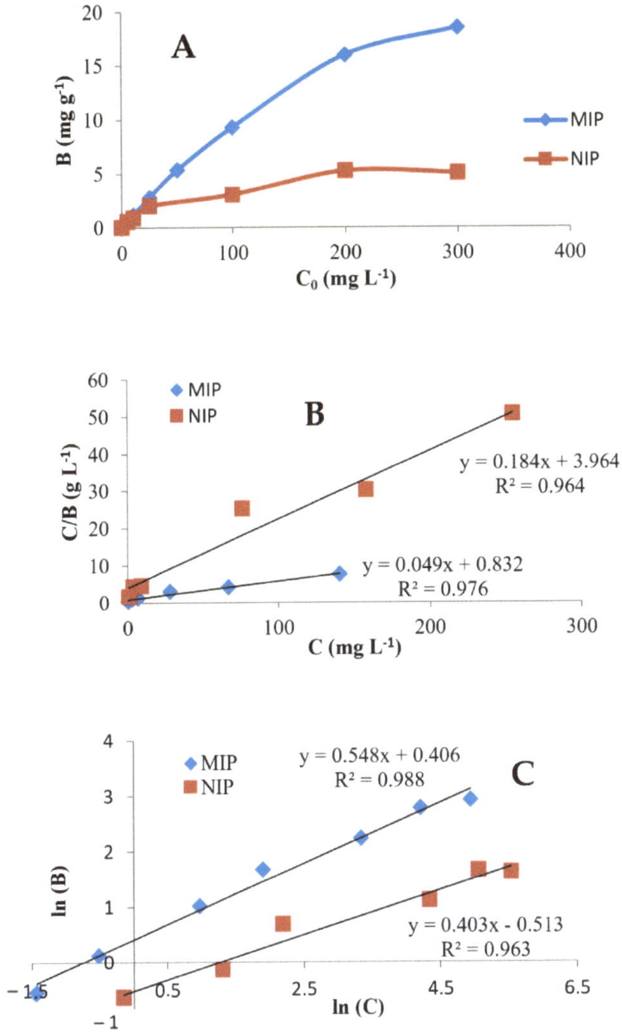

Figure 1. (**A**) Adsorption isotherm of MIP and NIP using batch adsorption assay. (B) is the binding of cloxacillin per g of polymer, (C_0) is the initial cloxacillin in the medium. (**B**) Langmuir isotherms linearized for cloxacillin adsorption onto MIP and NIP. (**C**) Freundlich isotherms linearized for cloxacillin adsorption onto MIP and NIP.

Different adsorption isotherm models have been applied to characterize the sorption behavior of MIPs. Langmuir and Freundlich isotherms are the most commonly used models and can be utilized to determine sorbent properties such as adsorption mechanism, surface area and affinities with the target, as these isotherms show the distribution of molecules between the liquid and solid phases at the equilibrium time. Experimental results fitted to Langmuir and Freundlich isotherms are included in Table 1.

Table 1. Langmuir and Freundlich isotherm data for adsorption of cloxacillin on ACN-MIP and ACN-NIP.

Isotherm Results	Langmuir				Freundlich		
	B_0 (mg g^{-1})	K_L (L mg^{-1})	R_L	R^2	KF (mg g^{-1})	n	R^2
ACN-MIP	0.25	80.2	0.001	0.964	1.5	1.8	0.988
ACN-NIP	1.21	4.51	0.004	0.976	0.5	2.5	0.963

The main premise assumed in the Langmuir isotherm model is that the sorbent is monolayer, where the sites only take one molecule of the target each, making the adsorption energy constant over all sites. When every adsorption site has sorbed one target molecule, the maximum surface adsorption is achieved and no more sorption can take place [38]. The obtained data on ACN-MIP and ACN-NIP were fitted to the Langmuir adsorption model using Equation (2):

$$\frac{C}{B} = \left(\frac{1}{B_0 \times K_L}\right) + \left(\frac{C}{B_o}\right) \quad (2)$$

In Equation (2), C (mg L^{-1}) refers to the remaining analyte concentration at equilibrium, B (mg g^{-1}) refers to the cloxacillin amount bound to ACN-MIP or ACN-NIP (g) at equilibrium, B_0 (mg g^{-1}) refers to the maximum amount of cloxacillin related to complete coverage of the monolayer on the surface and K_L (L mg $^{-1}$) refers to the Langmuir constant, which is associated with adsorption energy. The experimental data were plotted in Figure 1B, showing a linear relationship where the Langmuir constants B_0 and K_L can be calculated with Equation (2). The maximum monolayer capacities, B_0, were 0.25 mg g^{-1} and 1.21 mg g^{-1} for ACN-MIP and ACN-NIP, respectively, while the Langmuir constants K_L were 80.2 mg L^{-1} and 4.51 mg L^{-1} for MIP and NIP, respectively (Table 1).

The Langmuir model is restricted to a monolayer, while the Freundlich isotherm model corresponds to a heterogeneous adsorbent surface caused by multilayer adsorption and an exponential distribution of the active sites and their energies. The empirical equation that represents the Freundlich adsorption isotherm is as follows (3):

$$\ln(B) = \ln(K_F) + \frac{1}{n}\ln(C) \quad (3)$$

In Equation (3), K_F (mg g^{-1}) refers to a Freundlich constant corresponding to the binding energy and n (dimensionless) is a Freundlich constant representing the heterogeneity index according to this empirical model. The adsorption is linear when $n = 1$, a chemical process when $n < 1$ and a physical process when $n > 1$ [39]. Figure 1C shows the graph of $\ln(B)$ versus $\ln(C)$ using binding data from ACN-MIP and ACN-NIP, which give a straight line. The values of Freundlich constants K_F were 1.5 mg g^{-1} for MIP and 0.5 mg g^{-1} for ACN-NIP, while the n constants were 1.8 and 2.5 for ACN-MIP and ACN-NIP, respectively. The value of n for ACN-MIP was lower than for ACN-NIP, which means that ACN-MIP has a higher percentage of binding sites of high affinity.

To finish the evaluation of the binding characteristics between cloxacillin and polymer at equilibrium and predict the affinity between the solid phase and the analyte, the R_L parameter (dimensionless) was studied. R_L, which describes the nature of the adsorption process, is represented in Equation (4), where K_L (L mg $^{-1}$) refers to Langmuir constant and C_0 (mg L^{-1}) refers to the beginning analyte concentration.

$$R_L = \frac{1}{1 + (K_L \times C_0)} \quad (4)$$

The type of Langmuir isotherm is defined by the value of R_L: irreversible when $R_L = 0$, favorable when $R_L < 1$, linear when $R_L = 1$ or unfavorable when $R_L > 1$ [40]. The

R_L obtained for ACN-MIP and ACN-NIP were all positive and less than 1, between 0.001 and 0.04, indicating that the adsorption process of cloxacillin was favorable.

The Scatchard plot is used to determine the number and type of binding sites of the polymers. This estimated model is often used in solid-phase extraction to characterize the maximum amount of adsorption and the dissociation constant. The binding data were linearly transformed according to the Scatchard equation [39].

$$\frac{B}{C} = \frac{(B_{\max} - B)}{K_d} \quad (5)$$

In Equation (5), B (mg g^{-1}) refers to cloxacillin bound to 1 g of ACN-MIP or ACN-NIP at equilibrium, C (mg L^{-1}) refers to the remaining analyte concentration at equilibrium, $B_{\max}$ (mg g^{-1}) refers to the maximum amount of adsorption of cloxacillin, and K_d (mg L^{-1}) refers to the equilibrium dissociation constant of the binding sites. The slope and the intercept of the plot B/C versus B correspond to K_d and $B_{\max}$, respectively. The Scatchard plots of ACN-MIP and ACN-NIP (Figure 2A and 2B, respectively) are both nonlinear.

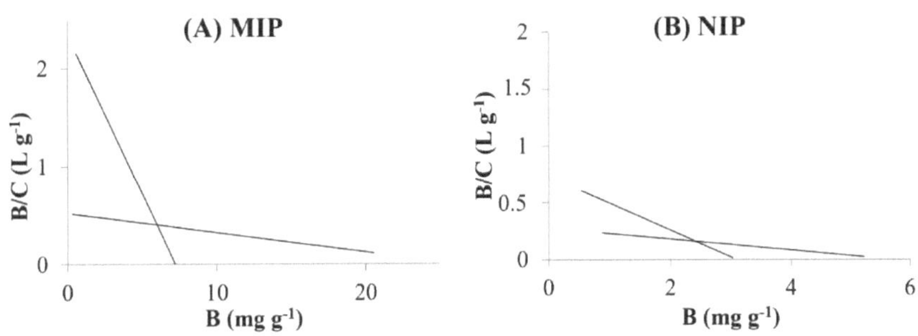

Figure 2. Scatchard plots of the binding of cloxacillin to the MIP (**A**) and to the NIP (**B**).

The data could be fitted to two straight lines with different slopes, which means the existence of heterogeneous binding sites, both in ACN-MIP and ACN-NIP, representing binding sites of high and low affinity. The dissociation constant of the binding sites (K_d) and the apparent maximum number of the binding sites ($B_{\max}$) could be obtained from the slope and the intercept, respectively. Table 2 shows the outcomes of Scatchard assays for both polymers ACN-MIP and ACN-NIP. In ACN-MIP, the values of $B_{\max}$ and K_d are 2.9 mg g^{-1} and 1.3 mg L^{-1}, respectively, for binding sites of high affinity and 22.6 mg g^{-1} and 57.2 mg L^{-1}, respectively, for the low-affinity binding sites. Meanwhile, the ACN-NIP results revealed values of 1.8 mg g^{-1} and 2.7 mg L^{-1} for $B_{\max}$ and K_d (high-affinity binding sites) and 6.1 mg g^{-1} and 181.6 mg L^{-1} (binding sites of low affinity), respectively. These results show that ACN-MIP had bigger $B_{\max}$ and lower K_d in ACN-MIP than ACN-NIP, due to the specific binding sites.

Table 2. Results of Scatchard analysis for ACN-MIP and ACN-NIP.

Binding Sites	**ACN-MIP**		**ACN-NIP**	
	$B_{\max}$ (mg g^{-1})	K_d (mg L^{-1})	$B_{\max}$ (mg g^{-1})	K_d (mg L^{-1})
High affinity	2.9	1.3	1.8	2.7
Low affinity	22.6	57.2	6.1	181.6

The synthesis of polymers via the non-covalent method could explain the heterogeneity of the surface of the MIP. Although this method generates specific binding sites, some interactions between the template and the functional monomers cannot be made properly. This incompleteness could generate non-specific binding sites, causing the heterogeneous surface.

3.3. MISPE Procedure Optimization

The MISPE procedure was optimized using ACN-MIP as the sorbent to obtain the best selectivity and recovery conditions for cloxacillin. Firstly, the conditioning stage was evaluated using MeOH (3 × 1 mL) or 2 mL of MeOH and 1 mL of Milli-Q water and, subsequently, a solution of cloxacillin in drinking water at 1 mg L^{-1} was loaded onto the SPE column. The higher retention of cloxacillin (96%) occurred after conditioning the column with 2 mL MeOH and 1 mL of Milli-Q water, being 76% when only MeOH was used. To optimize the washing step, different solvents (water, acetonitrile and toluene) were tested. Water and acetonitrile washed cloxacillin from the polymer. Given the low solubility of cloxacillin in toluene, this solvent was chosen to wash the non-specifically nonpolar bindings. To achieve a quantitative extraction of cloxacillin from the MISPE column, different volumes (1–5 mL) of MeOH and ACN or MeOH/ACN mixtures were examined as eluents. Recoveries were calculated from the fractions collected after the elution step, taking into account the amount of amoxicillin retained on the column in the loading step. Figure 3A shows the elution capacity of the solvents tested when 3 mL (2 × 1.5) was used. It is shown that recoveries increased with the increment of methanol percentage, reaching the maximum when 100% methanol was used. The effect of solvent volume on the elution process was tested (Figure 3B). The complete elution of cloxacillin was reached with MeOH (3 × 1.5 mL).

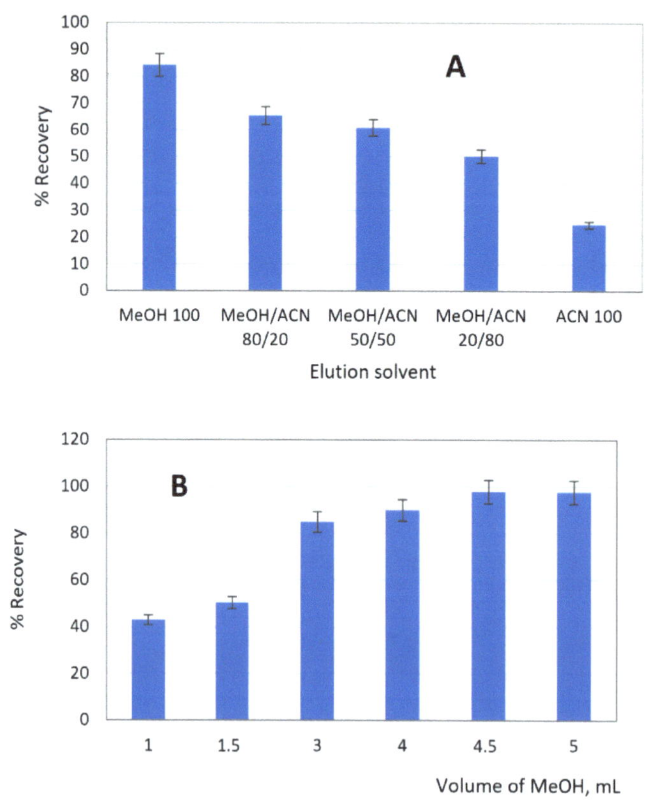

Figure 3. Optimization of MISPE. (**A**) Solvent elution. (**B**) Volume of MeOH as optimum solvent elution.

3.4. Validation and Applicability of the MISPE-HPLC Methodology

The MISPE method developed for cloxacillin was validated by testing analytical parameters such as linearity, accuracy, precision and limits of detection and quantification using spiked water samples. River water required a previous filtration step through a PTFE filter (0.2 µm) to remove suspended matter before being submitted to an optimized MIPSE procedure.

Linearity was evaluated using spiked water samples with increasing concentrations of cloxacillin in the range of 0.05–1.5 µg L^{-1}. Calibration curves were plotted by means of the cloxacillin concentration versus peak area. The obtained results showed good linearity in both drinking and river water ($R^2 > 0.998$). The accuracy and precision of the developed method were assessed via analysis of drinking and river water samples spiked at three concentration levels (0.1, 0.5, 1 µg L^{-1}). Samples were analyzed in quintuplicate for the intra-day study and in triplicate for the inter-day and accuracy studies. Standard deviations were calculated in all cases (Table 3). The % RSD values showed good precision of the developed method. The obtained values for the detection limit (LOD), calculated as a signal-to-noise ratio of 3, were 0.29 and 0.37 µg L^{-1} for drinking and river water, respectively. The quantification limits (LOQs), calculated as a signal-to-noise ratio of 10, were 0.8 and 0.98 µg L^{-1} for drinking and river water, respectively.

Table 3. Analytical characteristics for cloxacillin of the developed MISPE method for drinking and river water.

Sample	Linearity		Spiking Level µg/L	Recovery ± RSD %		LOD µg/L	LOQ µg/L
	Concentration Range µg/L	R^2		Inter-Day	Intra-Day		
Drinking water	0.05–1.50	0.999	0.10 0.50 1.00	96.9 ± 6.3 94.7 ± 4.2 93.8 ± 6.4	90.5 ± 7.5 92.5 ± 6.8 91.8 ± 7.9	0.29	0.8
River water	0.05–1.50	0.999	0.10 0.50 1.00	83.1 ± 2.2 84.3 ± 5.2 89.1 ± 4.2	89.9 ± 6.0 80.5 ± 5.0 81.7 ± 7.7	0.37	0.98

The comparison of the developed method with other reported techniques for the determination of cloxacillin, such as the detection method, extraction method, LOD, LOQ and recovery, is summarized in Table 4.

Table 4. Comparison of the proposed method with other published methods for determination of cloxacillin.

Sample Matrix	Extraction Method	Detection Method	Recovery (%)	LOD	LOQ	Ref.
Milk	SPE	Electrochemical sensor	98.6–101.8	36 nM		[31]
Shrimp	MIM	HPLC-UV	80.9–94.9	0.03 µg/g	0.10 µg/g	[33]
Pig plasma	MMIP	SERS	<80.0	7.80 pmol		[34]
Drinking water Tap water	MISPE	HPLC-DAD	93.8–96.9 83.1–89.1	0.29 µg/L 0.37 µg/L	0.80 µg/L 0.98 µg/L	Present work

MIM: molecularly imprinted membranes; MMIP: magnetic molecularly imprinted polymer; SERS: surface-enhanced Raman spectroscopy.

3.5. Selectivity of the MIP

Once the MISPE procedure was optimized, the selective adsorption of the ACN-MIP towards other ß-lactam antibiotics structurally analogous to cloxacillin such as amoxicillin, penicillin V, oxacillin and cefazoline was evaluated. The recovery percentages were calculated considering the fractions collected after the elution step with respect to the initial concentration of each compound loaded onto the SPE column. The results obtained are summarized in Figure 4. Recoveries for cloxacillin, penicillin V, oxacillin and amoxicillin were higher than 85%, while the recovery for cefazoline was approximately 74%. The results were in accordance with the chemical structures of the compounds assayed. This study revealed a high adsorption capacity of ACN-MIP towards the ß-lactam antibiotics with similar chemical structures to cloxacillin (molecule template) in comparison with cefazoline, the compound that most differs structurally from cloxacillin, which means that residues of five ß-lactam antibiotics could be extracted using the developed MISPE methodology. Furthermore, the adsorption capacity of the antibiotics onto ACN-NIP was evaluated. The obtained results showed a retention capacity lower than 22% for all the analytes.

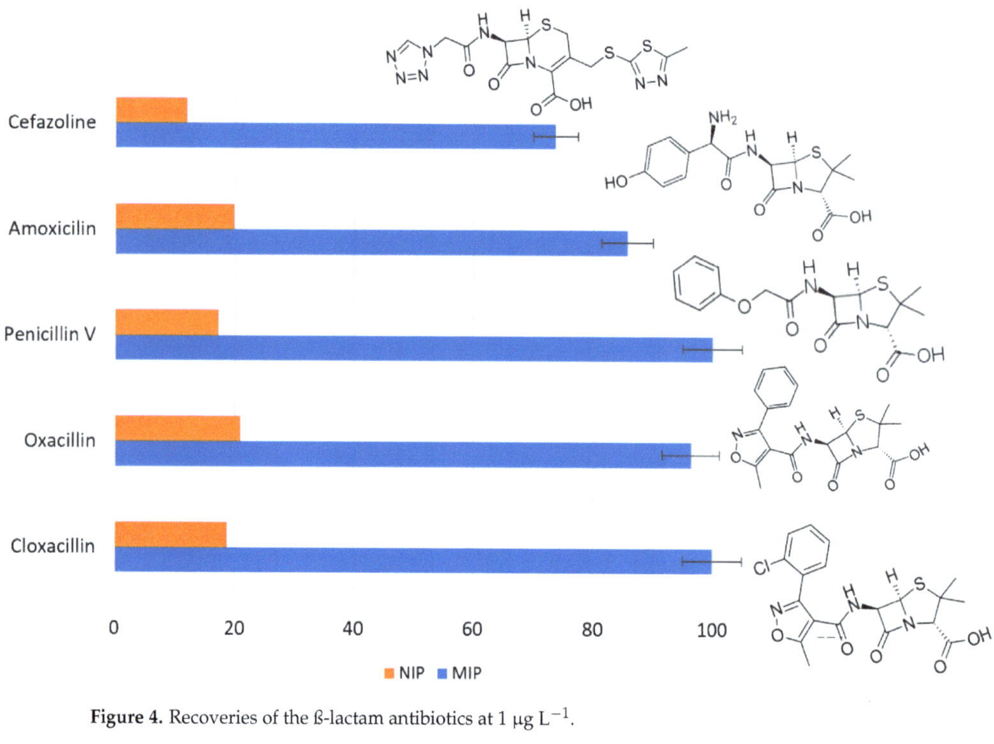

Figure 4. Recoveries of the ß-lactam antibiotics at 1 μg L^{-1}.

4. Conclusions

The use of a molecularly imprinting polymer was evaluated to develop an SPE method for the detection of cloxacillin in drinking and river water samples via HPLC-DAD. Binding properties of the polymer selected as optimum were studied using the Langmuir and Freundlich models and Scatchard analysis. Imprinted and non-imprinted polymers synthesized in ACN, ACN-MIP and ACN-NIP are well-described by these models with high- and low-affinity binding sites. Synthetized MIP exhibited a very high adsorption capacity for cloxacillin and good selectivity. The specific recognition of MIP towards other antibiotics belonging to the family of ß-lactams was tested. The obtained results showed successful recovery of antibiotics with a similar chemical structure (amoxicillin, penicillin V and oxacillin) and lower recovery for cefazoline because of lesser structural similarity. Hence, our work revealed that the obtained ACN-MIP has significant potential for use as specific sorbent in the SPE method for the analysis of five beta-lactams in drinking and river waters at trace levels. The developed MISPE-HPLC methodology offers a rapid, easy and reliable alternative procedure, consuming low volumes of solvents. This work thus makes a significant advance in water quality control with respect to other published methods due to the simplicity, low cost and sensitivity of our method.

Author Contributions: Conceptualization, R.M.G. and P.F.H.; methodology, R.M.G., P.F.H., G.P. and J.C.B.; investigation, E.J.C. and G.P.; data curation, E.J.C. and J.C.B.; writing—original draft preparation, E.J.C. and G.P.; writing—review and editing, R.M.G., P.F.H., J.C.B. and G.P.; supervision, R.M.G., P.F.H. and G.P.; project administration, P.F.H. and R.M.G.; funding acquisition, P.F.H. and R.M.G. All authors have read and agreed to the published version of the manuscript.

Funding: This work was supported by the Community of Madrid (Spain) and European funding from the FSE and FEDER programs (project S2018/BAA-4393, AVANSECAL-II-CM) and the National University of Distance Education (EUROPA INVESTIGACIÓN UNED—SANTANDER, ARTECAP).

Institutional Review Board Statement: Not applicable.

Data Availability Statement: Data are contained within this article.

Conflicts of Interest: The authors declare no conflict of interest.

References

1. Jia, W.L.; Song, C.; He, L.Y.; Wang, B.; Gao, F.Z.; Zhang, M.; Ying, G.G. Antibiotics in Soil and Water: Occurrence, Fate, and Risk. *Curr. Opin. Environ. Sci. Health* **2023**, *32*, 100437. [CrossRef]
2. Kovalakova, P.; Cizmas, L.; McDonald, T.J.; Marsalek, B.; Feng, M.; Sharma, V.K. Occurrence and Toxicity of Antibiotics in the Aquatic Environment: A Review. *Chemosphere* **2020**, *251*, 126351. [CrossRef] [PubMed]
3. Zuo, R.; Liu, X.; Zhang, Q.; Wang, J.; Yang, J.; Teng, Y.; Chen, X.; Zhai, Y. Sulfonamide Antibiotics in Groundwater and Their Migration in the Vadose Zone: A Case in a Drinking Water Resource. *Ecol. Eng.* **2021**, *162*, 106175. [CrossRef]
4. Shi, J.; Dong, Y.; Shi, Y.; Yin, T.; He, W.; An, T.; Tang, Y.; Hou, X.; Chong, S.; Chen, D.; et al. Groundwater Antibiotics and Microplastics in a Drinking-Water Source Area, Northern China: Occurrence, Spatial Distribution, Risk Assessment, and Correlation. *Environ. Res.* **2022**, *210*, 112855. [CrossRef] [PubMed]
5. Mishra, S.; Singh, A.K.; Cheng, L.; Hussain, A.; Maiti, A. Occurrence of Antibiotics in Wastewater: Potential Ecological Risk and Removal through Anaerobic–Aerobic Systems. *Environ. Res.* **2023**, *226*, 115678. [CrossRef]
6. Bilal, M.; Mehmood, S.; Rasheed, T.; Iqbal, H.M.N. Antibiotics Traces in the Aquatic Environment: Persistence and Adverse Environmental Impact. *Curr. Opin. Environ. Sci. Health* **2020**, *13*, 68–74. [CrossRef]
7. Ding, D.; Wang, B.; Zhang, X.; Zhang, J.; Zhang, H.; Liu, X.; Gao, Z.; Yu, Z. The Spread of Antibiotic Resistance to Humans and Potential Protection Strategies. *Ecotoxicol. Environ. Saf.* **2023**, *254*, 114734. [CrossRef]
8. Yu, C.; Pang, H.; Wang, J.H.; Chi, Z.Y.; Zhang, Q.; Kong, F.T.; Xu, Y.P.; Li, S.Y.; Che, J. Occurrence of Antibiotics in Waters, Removal by Microalgae-Based Systems, and Their Toxicological Effects: A Review. *Sci. Total Environ.* **2022**, *813*, 151891. [CrossRef]
9. Takács, E.; Wang, J.; Chu, L.; Tóth, T.; Kovács, K.; Bezsenyi, A.; Szabó, L.; Homlok, R.; Wojnárovits, L. Elimination of Oxacillin, Its Toxicity and Antibacterial Activity by Using Ionizing Radiation. *Chemosphere* **2022**, *286*, 131467. [CrossRef]
10. Ahmed, S.; Ning, J.; Peng, D.; Chen, T.; Ahmad, I.; Ali, A.; Lei, Z.; Abu bakr Shabbir, M.; Cheng, G.; Yuan, Z. Current Advances in Immunoassays for the Detection of Antibiotics Residues: A Review. *Food Agric. Immunol.* **2020**, *31*, 268–290. [CrossRef]
11. Merlo, F.; Montagna, J.; Maraschi, F.; Profumo, A.; Baldanti, F.; Speltini, A. A Versatile Method for Multiclass Determination of β-Lactam Drugs in Urine by Solid-Phase Extraction Followed by HILIC-UV. *J. Chromatogr. Open* **2022**, *2*, 100048. [CrossRef]
12. Cámara, M.; Gallego-Picó, A.; Garcinuño, R.M.; Fernández-Hernando, P.; Durand-Alegría, J.S.; Sánchez, P.J. An HPLC-DAD Method for the Simultaneous Determination of Nine β-Lactam Antibiotics in Ewe Milk. *Food Chem.* **2013**, *141*, 829–834. [CrossRef] [PubMed]
13. El-Kimary, E.I.; Korany, M.A.; Issa, A.E.; Basuny, M.G. Simultaneous Microdetermination of Different Penicillin Antibiotics Residues for Cross-Contamination Study in Non-Penicillin Dosage Forms. *Microchem. J.* **2023**, *185*, 108291. [CrossRef]
14. Karci, A.; Balcioğlu, I.A. Investigation of the Tetracycline, Sulfonamide, and Fluoroquinolone Antimicrobial Compounds in Animal Manure and Agricultural Soils in Turkey. *Sci. Total Environ.* **2009**, *407*, 4652–4664. [CrossRef]
15. Igualada, C.; Giraldo, J.; Font, G.; Yusà, V. Validation of a Multi-Residue UHPLC-HRMS Method for Antibiotics Screening in Milk, Fresh Cheese, and Whey. *J. Food Compos. Anal.* **2022**, *106*, 104265. [CrossRef]
16. Peris-Vicente, J.; Peris-García, E.; Albiol-Chiva, J.; Durgbanshi, A.; Ochoa-Aranda, E.; Carda-Broch, S.; Bose, D.; Esteve-Romero, J. Liquid Chromatography, a Valuable Tool in the Determination of Antibiotics in Biological, Food and Environmental Samples. *Microchem. J.* **2022**, *177*, 107309. [CrossRef]
17. Guironnet, A.; Wiest, L.; Vulliet, E. Improvement of the QuEChERS Extraction Step by Matrix-Dispersion Effect and Application on Beta-Lactams Analysis in Wastewater Sludge by LC-MS/MS. *Talanta* **2022**, *237*, 122923. [CrossRef]

18. Melekhin, A.O.; Tolmacheva, V.V.; Goncharov, N.O.; Apyari, V.V.; Dmitrienko, S.G.; Shubina, E.G.; Grudev, A.I. Multi-Class, Multi-Residue Determination of 132 Veterinary Drugs in Milk by Magnetic Solid-Phase Extraction Based on Magnetic Hypercrosslinked Polystyrene Prior to Their Determination by High-Performance Liquid Chromatography–Tandem Mass Spectrometry. *Food Chem.* **2022**, *387*, 132866. [CrossRef]
19. Wu, J.; Guo, E.L.; Wang, M.L.; Wang, K.; Ma, L.; Lian, K. Determination of β-Lactam Antibiotics in Animal Derived Foods by Modified QuEChERS Coupled with Ultra Performance Liquid Chromatography-Tandem Mass Spectrometry. *J. Food Compos. Anal.* **2023**, *122*, 105437. [CrossRef]
20. Chiesa, L.M.; Nobile, M.; Panseri, S.; Arioli, F. Antibiotic Use in Heavy Pigs: Comparison between Urine and Muscle Samples from Food Chain Animals Analysed by HPLC-MS/MS. *Food Chem.* **2017**, *235*, 111–118. [CrossRef]
21. Fabregat-Safont, D.; Pitarch, E.; Bijlsma, L.; Matei, I.; Hernández, F. Rapid and Sensitive Analytical Method for the Determination of Amoxicillin and Related Compounds in Water Meeting the Requirements of the European Union Watch List. *J. Chromatogr. A* **2021**, *1658*, 462605. [CrossRef]
22. Goessens, T.; Huysman, S.; De Troyer, N.; Deknock, A.; Goethals, P.; Lens, L.; Vanhaecke, L.; Croubels, S. Multi-Class Analysis of 46 Antimicrobial Drug Residues in Pond Water Using UHPLC-Orbitrap-HRMS and Application to Freshwater Ponds in Flanders, Belgium. *Talanta* **2020**, *220*, 121326. [CrossRef]
23. Suseela, M.N.L.; Viswanadh, M.K.; Mehata, A.K.; Priya, V.; Vikas; Setia, A.; Malik, A.K.; Gokul, P.; Selvin, J.; Muthu, M.S. Advances in Solid-Phase Extraction Techniques: Role of Nanosorbents for the Enrichment of Antibiotics for Analytical Quantification. *J. Chromatogr. A* **2023**, *1695*, 463937. [CrossRef]
24. Wang, C.; Ye, D.; Li, X.; Jia, Y.; Zhao, L.; Liu, S.; Xu, J.; Du, J.; Tian, L.; Li, J.; et al. Occurrence of Pharmaceuticals and Personal Care Products in Bottled Water and Assessment of the Associated Risks. *Environ. Int.* **2021**, *155*, 106651. [CrossRef] [PubMed]
25. Cha, J.; Carlson, K.H. Occurrence of β-Lactam and Polyether Ionophore Antibiotics in Lagoon Water and Animal Manure. *Sci. Total Environ.* **2018**, *640–641*, 1346–1353. [CrossRef] [PubMed]
26. Singhal, A.; Sadique, M.A.; Kumar, N.; Yadav, S.; Ranjan, P.; Parihar, A.; Khan, R.; Kaushik, A.K. Multifunctional Carbon Nanomaterials Decorated Molecularly Imprinted Hybrid Polymers for Efficient Electrochemical Antibiotics Sensing. *J. Environ. Chem. Eng.* **2022**, *10*, 107703. [CrossRef]
27. Neng, J.; Wang, J.; Wang, Y.; Zhang, Y.; Chen, P. Trace Analysis of Food by Surface-Enhanced Raman Spectroscopy Combined with Molecular Imprinting Technology: Principle, Application, Challenges, and Prospects. *Food Chem.* **2023**, *429*, 136883. [CrossRef] [PubMed]
28. Rahman, S.; Bozal-Palabiyik, B.; Unal, D.N.; Erkmen, C.; Siddiq, M.; Shah, A.; Uslu, B. Molecularly Imprinted Polymers (MIPs) Combined with Nanomaterials as Electrochemical Sensing Applications for Environmental Pollutants. *Trends Environ. Anal. Chem.* **2022**, *36*, e00176. [CrossRef]
29. Basak, S.; Venkatram, R.; Singhal, R.S. Recent Advances in the Application of Molecularly Imprinted Polymers (MIPs) in Food Analysis. *Food Control* **2022**, *139*, 109074. [CrossRef]
30. Benito-Peña, E.; Partal-Rodera, A.I.; León-González, M.E.; Moreno-Bondi, M.C. Evaluation of Mixed Mode Solid Phase Extraction Cartridges for the Preconcentration of Beta-Lactam Antibiotics in Wastewater Using Liquid Chromatography with UV-DAD Detection. *Anal. Chim. Acta* **2006**, *556*, 415–422. [CrossRef]
31. Jafari, S.; Dehghani, M.; Nasirizadeh, N.; Baghersad, M.H.; Azimzadeh, M. Label-Free Electrochemical Detection of Cloxacillin Antibiotic in Milk Samples Based on Molecularly Imprinted Polymer and Graphene Oxide-Gold Nanocomposite. *Measurement* **2019**, *145*, 22–29. [CrossRef]
32. Cheng, G.; Zhao, J.; Wang, X.; Yang, C.; Li, S.; Lu, T.; Li, X.; Wang, X.; Zhu, G. A Highly Sensitive and Selective Method for the Determination of Ceftiofur Sodium in Milk and Animal-Origin Food Based on Molecularly Imprinted Solid-Phase Extraction Coupled with HPLC-UV. *Food Chem.* **2021**, *347*, 129013. [CrossRef] [PubMed]
33. Du, W.; Sun, M.; Guo, P.; Chang, C.; Fu, Q. Molecularly Imprinted Membrane Extraction Combined with High-Performance Liquid Chromatography for Selective Analysis of Cloxacillin from Shrimp Samples. *Food Chem.* **2018**, *259*, 73–80. [CrossRef]
34. Ashley, J.; Wu, K.; Hansen, M.F.; Schmidt, M.S.; Boisen, A.; Sun, Y. Quantitative Detection of Trace Level Cloxacillin in Food Samples Using Magnetic Molecularly Imprinted Polymer Extraction and Surface-Enhanced Raman Spectroscopy Nanopillars. *Anal. Chem.* **2017**, *89*, 11484–11490. [CrossRef]
35. Camel, V. Microwave-Assisted Solvent Extraction of Environmental Samples. *TrAC Trends Anal. Chem.* **2000**, *19*, 229–248. [CrossRef]
36. Bravo, J.C.; Garcinuño, R.M.; Fernández, P.; Durand, J.S. A New Molecularly Imprinted Polymer for the On-Column Solid-Phase Extraction of Diethylstilbestrol from Aqueous Samples. *Anal. Bioanal. Chem.* **2007**, *388*, 1039–1045. [CrossRef]
37. Yilmaz, E.; Mosbach, K.; Haupt, K. Influence of Functional and Cross-Linking Monomers and the Amount of Template on the Performance of Molecularly Imprinted Polymers in Binding Assays. *Anal. Commun.* **1999**, *36*, 167–170. [CrossRef]
38. Aly, Z.; Luca, V. Uranium Extraction from Aqueous Solution Using Dried and Pyrolyzed Tea and Coffee Wastes. *J. Radioanal. Nucl. Chem.* **2013**, *295*, 889–900. [CrossRef]

39. Pourfarzib, M.; Shekarchi, M.; Rastegar, H.; Akbari-Adergani, B.; Mehramizi, A.; Dinarvand, R. Molecularly Imprinted Nanoparticles Prepared by Miniemulsion Polymerization as a Sorbent for Selective Extraction and Purification of Efavirenz from Human Serum and Urine. *J. Chromatogr. B* **2015**, *974*, 1–8. [CrossRef]
40. Aly, Z.; Graulet, A.; Scales, N.; Hanley, T. Removal of Aluminium from Aqueous Solutions Using PAN-Based Adsorbents: Characterisation, Kinetics, Equilibrium and Thermodynamic Studies. *Environ. Sci. Pollut. Res.* **2014**, *21*, 3972–3986. [CrossRef]

polymers

Article

Investigation of the Release Rate of Biocide and Corrosion Resistance of Vinyl-, Acrylic-, and Epoxy-Based Antifouling Paints on Steel in Marine Infrastructures

Adel Jalaie [1], Abdolah Afshaar [1,*], Seyed Borhan Mousavi [2,*] and Mohammad Heidari [3]

1 Department of Materials Science and Engineering, Sharif University of Technology, Tehran 1458889694, Iran
2 J. Mike Walker '66 Mechanical Engineering Department, Texas A&M University, College Station, TX 77843, USA
3 Faculty of Chemical & Petroleum Engineering, University of Tabriz, Tabriz 5166616471, Iran
* Correspondence: abdolahafshaar@gmail.com (A.A.); seyedborhanm@gmail.com or borhanm@tamu.edu (S.B.M.)

Abstract: This study comprehensively assesses the release rate of biocides, corrosion effects related to antifouling, and the physical properties of different paint types. Tests were conducted to measure thickness, viscosity, hardness, bending, adhesion, gloss, impact resistance, abrasion resistance, scratch resistance, polarization, and salt spray. The paints evaluated include resin-based, acrylic-based, epoxy-based, and vinyl-based formulations. The study investigates the influence of biocide content, biocide particle size, and immersion time on release rate using a lab-scale setup. Results showed that acrylic-based paints had a higher biocide release rate due to faster hydrolysis, while smaller biocide particle sizes led to higher release rates in resin-based paints. Optimal total biocide contents were determined to be 30% for acrylic-based, 60% for epoxy-based, and 50% for vinyl-based paints. Antifouling corrosion analysis demonstrated that sample with an optimal release rate effectively prevent algae growth and fouling. Acrylic-based paint with 30 wt.% biocide content exhibited superior adhesion with a dolly separation force of 4.12 MPa. Evaluating the impact of synthesized polyaniline on 30 wt.% epoxy-based paint, a sample coated with 10 wt.% polyaniline represented a low corrosion rate of 0.35 μm/year and a high impedance value of approximately 37,000 Ohm·cm^{-2}.

Keywords: antifouling (AF); Cu_2O/ZnO biocide; polyaniline; vinyl paint; acrylic paint; epoxy paint

1. Introduction

Accumulation and adherence of marine organisms on submerged surfaces, known as biofouling, can cause drastic harmful influences on manufactured marine infrastructure during its service time, leading to greater fuel consumption, arduous hull cleaning processes, expedited corrosion, and a need for more maintenance [1–4]. Hence, developing practical and durable antifouling (AF) approaches is vital to preserving marine infrastructure from biofouling organisms. To overcome the mentioned drawbacks, antifouling paints are usually used to constrain the organisms' growth on the ship's body [5]. Recently, while numerous investigations have experimented with utilizing biocide-free antifouling methods, namely bio-inspired coatings, fouling-resistant coatings, and fouling-release coatings, biocides have been mostly used as the principal effective components in commercial antifouling coatings [6–9]. Microbial biofilms have been observed to perform an essential part in interfering with the establishment of invertebrate larvae and microorganisms from fungi, which is the primary form of fouling. Recent research has proposed that the relationship between microbial biofilms and the growth of consequent macrofouling areas could be notable [10–12]. Tributyltin self-polishing copolymer (TBT-SPC) paints have been used widely for more than two decades due to their remarkable antifouling features; however, tributyltin-based antifouling coatings have been forbidden due to environmental issues. Since the complete prohibition of the usage of tributyltin (TBT) as a biocide in the marine ecosystem, copper oxide (Cu_2O) has attracted attention as an antifouling biocide in marine paints due to its substantial characteristics such as durability, performance, and effectiveness; in addition, antifouling paints including copper oxide can provide sufficient fouling control to a great number of steels [13–16]. Lately, considerably conducting polymers (ICPs) such as polyaniline (PANI) and polypyrrole (PPy) and their derivatives have been investigated due to their outstanding features with regard to reversible electrochemical performance, electrical conductivity, optical and electrical traits, etc. [17,18]. In addition, these materials have been used extensively in industrial applications, namely batteries, electronic devices, and sensors, and as anticorrosive additives due to their promising features such as excellent mechanical and physicochemical attributes, low costs, and safety [19–22]. Since a thin film of conducting polymers could preserve the substrate only for a short period, organic coatings and conducting polymers are usually combined to achieve effective and durable corrosion resistance [23–25]. Numerous studies have focused on examining the corrosive characteristics of PANI coatings, proving that with the addition of nanoparticles (NPs) to the coatings, better anticorrosive properties will be gained due to the formation of a stable passive film on the surface of the substrate. These nanoparticles include Fe_2O_3 [26–28], CNT, ZnO, ZrO_2 [29], activated carbon [30,31], and TiO_2 [32–39]. Zinc oxide (ZnO), an inorganic material, has been widely used in many applications such as photocatalysis [40], optics, and heat transfer [41] and especially for antibacterial purposes [42,43].

Yong et al. [42] utilized ZnO NPs in an alkyd resin to prepare a biocide antifouling paint. The results confirmed that ZnO NPs could potentially coat infrastructure in marine industries. Al-Belushi et al. [44] first prepared ZnO nanoparticles to fabricate chitosan (CHT)-ZnO NPs coatings at two different chitosan solutions, CHT 1% and CHT 2%. Then, they assessed the antifouling features of the prepared coatings using Gram-negative and-positive bacteria. The outcomes revealed that the prepared specimens had higher antifouling features compared with the pure nano-ZnO rods or CHT coatings than those of commercially available paints in the dark and under light irradiation. They further asserted that the fabricated coatings had better efficacy in the removal of Gram-negative bacteria rather than Gram-positive ones. Rajan et al. [45] conducted a study to investigate epoxy-based paint's antifouling and anticorrosive characteristics by applying the disc diffusion method. The results showed that the investigated properties of the measured paint could maintain antifouling and anticorrosive characteristics in natural seawater for about six months.

Moreover, the effectiveness and durability of the coating can be improved by raising the G. edulis extract contents. Ni et al. [46] evaluated the preparation, antifouling, and self-polishing attributes of the multifunctional resins by hindering algae addition analysis, anti-protein adsorption, and antifouling test. They concluded that the acrylic metal salt resins, including the indole derivative, had substantial antifouling and self-polishing attributes compared to pure acrylic metal salt resins. Adeleye et al. [47] studied the chemical state of copper particles in water and the effect of drying time, surface type, and water salinity on the release rate of different copper particles. They asserted that the release rate of copper increased with water salinity. They further stated that drying paints for longer than what is suggested by the manufacturer will cause the additional release of biocides into water. Palanichamy et al. [48] conducted a study to evaluate the antifouling behavior of five different marine bacteriocins using epoxy-based paint. The outcomes demonstrated that the prepared coatings could maintain antifouling behavior for three months in seawater, and the life expectancy could be improved as biocide content increased. Quan et al. [49] conducted a study to evaluate the antifouling, antibacterial, and anticorrosion behavior of epoxy coatings containing prepared poly(m-aminophenol) (PmAP) at four different concentrations. The antibacterial characteristics were measured via the plate count determination method, showing that sufficient cationic moieties and phenolic hydroxyl groups available in PmAP had effective antibacterial activity toward B. subtilis and E. coli. It was further asserted that the antifouling feature of the prepared coatings increased with the increase in concentration; thus, the coatings were preserved from failure affected by microbes. Quan et al. [50] investigated the antifouling and antibacterial execution of a binary epoxy coating comprising bromine-benzyl-disubstituted polyaniline (BBP). The comprehensive analyses confirmed that the prepared samples had better antifouling execution than the normal PANI coating, and the optimal concentration was 5%. Moreover, it was found that the antibacterial properties of the coatings were enhanced with the addition of BBP nanoparticles. Tonelli et al. [51] analyzed the efficacy of utilizing halloysite clay nanotubes (HNTs), which are nanocontainers, on the antifouling behaviors of epoxy coatings, proving that the measured properties of the coatings were enhanced with the incorporation of used nano-containers. The antifouling capability of biodegradable polyurethane was examined by Ma et al. [52]. They reported that the composition of the polyurethane could control the antibiofouling capacity.

Based on the performed literature review, thorough examinations have been carried out on the release rate of biocides and the corrosion rate of resin-based paints, while limited considerations have been paid to the release rate of disparate biocides and the corrosion rate of different resin-based paints using wide weight ratio range of additives to acquire the optimum concentrations. Moreover, a novel bimetallic biocide containing two different biocides, Cu_2O and ZnO, was utilized to prepare all resin-based paints. In addition, a novel lab-scale setup using artificial seawater was designed and built with a high similarity in composition and concentration to realistic seawater to measure the release rate.

In this investigation, we aimed to consider all affecting parameters of the antifouling properties of different biocides, environmental issues, and physical features of the various coatings. Moreover, all essentially required industrial aspects were comprehensively considered to assess the industrialization facilities, confirming that the mentioned coatings could remarkably ease industrialized capabilities, which, to the best of our knowledge, has not been studied so far.

In this study, the biocidal release rate should be examined to achieve a suitable formulation for antifouling paint, which is the most important component in the paint formulation to combat growth fouling. For this reason, first, the basic formulation of paints was extracted from the articles, and then by changing the effective parameters, such as the type of resin, the content of the biocide in the paint, and the grain size of the pigments on the release rate, the most appropriate formulation was selected. Mechanical, physical, corrosion resistance, and paint quality tests were performed on St37 steel substrate for samples at each optimal formulation. In order to guarantee the consistency and replicability of outcomes, every experiment was performed a minimum of three times.

2. Experimental Materials and Methods

2.1. Materials

Commercial Cu_2O and ZnO were purchased from Sigma-Aldrich and Merck, respectively. To formulate antifouling paints, acrylic resin No. 2378, EPON 828 resin, and Mowital B 30 H polyvinyl butyral (PVB) were obtained from Taak Resin Kaveh Chemical Co. (Tehran, Iran), Shell Chemical Co. (Deer Park, TX, USA), and Kuraray (Merrillville, IN, USA), respectively. Zinc phosphate epoxy No. 15300 and polyamine epoxy No. 45180 to be used as primers were obtained from Bajak Paint Co. (Tehran, Iran). To prepare the specimens, St-37 steel was used as the base metal with a thickness of 2 mm. The chemical composition of the used steel is listed in Table 1.

Table 1. Chemical composition of St-37 steel.

Element	Mn	P	S	C	Fe
Percent (wt.%)	0.6	0.03	0.02	0.17	0.18

After preparing the samples, abrasive blasting was conducted to increase the surface roughness and adhesion of the coating and remove possible oxide layers on the surface. Then, the specimens were cleaned using acetone to remove surface contaminants such as grease. To examine the biocide release rate, glass plates with dimensions of 10 × 15 cm^2 were used.

2.2. Paints Preparation

To prepare the acrylic-based paint, first xylene and *n*-butanol were mixed, at an appropriate ratio of each (1:1), using a stirrer at 2000 rpm for 30 min, then acrylic resin powder with other additives was added in the ratio listed in Table 2. The mixture was stirred at 2000 rpm for 1 h; last, Cu_2O and ZnO were added to the mixture at different contents. The particle size was measured using a grindometer which was set to 50 μm. By adding glass beads at a ratio of 1:2 to the prepared sample and mixing for about 30 min, the particle size decreased to 20 μm. The paint particles were separated from glass beads using a metal mesh [53].

Table 2. Composition of Cu_2O and ZnO biocide paints at different contents and the role of used additives.

Material	Role	wt.%
Resin (binder)	Film-forming component of paint	10, 20, 30, 40, 50, 60, and 70 ± 0.1
Cu_2O particles	Biocide activity	0, 7, 14, 20, 27, 33, and 40 ± 0.1
ZnO particles	Biocide activity	0, 3, 6, 10, 13, 17, and 20 ± 0.1
Titanium dioxide (pigment)	Providing the color of the paint	5 ± 0.1
Iron (III) oxide red (pigment)	Providing the color of the paint	5 ± 0.1
Dioctyl phthalate	Providing better pigment dispersion	5 ± 0.1
Talc	Providing the color of the paint	5 ± 0.1
Efka FA 4644	Dispersing agent	2 ± 0.1
BYK310	Providing uniform dispersion	2 ± 0.1

To prepare an epoxy-based paint, liquid resin and the additives listed in Table 2 as well as Cu_2O and ZnO biocides, were mixed at 2000 rpm for 1 hr. The glass beads were utilized to decrease the particle size. One hour prior to the application, the paint was mixed with a Triethylenetetramine (TETA)-based amine hardener (Triethylenetetramine, Aradur® HY951, Huntsman Corporation, Mumbai, India) at a ratio of 100 to 15.

Rosin particles must be dissolved to be used in surface coating. Thus, the rosin particles (CAS No: 63449-39-8, Rosin Factory, Clifton, NJ, USA) were powdered well using a mortar. The rosin powders were mixed with xylene at a ratio of 1:2. Mowital B 30 H polyvinyl butyral (PVB) was dissolved into cyclohexane using a stirrer at 2000 rpm for about 30 min. The listed additives in Table 2 and biocides were also mixed at the same speed and time. Eventually, the prepared rosin was mixed with the PVB mixture with a ratio of 1:2 for 30 min. Particle size was decreased using the glass beads.

2.3. Curing Time

The paints were applied to the plates using a 60 μm applicator at 22–25 °C, and the curing time was determined based on the ISIRI6456 standard [54,55]. All tests were carried out three times to minimize error and ensure the repeatability of data.

2.4. Biocide Release Rate Measurement

A laboratory setup was sketched and built based on the ASTM D6442-06 [56,57]. A layer of the prepared paints with a thickness of 120 μm was applied to the glass plates to an area of 10 × 15 cm^2 using an applicator. The laboratory system consisted of a container that was filled with artificial seawater. The artificial seawater was prepared according to the ASTM D1141 standard [58,59], and its specifications are listed in Table S1. The temperature and pH of the water were maintained at 23 ± 2 °C and 7.9–8.1, respectively, and they were monitored continuously during the tests. The size ratio of the painted plates in the water and the total amount of water in each chamber were maintained at between approximately 1:3–1:6. A 1000 cm^3 quantity of artificial water was poured into each chamber. To create dynamic and wave-like seawater conditions, blowing was carried out in each chamber by channeling the air pump, which caused turbulence in the water. The outlet air flow rate for each chamber was determined via a flowmeter and kept at 60 ± 5 mL/min. Figures S1 and S2 present the built laboratory setup and its corresponding schematic, respectively.

Cu concentration was evaluated at 5-, 10- and 15-day intervals by sampling 20 cm^3 of water in each chamber and applying atomic absorption spectroscopy (AAS). The reason for choosing 15 days is that the release rate was investigated when the water had completely penetrated the paint layer, breaking and reforming the penetration barriers of the paint. It was demonstrated that penetration and biocidal dissolution occurred in 7 days [60,61]. Although the water evaporated during the tests, the initial water level was maintained by adding distilled water to the chambers. The release rate was calculated by Equation (1) [56].

$$R = \frac{CVD}{AT} \tag{1}$$

where R, C, V, D, A, and T are the release rate of biocide ($\frac{\mu g}{cm^2.day}$), the content of biocide (($\frac{\mu g}{L}$), the volume of seawater in the measuring container (L), the number of hours per day (24), the surface area of the paint film (cm^2), and the time the flat panel was immersed in the measuring container (h), respectively.

If the calculated release rate is $\geq 15\frac{\mu g}{cm^2.day}$, it means that the paint has antifouling activity, but for $R < 15\frac{\mu g}{cm^2.day}$, the antifouling performance of the paint is not satisfying [62].

2.5. Antifouling Resistance Analysis

A laboratory setup was built to examine the antifouling features of the samples. The setup consisted of two glass tanks under the radiation of artificial light with UV lamps with an intensity of about 90 lux that provided necessary light to the algae at night and accelerated their growth. An air and CO_2 gas blower pump were also provided in the setup. A photo of the setup is presented in Figure S3. The water temperature was maintained at 27–30 °C. The chemical composition of the water used in the tanks is listed in Table S2. The samples with satisfying biocide release rates were considered in this test, and the algae formation process was analyzed. The samples contained substrates, zinc phosphate epoxy No. 15300, polyamine epoxy No. 45180, and antifouling paint. A thickness of 150 µm of each layer was applied to the steel plate, and the painted samples were compared to the pure ones. Each sample was kept in the system for 3 months.

2.6. Thickness Measurement

The thickness of the samples was studied using the PosiTector 6000 (DeFelsko, The Woodlands, TX, USA) instrument based on ASTM D1186 [63,64]. The tests were conducted at 5 points for each sample, and the mean value was reported.

2.7. Viscosity Measurement

Since the viscosity of paints plays an influential role in their properties, to remove this parameter, the tests had to be conducted at fixed viscosity; thus, the viscosity of all samples was kept at 110 Krebs by employing a Brookfield (Ametek, Middleborough, MA, USA) viscometer at 25 °C temperature based on ASTM D562 [65,66].

2.8. Hardness Test

To measure the hardness of the samples, the pencil hardness and König pendulum hardness tests were performed according to ASTM D3363 and D4366 standards [67,68], respectively [69,70]. The pencil hardness and König pendulum hardness tests were conducted employing Wolff-Wilborn (BYK, Germany) and byko-swing König (BYK, Geretsried, Germany) instruments, respectively. The samples were prepared by applying a layer of the formulated paints at a thickness of 120 μm.

2.9. Bending Test

The analyses were performed by spraying a layer of the prepared paints at a thickness of 120 μm on the 10 × 15 cm^2 plates after 1 week according to ASTM D522 standard [69,71]. The tests were conducted using a Mandrel bender machine (Ercolina, Davenport, IA, USA).

2.10. Paint Adhesion Test

The adhesion tests were performed using a Pull-off adhesion tester (DeFelsko PosiTest AT, The Woodlands, TX, USA) according to ASTM D4541 standard [72] for a layer of the prepared paints with a thickness of 120 μm [73].

2.11. Gloss Test

The gloss tests were conducted for a layer of the prepared paints with a thickness of 120 μm at 60° angle utilizing a gloss meter apparatus (3NH NHG60, Ramsey, NJ, USA) based on ASTM D523 standard [74,75].

2.12. Impact Resistance Test

The impact resistance tests were conducted utilizing an impact resistance machine (HD-A520, Haida international, Dongguan, China) based on the ASTM D1794 standard [76]. In this test, the deformation of the coating was evaluated with a 2-pound ball released from different heights. Three samples were prepared for each test.

2.13. Abrasion Resistance Test

This test was conducted to measure the life of the paints against abrasion. A layer of the prepared paints with a thickness of 200 μm was applied to an area of the plate with dimensions of 100 × 100 × 1 mm^3, having an 8 mm hole in its center. A 1 mg quantity of paint was placed on the machine's rotational surface, the abrasive wheels were placed tangentially to the sample's surface, and a 250 g loading bar was applied to each. A suction device was installed close to the abrasion site, at a distance of 1 to 2 mm from the sample's surface, to remove dust caused by abrasion to the device from the test site. The mass loss of the samples was measured after 200 cycles of wear compared with the initial weight. The tests were conducted using an abrasion resistance test machine (GT-C14A, Nanjing, China) according to ASTM D4060 standard [77,78].

2.14. Scratch Resistance Test

Scratch resistance tests of the samples were performed at different scratch forces, 5–30 N, employing a scratch resistance tester (SMT-500, Rtec, San Jose, CA, USA) according to ASTM D7027 standard [79,80].

2.15. Polarization Test

To study the corrosion behavior of the prepared resins and the effect of polyaniline as an additive in increasing the corrosion resistance of antifouling paints, the polarization curves by Autolab in the range of −0.6 to 0.6 V relative to the open circle potential with a sweeping rate of 1 $\frac{mV}{s}$ in the formulated 3.5% NaCl solution was used. Prior to the tests, the samples were placed in the solution for 10 min to achieve equilibrium conditions. According to ASTM G102-89 standard [81], the extrapolation method was used to calculate samples' corrosion rate from polarization curves [82].

2.16. Salt Spray Test

Salt spray tests were performed to study the corrosion behavior of the samples according to ASTM B117 standard [83,84]. The samples were placed inside the chamber so that they were at an angle of about 15 to 30 degrees to the vertical axis and had no contact with each other. To speed up the conclusion time, scratches were applied to the samples up to the depth of the base metal surface and along the diameter. The dimensions of the salt spray test samples in the study were 2×5 cm^2 and included three coats of primer, middle paint, and antifouling paint. The duration of the test was 240 h. To prevent the penetration of corrosion from the back and edges of the sample, the back of all the samples was protected with varnish and adhesive tape.

2.17. Polyaniline Synthesis

In this study, polyaniline was synthesized, and its antifouling property and efficacy against corrosion were investigated. A 20 g (0.1 mol) quantity of ammonium persulfate was dissolved as oxidizer and initiator of the polymerization reaction in 100 cm^3 of an aqueous 1 M perchlorate acid solution. A 0.1 mol quantity of double-distilled aniline (10 g) in 100 cm^3 of an aqueous 1 M perchlorate acid solution was dissolved. The initiator solution was added dropwise to the monomer solution, and the reaction temperature was stabilized at 0 °C. The primer solution was added to the polymerization container within 1 h, and the reaction mixture was stirred with a magnetic stirrer. After this time, the contents of the container, which contained a blackish-green precipitate, were filtered through a Buchner funnel. The resulting polymer was washed under filtration with methanol. The polymer was then washed with distilled water to remove residual methanol and dried in an oven for one day. After the synthesis of polyaniline powder, it was mixed with dimethylformamide as a solvent in a ratio of 1:1. Previous studies added different amounts of polyaniline/acid to the paint [85,86]; however, the best performance in increasing the corrosion resistance and antifouling qualities has been obtained in the range of 10–20 wt.% [87]. Therefore, 30 wt.% epoxy-based antifouling paints were prepared with the addition of 10, 15, 20, or 30 wt.% of polyaniline/acid. The prepared antifouling paints were applied to the surface of the steel with a thickness of 150 μm, and the results of the polarization and impedance tests were compared with the sample without polyaniline. Then, the optimal content of polyaniline added to the paint was determined.

3. Results and Discussion

3.1. Biocide Release Rate Measurement

3.1.1. Effect of Resin and Immersion Time on the Release Rate

Release rates of Cu_2O and ZnO biocides for paints with a grain size of 20 μm at 5-, 10-, and 15-day intervals are listed in Table 3. Based on the release rates of Cu_2O and ZnO biocides, it can be seen that in the case of paints in which acrylic resin was used in their preparation, the release rate was higher than that of the other two resins due to its polishing property. Binders and polymer chains can be hydrolyzed in acrylic copolymers. Because of the hydrolysis of their ester groups when in contact with water, acrylic resins weaken the bond between the binder and the biocides, releasing them more easily [88]. This reaction occurs in the highest layer in the presence of Cu_2O and ZnO biocides, which have partial solubility in seawater. Below this top layer, which is the paint's surface, the unhydrolyzed polymer remains completely hydrophobic until the bottom layer is exposed to seawater due to the dissolution of the top hydrolyzed layer. Due to the hydrolysis of the acrylic paint layer, the Cu_2O and ZnO biocides are also separated from the polymer and are available to act as antifouling agents. Once the hydrolyzed polymer is removed from the surface, the lower surfaces of the layer encounter water, and hydrolysis continues at the new binder surface. The biocides obtained from layer hydrolysis increase the release rate, resulting in higher release rates of Cu_2O and ZnO in paints with higher contents than the other two resins. Due to its intrinsic physical properties, the biocide release rates of epoxy resin paint were lower than the obtained values for acrylic and vinyl resin paints. Epoxy resin is used for insoluble matrices in antifouling paints. Since epoxy resin has a high crosslinking density, water can only fill the created pores by the dissolution of biocides and pigments on the paint surface; thus, seawater penetration paths to wash biocides out of the paint are low. Over time, the seawater gradually dissolves more biocides, expanding the infiltration layer. When a certain amount of biocide is removed from the surface of the paint layer, the movement of seawater easily erodes the brittle polymer chain, and the layer with less soluble biocides is exposed to seawater, which causes the lower biocide release rate of epoxy-based paints. At high contents of biocides, better antifouling performance was achieved.

Table 3. Release rates of Cu_2O and ZnO biocides at different contents and 5-, 10-, and 15-day intervals.

Resin	Total Biocide Content (wt.%)	Cu_2O Content in Paint (wt.%)	ZnO Content in Paint (wt.%)	Release Rate ($\frac{\mu g}{cm^2.day}$) 5 Days		5–10 Days		10–15 Days	
				Zn^+	Cu^{2+}	Zn^+	Cu^{2+}	Zn^+	Cu^{2+}
Acrylic	0	0	0	0	0	0	0	0	0
	10	7	3	7	9	3	3	3	2
	20	14	6	20	19	10	9	11	9
	30	20	10	31	29	13	14	12	14
	40	27	13	49	41	18	15	22	23
	50	33	17	53	60	25	18	25	20
	60	40	20	70	64	29	21	27	24
Epoxy	0	0	0	0	0	0	0	0	0
	10	7	3	5	0	5	3	5	0
	20	14	6	12	7	10	4	8	5
	30	20	10	13	6	9	6	7	5
	40	27	13	18	10	9	9	9	7
	50	33	17	25	14	12	10	10	8
	60	40	20	34	20	15	11	16	12
Vinyl	0	0	0	0	0	0	0	0	0
	10	7	3	5	3	2	0	2	2
	20	14	6	10	7	4	2	4	0
	30	20	10	16	9	9	3	8	3
	40	27	13	23	14	12	8	10	10
	50	33	17	35	21	18	14	16	15
	60	40	20	47	26	23	16	26	15

Vinyl/rosin-based paints had release rates that were lower than acrylic-based paints but higher than epoxy ones. This behavior could be due to the simultaneous presence of two types of resins with different properties in these paints. Vinyl resins have a high degree of polymerization and are resistant to seawater, which causes the pigments to be locked in this type of resin, and the porosity is greatly reduced. On the other hand, rosin is a natural substance obtained from pine gum, which is chemically very sensitive to alkaline conditions. If the solubility rate is not controlled in the alkaline environment, the seawater dissolves the paint rapidly, and the paint disappears. The antifouling coating quickly loses its effectiveness. For this purpose, hydrophobic or non-polar compounds should be used to control the rate of paint solubility. Mixing rosin with hydrophobic vinyl hydrolyzes the binder and increases the biocide release rate. According to the results of the release rate test, it can be said that in general, the release rate of biocides depends on the rate of hydrolysis of the copolymers from which the paint is made. Therefore, acrylic copolymers have a higher hydrolysis rate than epoxy, and vinyl copolymers have a higher biocide release rate. In epoxy copolymers, the rate of hydrolysis and release decreases due to the increase in the side chain length of the monomers. In addition, with an increase in the chain length and the spatial barrier of the lateral branches, the polarity of the resin decreases, and its hydrophobicity increases; moreover, the rate of water penetration and water contact with the resin decreases. In vinyl resin, hydrolysis and release rates increased with the rise of the reactive rosin group.

Table 3 shows that the release rates after 5 days were higher than those after 10 and 15 days for all paints. In the first days of the measurement, a sharp increase in release rates of biocides was seen for all prepared resins due to the high content gradient created between the paint and the aqueous environment because of the high biocide content in the paint's surface layer. The biocides were transferred out of the coating by forming a biocide content gradient, and a relative density was created in the boundary layer adjacent to the coating. The interesting point here is that after the first few days, the release rate of all prepared resins reached a steady state, and no remarkable change was seen. In the case of acrylic resin, due to the nature of the binder, which could be hydrolyzed, it always caused a new layer to be replaced as the biocides were eliminated in each layer, and therefore, the release rate decreased after the fifth day.

In the case of epoxy and vinyl/rosin resins, as mentioned, in the first five days, due to the high content of biocides on the paint surface, the release rate was high. However, after that, the release rate decreased, and the reason for this could be related to the formation of penetration barriers in the resin. Since epoxy resin is insoluble, water must penetrate via the pores created by the dissolution of the initial biocides on the surface into the paint. Over time, the penetration path became longer as the surface biocides were eliminated. After a certain time, the expansion rate of the depth of the pores due to the dissolution of pigments and the penetration of ions from the penetration layer was less than the biocide release rate and created a barrier against the penetration and release of biocides. After a thickness of 10 to 20 μm, these created barriers in the paint, causing the release rate to remain relatively constant until the end of the paint life.

3.1.2. Effect of the Content of the Biocides on the Release Rate

Figures 1 and 2 show the relationship between the content of the biocides and the mean release rate of Cu_2O and ZnO in a steady-state condition; steady-state refers to the biocide release rate after five days, when the release rate reached an almost constant rate for all prepared resins. Considering Figures 1 and 2 and Table 4, it can be seen that the release rate increased linearly with the increase of the content of biocides; however, at higher contents, the release rate almost reached a fixed state, and no remarkable change was seen. This is highlighted more for acrylic resin, since the amount of biocide available in each layer of paint remained fixed after 40 wt.% content of biocides. The mechanism of this type of paint is the simultaneous dissolution of the paint layer and biocides; thus, increasing the content of biocides did not affect the available amount of biocides in the most superficial layer of the dissolving paint. In the case of epoxy and vinyl paints, the release rate was fixed at higher contents since, in this type of insoluble resin, the particles are in physical contact with each other, and as the particle leaves the paint and enters seawater, a path for another particle to be exposed to the aquatic environment is created, so the release rate at high contents continues to increase slightly [60].

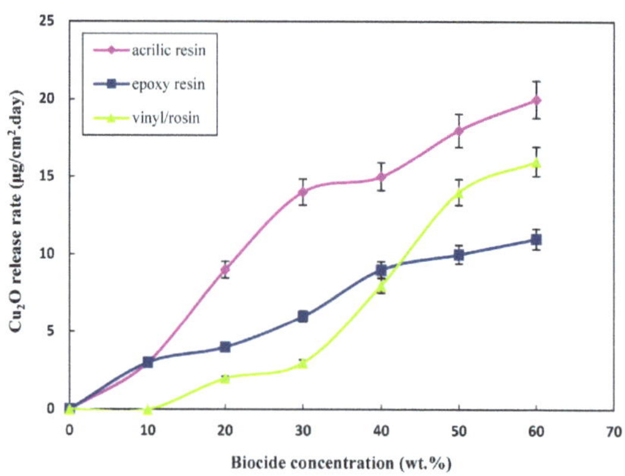

Figure 1. Variation of Cu_2O release rate at different contents in steady-state condition after 5 days.

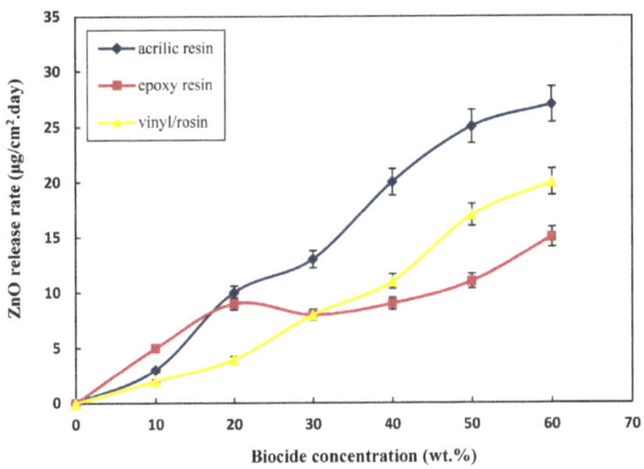

Figure 2. Variation of ZnO release rate at different contents in steady-state condition after 5 days.

Table 4. Mean release rates of Cu_2O and ZnO biocides at optimal contents after 5 days.

Resin	Total Biocide Content (wt.%)	Release Rate ($\frac{\mu g}{cm^2.day}$)	
		Zn^+	Cu^{2+}
Acrylic	30	31	29
Epoxy	60	34	20
Vinyl	50	35	21

Another point to consider here is the comparison of the release rates of Cu_2O and ZnO. The main biocide in antifouling paint is Cu_2O, and ZnO is used as an additional biocide in paint. Additionally, due to the high release rate of ZnO compared to Cu_2O, the concentration of ZnO in the formulation was half that of the Cu_2O (1:2). This can be well demonstrated by comparing the release rates of the two biocides. However, when the concentration of ZnO in the formulation was half that of Cu_2O, the release rates of the two were almost close in all cases. In addition to the higher partial dissolution of ZnO relative to Cu_2O in seawater, the lower density of ZnO also helps it to exit faster and penetrate more easily [89].

Moreover, due to the lower dissolution rate of Cu_2O, the dissolution of Cu_2O biocides controls the release rate kinetics. However, it is worth stating that with the faster dissolution of ZnO, pores and penetration paths for water entry and Cu_2O penetration increase. Accordingly, the Cu_2O release rate increased, especially in insoluble epoxy resin-based paints. Another equally important point is that, since ZnO biocides are lighter than Cu_2O biocides, after spraying the paint, its concentration on the surface of the paint will be higher than that of the inner layers; after the dissolution of ZnO on the surface, the water penetration path will be deeper.

ASTM D6442-06 states that the release rate of paint should be in the range of 10–15 ($\frac{\mu g}{cm^2.day}$) to have a good antifouling property. If the paint has a higher release rate than 15 ($\frac{\mu g}{cm^2.day}$), it poses a danger to other organisms in the marine environment, and using this type of paint is inappropriate. Accordingly, Table 4 lists the optimal content of biocides showing antifouling activity.

3.1.3. Effect of Biocide Particle Size on the Release Rate

To investigate the effect of the biocide particle size on the release rate, two different sizes were considered, 20 µm and 50 µm, and the tests were conducted at three intervals, as before. Table 5 lists the computed release rates for three prepared resins at different contents, test duration, and particle sizes. Based on the release rate values listed in Table 5, lower release rates were acquired for 50 µm particles in comparison with 20 µm ones for all samples, since the smaller biocide particles had a better chance of penetrating rather than the larger ones. Furthermore, as surface-to-volume ratio increased, after dissolution, the 20 μm particles created more pores in the paint surface for water to penetrate, which in turn increased the release rate. Smaller biocide particles also had uniform distribution, leading to a higher release rate. However, according to the results, with increasing the biocide particle size, the release rate reduction for epoxy and vinyl/rosin resins was greater than that of acrylic resins, which confirms that these resins had a penetration release mechanism.

Table 5. Release rates of Cu_2O and ZnO biocides at different contents, particle sizes, and 5-, 10-, and 15-day intervals.

Resin	Total Biocide Content (wt.%)	Particle Size (μm)	Release Rate ($\frac{\mu g}{cm^2.day}$) 5 Days		5–10 Days		10–15 Days	
			Zn^+	Cu^{2+}	Zn^+	Cu^{2+}	Zn^+	Cu^{2+}
Acrylic	20	20	9	11	9	10	19	20
		50	4	5	8	8	16	20
	40	20	23	22	15	18	41	49
		50	19	15	12	14	29	40
Epoxy	20	20	5	8	4	10	7	12
		50	2	5	2	4	5	5
	40	20	7	9	9	9	10	18
		50	3	4	3	3	4	8
Vinyl/rosin	20	20	0	4	2	4	7	10
		50	0	3	0	3	5	6
	40	20	10	10	8	12	14	23
		50	3	6	5	7	10	19

3.2. Effect of Biocide Release Rate on the Paint Thickness

To investigate the effect of biocide release rate on the thickness of the paint layer, first, steel samples were coated with two paint substrates with a total thickness of 150 μm. Then, a layer of antifouling paint with the optimal content of each prepared resin was applied (Table 5). The backs of the samples were covered entirely with a chlorinated rubber resin-based swimming pool coating (Pars Pamchal Co.) to prevent corrosion and errors in the release rate calculations. The release rate experiments were carried out, and the thickness of the paints was measured for 15 days (Figure 3). As shown in Figure 3, the largest reduction in thickness over time was related to acrylic resin, which was due to the hydrolyzable nature of the resin.

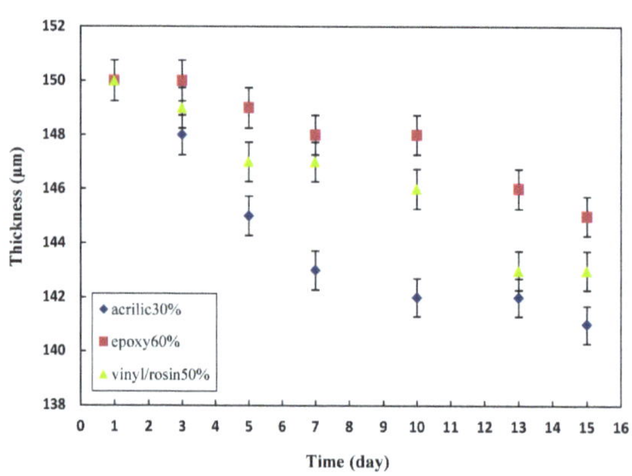

Figure 3. Thickness of the samples as a function of time at optimal content.

3.3. Antifouling Resistance Analysis

The antifouling resistance analyses were performed considering the optimal content of each resin. Some samples with a release rate lower than 10 ($\frac{\mu g}{cm^2.day}$) were randomly selected to investigate the correct functioning of the designed setup. The results were shown as a real photo after 3 months Figure 4. Sample (a) was without antifouling coating; as a result, the effect of fouling was seen after 3 months on the surface of the sample. Algae and fouling larvae were observed on the surface of sample (b). The release rate of sample (c) was less than the amount required to deal with fouling growth based on the standard. In practice, however, it could not resist the growth of fouling agents. The release rate in sample (d) was optimal and resisted fouling growth, so no effect from algae or fouling was observed after 3 months. Although the release rate in sample (e) was lower than the counteracting fouling growth, the rate of fouling formation was not high. The brown particles present were related to the corrosion products of the substrate steel. The release rate in the sample (f) was such that it could resist fouling growth. The release rate of the sample (g) was less than the amount required to withstand the growth of fouling mentioned in the standard and practically could not resist fouling growth. Sample (h) had an optimal release rate, and no algae formed, but due to corrosion products on the edge of the sheet, brown particles were observed on the paint surface. Corrosion kinetics also intensified due to the lack of fouling formation.

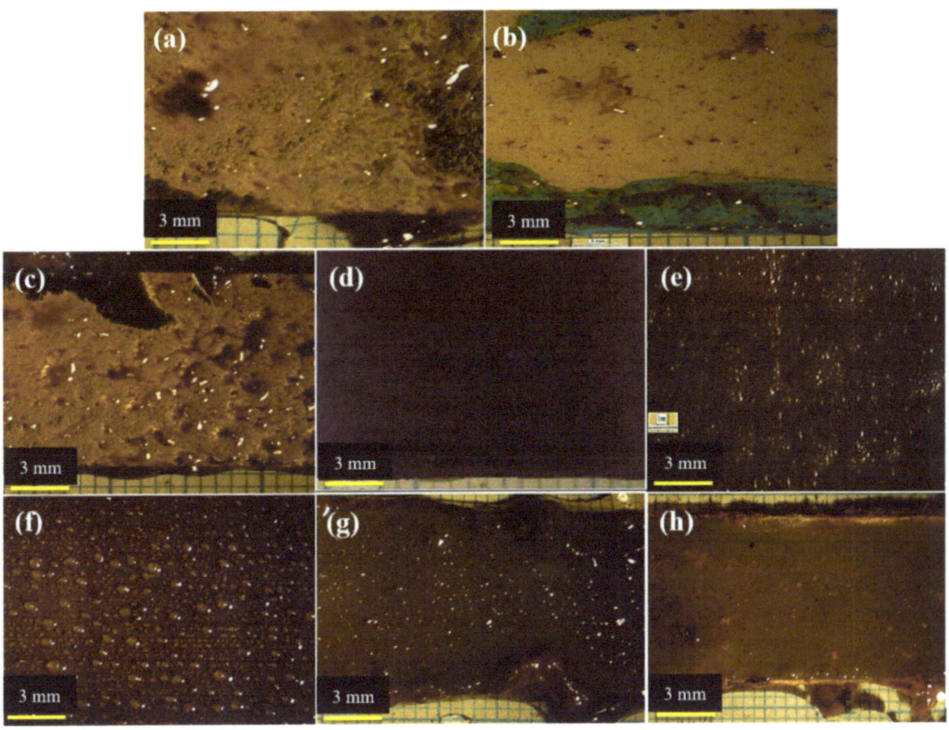

Figure 4. Real photos after 3 months of sample (**a**) without antifouling coating, (**b**) with substrate and without antifouling coating, (**c**) 20 wt.% epoxy-based antifouling coating, (**d**) 60 wt.% epoxy-based antifouling coating, (**e**) 10 wt.% acrylic-based antifouling coating, (**f**) 30 wt.% acrylic-based antifouling coating, (**g**) 20 wt.% vinyl/rosin-based antifouling coating, (**h**) 50 wt.% vinyl/rosin-based antifouling coating.

3.4. Curing Time

The curing time for acrylic-based and vinyl/rosin-based paints was in the range of 25–30 min, and no fingerprint effects were observed after this time. Epoxy-based paints had a longer curing time than the other prepared samples, and the samples were dried after 8 h, but a fingerprint effect was observed. One of the important parameters affecting the initial curing time of a coating is the type of solvent in it. In the case of vinyl and acrylic resins, the curing time was almost the same for both due to the similar evaporation time of the solvent, regardless of the nature of their chemical structure. On the other hand, the curing time of epoxy resins was longer due to the chemical reaction between the resin and the hardener. It has been previously stated [60] that epoxy-based paint's curing time is when the polymerization process is completed and the paint is at maximum hardness. In this study, the epoxy-based paints were completely dried after one week of application.

3.5. Hardness

The obtained results of the pencil hardness and König pendulum hardness tests at optimal contents of biocides are listed in Table 6. The graphite pencil hardness range is 8−1B, HB, F, and 1−6H, of which 6H was the hardest and 8B was the least hard. Based on the data listed in Table 6, it can be seen that the epoxy resin was the hardest, and the acrylic resin was the least hard of all. In the pencil hardness test, the sample must have a 2B degree of hardness, and in the König pendulum hardness test, the oscillation time must be >60 s to meet the standards. Variation of hardness depends on the type and nature of the resin. As the biocide content increased, the hardness degree of the paints increased, since the presence of a mineral compound in an organic resin increases the hardness of that resin. By comparing the obtained results of the release rate and hardness tests, it can be concluded that lower hardness was seen in the acrylic resin, which had a higher release rate at the optimal content of biocides. Therefore, the harder the resin, the lower the release rate, and in hard resins, such as epoxy, with high contents of biocides, optimal release can be achieved. The low release rate in harder resins is also due to more complex networks and higher crosslinking density, which increases the hardness of these resins and makes biocide release difficult. Furthermore, acrylic monomers involved in the polymerization process of acrylic resins, including ethyl acrylate, ethylhexyl acrylate, and acrylic acid, have low T_g temperature, which makes them less hard than the other two resins [90,91].

Table 6. The pencil hardness and König pendulum hardness results of prepared antifouling paints.

Resin	Total Biocide Content (wt.%)	Pencil Hardness Degree	König Pendulum Oscillation Time (s)
Acrylic	10	B	65
	30	H	67
	40	3H	70
Epoxy	20	5H	91
	40	5H	127
	60	6H	165
Vinyl/rosin	20	4H	78
	40	4H	109
	50	B	65

3.6. Bending Test

ASTM D522 states that after bending painted specimens around cylinders with a maximum diameter of 6 mm, if the effect of destruction, cracking, and separation is not seen in the paint, as the paint's flexibility is very high and desirable. If bending around cylinders with a diameter of 6 to 10 mm does not cause rupture or cracking, its flexibility is acceptable; otherwise, the paint will be poor in terms of flexibility. It is worth stating that the smaller the diameter of the cylinder or mandrel, the more intense the bending of the paint. Figure 5 presents photos of the prepared antifouling paints after the bending test. In epoxy resin-based paints, a 6 mm-diameter rupture was created due to bending around the mandrel. The paint was not damaged at a cylinder diameter of 8 mm, which is classified as suitable flexibility. The reason for this phenomenon is the epoxy resin's high molecular weight. However, in the acrylic and vinyl/rosin-based paints, no bending effects were seen after bending around mandrels with a diameter of 3 and 5 mm, respectively, so they can be considered high-flexibility paints. The flexibility of the acrylic resin was due to its low molecular weight and the lower content of biocides in the optimal formulation of this resin.

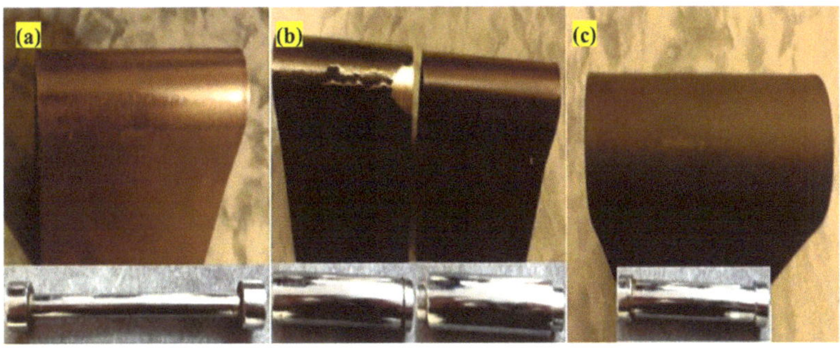

Figure 5. Photos of (**a**) acrylic resin, (**b**) epoxy resin, (**c**) vinyl/rosin resin after bending test.

3.7. Paint Adhesion Test

In this test, all affecting factors were considered to be the same to compare the adhesion of the samples. The results of this test are reported in MPa based on the force required to separate the pin (dolly) attached to the coating from the substrate. Table 7 lists the results of the paint adhesion test of the prepared antifouling paints at different biocide contents. Figure 6 displays the photo of the pins and samples after running the tests. Acrylic resins had good adhesion to their substrate due to the polarity of the functional groups in the structure of the resins.

On the other hand, the adhesion of epoxy and vinyl/rosin coatings was much higher than that of the acrylic resin. The reasons for this were higher chemical and surface absorption and mechanical interlocking. The high polarity of the epoxy resin chain network, the presence of hydroxyl and ether aliphatic groups, and the presence of metal oxide on the steel surface caused electromagnetic absorption between the two materials. The adhesion of epoxy resin paints is related to their hydroxyl group [92]. The formation of chemical bonds between active hydrogen on the steel surface and the epoxide group distributed in the resin causes adhesion. In addition, the polymerization of the epoxy resin occurs after adding hardener and coating on the steel surface; performing this reaction in the cavities and pores created in the sample preparation increases the mechanical adhesion.

Table 7. Obtained dolly separation force at diverse biocides content of acrylic, epoxy, and vinyl/rosin resins.

Resin	Total Biocide Content (wt.%)	Pin (Dolly) Separation Force (MPa)
Acrylic	30	12/4
	40	61/4
Epoxy	40	58/9
	60	41/8
Vinyl/rosin	40	61/6
	50	17/7

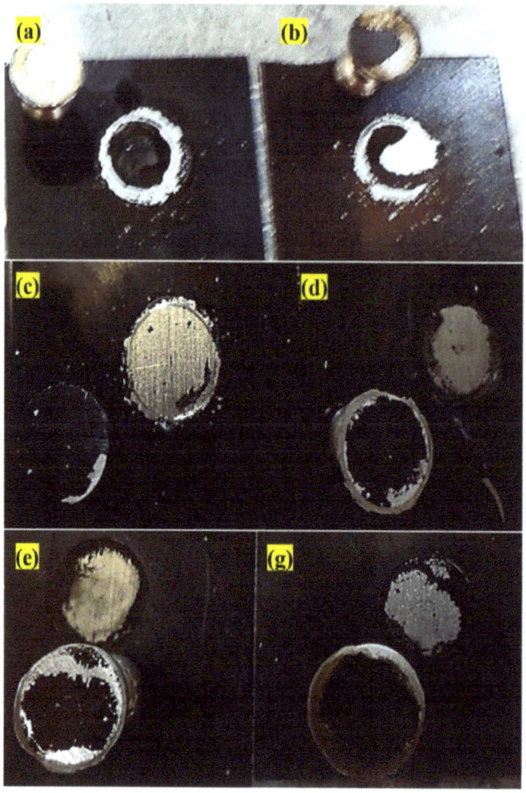

Figure 6. Photos of (**a**) 30 wt.% acrylic-based antifouling coating, (**b**) 40 wt.% acrylic-based antifouling coating, (**c**) 40 wt.% epoxy-based antifouling coating, (**d**) 60 wt.% epoxy-based antifouling coating, (**e**) 40 wt.% vinyl/rosin-based antifouling coating, (**g**) 50 wt.% vinyl/rosin-based antifouling coating after pull-off test.

3.8. Gloss Test

The outcomes of the gloss test for prepared resins with diverse biocide contents are summarized in Table 8. The larger the gloss number, the more gloss the sample has. The gloss decreased as the biocide content increased, and the paint became dull. The reason is that the surface roughness increased with increased biocide content. As a result, when light radiation struck the surface, the surface reflected less radiation, scattering occurred, and the gloss decreased. In vinyl/rosin resin, due to the high oil absorption by rosin, which caused better miscibility of the biocides with the resin, it had less roughness and more gloss.

Table 8. Gloss test results at diverse biocides content of acrylic, epoxy, and vinyl/rosin resins.

Resin	Total Biocide Content (wt.%)	Lux
Acrylic	30	26
	40	20
Epoxy	40	42
	60	40
Vinyl/rosin	40	45
	50	38

3.9. Impact Resistance Test

The photos of the samples after running the tests at each optimal content of biocides are shown in Figure 7. Tests were carried out 3 times for each sample. In all samples, the paint cracked after the ball was released from a height of 60 cm. In vinyl/rosin paint, the number of cracks was smaller compared with the other prepared samples due to the presence of rosin, and after the ball was thrown from a height of 50 cm, no cracks were seen. Due to its oily nature, rosin could reduce the tensile stress applied to the paint at the tip of the ball and increase the impact resistance. Epoxy and acrylic samples had lower impact resistance.

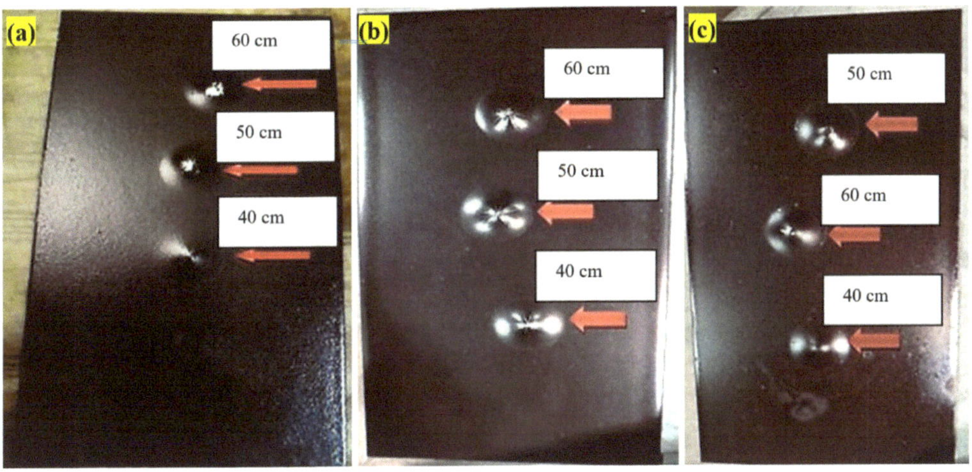

Figure 7. Photos of (**a**) acrylic-based antifouling coating, (**b**) epoxy-based antifouling coating, (**c**) vinyl/rosin-based antifouling coating after impact resistance test from different heights.

3.10. Abrasion Resistance Test

Since these tests are time-consuming, abrasion resistance tests were performed on only three samples of antifouling paint with optimal content of biocides to control fouling, and the results are given in Table 9 and Figure 8 shows the samples after the tests. The mass losses of acrylic resin, epoxy resin, and vinyl/rosin resin were 0.017, 0.016, and 0.001, respectively. As is obvious, most of the wear was seen in the acrylic sample because, in this resin, weak longitudinal bonds between monomeric chains caused the paint to have less resistance to abrasion and shear forces, but in epoxy and vinyl resins, bonds are established in all directions.

Table 9. Mass loss of the samples after abrasion resistance test.

Resin	Total Biocide Content (wt.%)	Initial Weight (g)	Final Weight (g)	Mass Loss (g)
Acrylic	30	62.847	62.83	0.017
Epoxy	60	64.196	64.18	0.016
Vinyl/rosin	50	63.301	63.30	0.001

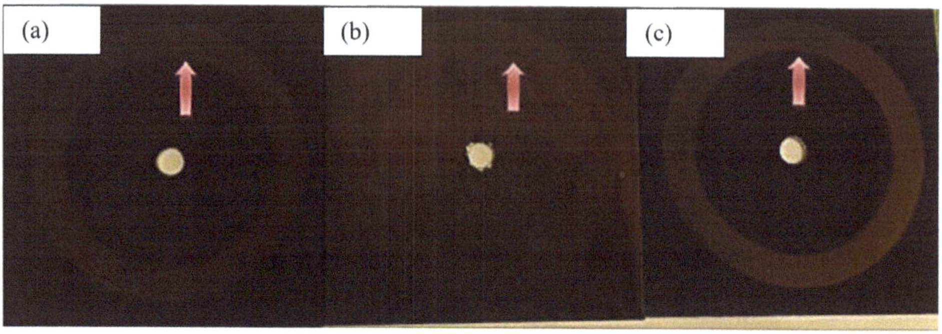

Figure 8. Photos of (**a**) acrylic-based antifouling coating, (**b**) epoxy-based antifouling coating, (**c**) vinyl/rosin-based antifouling coating after abrasion resistance test.

3.11. Scratch Resistance Test

Table 10 reports the force applied to the samples that caused the first scratch on the paint (g). According to the results, the lowest force required in order to create scratches was for acrylic antifouling paint with 30 wt.% content of biocides due to their lower modulus of elasticity. Moreover, the lower biocide content in acrylic-based paint reduced scratch resistance, since biocides can act as fillers and increase scratch resistance in paint. Biocides increase scratch resistance due to the hardness of resin, which has a lower elastic modulus. The highest scratch resistance was related to the epoxy resin, in which no scratch was seen under the weight of 3000 g because the occupation of intermolecular space between polymer chains by more biocides caused the stress applied by the needle of the device via hydrodynamic changes to be distributed between the polymer chains, and as a result, the paint resisted scratching.

Table 10. Required force for the scratch resistance of the prepared paints.

Resin	Total Biocide Content (wt.%)	Applied Weight (g)	Photo of Formed Scratch
Acrylic	30	1500	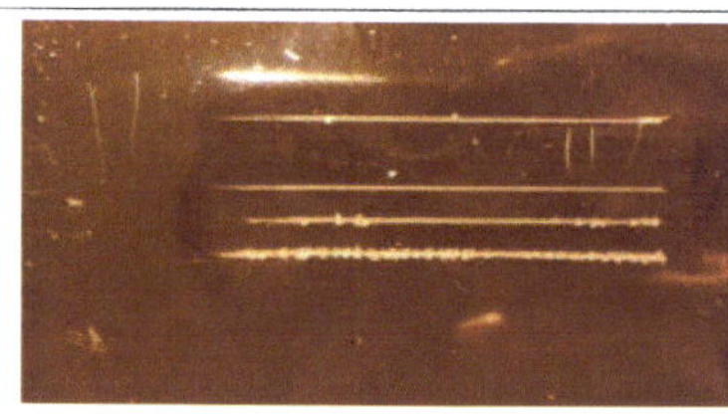
Epoxy	60	3000	No scratch
Vinyl/rosin	50	2000	

3.12. Polarization TEST

Figure 9 shows the polarization curves of different antifouling paints and the primer coating. The corrosion current density, potential, and corrosion rate of the samples are listed in Table 11. The results of the prepared resins were compared with a sample having only one primer coating to assess the effect of the antifouling paints. According to the results, it can be seen that the corrosion rate decreased, and the potential went toward positive values. The corrosion rate of the three paints was close to each other since their anticorrosion mechanism was the same. All three resins protected the substrate against corrosion by providing the barrier layer with high adhesion. Since the motion of ions in the interface between the metal and the coating depends on the amount of gravitational force between the metal and the coating, a stronger interaction between the coating and the metal and the presence of more gravitational forces in the interface increases the bond strength of the interface. This reduces the movement of ions in this area. In other words, the high adhesion of the coating to the metal surface causes the coating to act as an electrical insulator to prevent the formation of anode–cathode microcells on the metal surface or to minimize their formation.

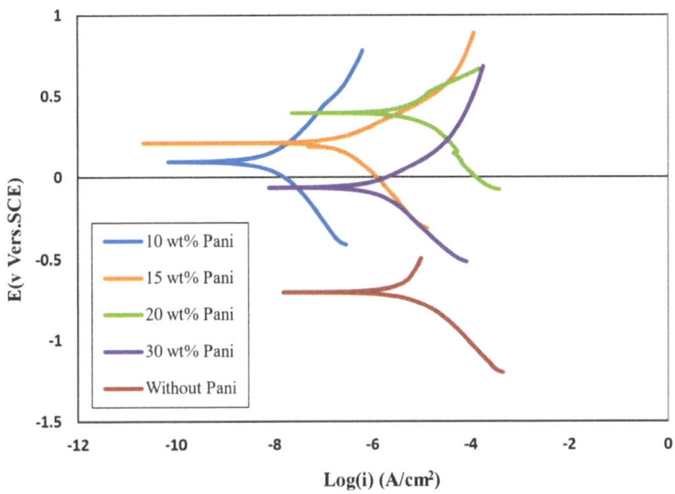

Figure 9. Polarization curves of different antifouling paints and sample with primer coating in 3.5% NaCl solution.

Table 11. Obtained data from the polarization curve of antifouling paints and primer coating.

Resin	E_{corr} (V)	$I_{corr} \times 10^{-6}$ $(\frac{A}{cm^2})$	Corrosion Rate $(\frac{mm}{year})$
Primer coating	−0.83	79	3.61
Acrylic 30 wt.%	−0.39	3.2	0.128
Epoxy 60 wt.%	−0.27	0.082	0.013
Vinyl/rosin 50 wt.%	−0.58	0.47	0.0188

3.13. Salt Spray Test

Figure 10 shows the impact of corrosion on the surface of the samples after salt spray tests. According to Figure 10, corrosion products can be seen in all samples in the scratch zone where the metal was exposed to a corrosive environment. The amount of corrosion in each sample depends on the resin's nature and the biocides' content. With the 40 wt.% acrylic-based antifouling coating, due to the incomplete distribution of pigments in the paint, non-uniformity occurred in the coating, which reduced the corrosion resistance. With the 60 wt.% acrylic-based antifouling coating, a large amount of corrosion products were formed due to small cracks in the paint. In general, the corrosion resistance of all the prepared paints was good. However, since they were denser than the acrylic-based paints, the epoxy- and vinyl/rosin-based paints had better anticorrosion performance than the acrylic ones. Furthermore, increasing the content of biocides reduced the corrosion resistance due to the increase in surface roughness.

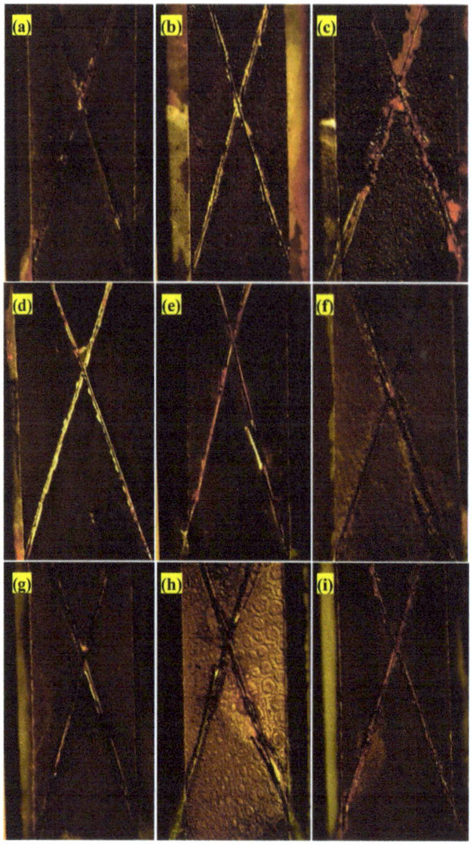

Figure 10. Corrosion impact after salt spray test of sample (**a**) 30 wt.% acrylic-based antifouling coating, (**b**) 40 wt.% acrylic-based antifouling coating, (**c**) 60 wt.% acrylic-based antifouling coating, (**d**) 20 wt.% epoxy-based antifouling coating, (**e**) 40 wt.% epoxy-based antifouling coating, (**f**) 60 wt.% epoxy-based antifouling coating, (**g**) 20 wt.% vinyl/rosin-based antifouling coating, (**h**) 40 wt.% vinyl/rosin-based antifouling coating, (**i**) 50 wt.% vinyl/rosin-based antifouling coating.

The criterion for destroying organic coatings is swelling and lifting of the paint layer from the scratch area. The rate of adhesion loss of the coating layer increases with time and the penetration of corrosive agents into the coating layer. Paint layers applied to the steel surface can protect the surface from corrosion by forming a layer. A layer of paint on the surface containing zinc limits the activity of corrosive agents on the surface. For example, this layer prevents oxygen and water from entering the steel surface, stopping the cathodic reaction, which will reduce the corrosion rate. The most common failure mechanism of epoxy coatings against corrosion involves the gradual loss of adhesion of the coating to the steel substrate. In fact, due to the presence of scratches in the coating, the resin cannot completely prevent corrosive agents from reaching the metal surface. The interface resistance determines the corrosion resistance after corrosive agents reach the metal surface. The important thing is to maintain corrosion resistance over time. In fact, it is important to maintain adhesion over time for coatings used in corrosive environments. However, due to environmental conditions and mechanical and chemical damage, these coatings cannot provide complete protection over time.

In some cases, the damaged surface performs worse than a surface without coating. Through the interfacial layer of steel and paint, water can penetrate through the scratches.

Water molecules form new hydrogen bonds with the hydrated surface of the metal, replacing the resin's bonds with the metal, reducing adhesion, and thus detaching the paint coating. An anodic reaction at the site of the coating defect is usually accompanied by a cathodic reaction near the coating substrate, which results in further separation of the coating from the metal and, ultimately, more corrosion.

3.14. Investigation the Antifouling Property of Polyaniline in Paint and Its Efficacy on Corrosion Rate

Polyaniline was added to the epoxy-based paint after it was dissolved in dimethylformamide as an additive in proportions of 10, 15, 20, or 30 wt.%. Then, its effect on electrochemical properties was evaluated via Tafel analysis and electrochemical impedance, and its antifouling property was investigated. Epoxy-based antifouling paints containing 30 wt.% of biocide and 10, 15, 20, or 30 wt.% of polyaniline/acid were prepared and applied to steel. Then, a polarization test was conducted in a 3.5% NaCl solution. Polarization curves and data are shown in Figure 11 and Table 12, respectively. By comparing 4 paints containing polyaniline with a paint without polyaniline, it can be concluded that the corrosion potential was more positive, and the corrosion current density was significantly reduced with the addition of polyaniline. The corrosion potential increased by 0.75, 0.85, and 0.93 for paints containing 10, 15, and 20 wt.% polyaniline, respectively.

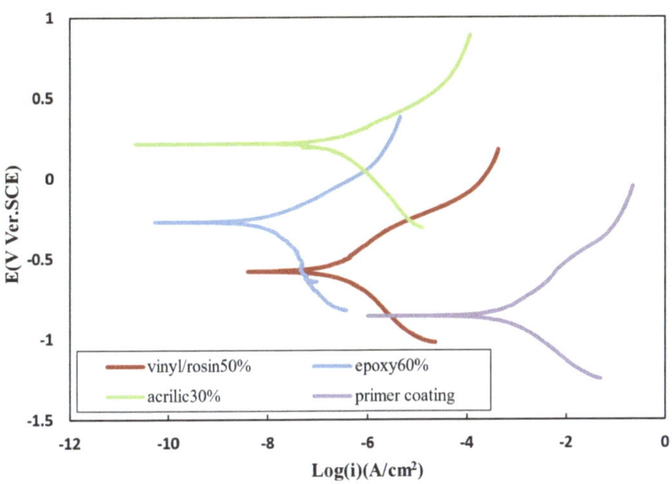

Figure 11. Polarization curves of antifouling paints at different contents of polyaniline in 3.5% NaCl solution.

Table 12. Obtained data from the polarization curve of antifouling paints containing polyaniline.

Polyaniline Concentration (wt.%)	E_{corr} (V)	$I_{corr} \times 10^{-6}$ ($\frac{A}{cm^2}$)	Corrosion Rate ($\frac{mm}{year}$)
0	−0.73	6.3	0.24
10	0.11	0.0093	0.00035
15	0.23	0.37	0.0148
20	0.47	8.7	0.348
30	−0.07	3.1	0.11

It should be noted that metals with coatings containing polyaniline are placed in the passive zone because a dense oxide layer is formed on the metal and polyaniline joint, protecting the metal against corrosion. An electric field is formed in the direction opposite to the electron transfer between the metal surface and the conductive coating surface (containing polyaniline), which can act as a charge transfer barrier layer to prevent the transfer of electrons from the metal to the oxidant and thus reduce corrosion. The corrosion potential decreased in the 30 wt.% sample, and the corrosion current density increased. The reason for this phenomenon is that at low concentrations of polyaniline (less than 20 wt.%), the distance between the conducting particles of polyaniline is large and fewer components are in contact, resulting in the formation of a chain of polyaniline, which can create charge conduction paths which reduce the strength of the coating. However, the conduction pathways are completed at higher concentrations, and adding polyaniline does not contribute to forming new conduction pathways in the paint. Even acid mixed with excess polyaniline may block the conduction channels in the paint.

Moreover, we observed that with an excessive increase of polyaniline, surface roughness increased. Density decreased, which increased surface defects, and the corrosive medium's penetration rate into the paint coating was faster than the passive layer's formation rate. As a result, the passive layer was destroyed by the corrosive medium before formation; thus, the anticorrosion effect of the coating was reduced. The impedance criterion is an important parameter for investigating the corrosion resistance behavior of polyaniline-containing coatings, which is directly related to the degree of penetration of the corrosive medium into the coating. Figure 12 shows the electrochemical impedance spectroscopy (EIS) results of antifouling polyaniline paints. As can be seen in Figure 12, the high-frequency curves are circular. The low-frequency curves are straight lines, which correspond to the Warburg region and indicate that the presence of polyaniline in the coating had a high protective effect and corrosion occurred only at the metal/environment interface, which was caused by the penetration of the corrosive medium into the pores [93]. Hence, the corrosion process largely depended on the mass transfer phenomenon and the electrochemical reaction. The sample curve of 10 wt.% polyanilines is only a semicircle with the highest impedance compared to other samples. In this sample, the process was often controlled by an electrochemical reaction, and the conductive coating effectively prevented the corrosive medium from penetrating the metal's surface; this is also consistent with the polarization diagram. As the concentration of polyaniline and the surface roughness increased, and the density of the coating decreased, water penetrated the metal–coating interface, and corrosion began.

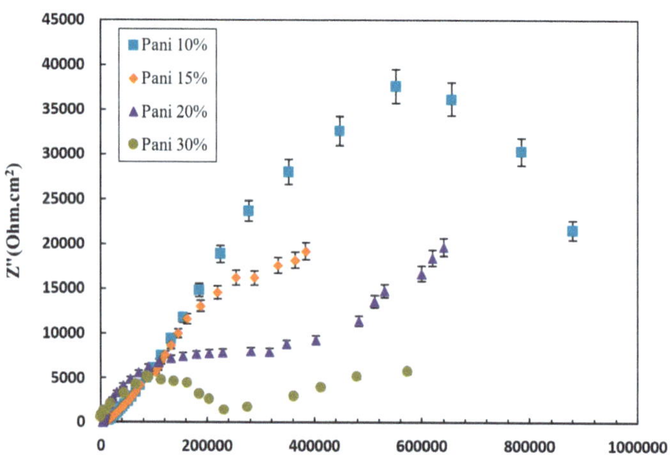

Figure 12. Impedance curves of antifouling paints at different contents of polyaniline in 3.5% NaCl solution.

Since polyaniline is conductive, when paint containing polyaniline is applied to a layer of insulating paint on a metal substrate, the antifouling paint layer containing polyaniline plays the role of the anode, and the electrolysis of seawater produces Hyspochlorous acid (HOCl). HOCl covers the surface of the paint coating and prevents the formation and adhesion of marine organisms.

In traditional conductive coatings, which consist of the distribution of additives such as graphite, carbon black, and metal powders in organic or inorganic resins, the additive consumption is high, and it is unevenly distributed in the resin, so these coatings are very expensive, and their conductivity properties are not efficient. However, developing conductive polymers such as polyaniline, polythiophene, and pyrrole and mixing them with an acid improves conductivity. However, among these, polyaniline has the best miscibility with resin after mixing with acid. Therefore, a sample containing 10 wt.% of polyaniline was placed in an algae growth system for 3 months to evaluate its antifouling property, and Figure 13 shows photos of the samples after the test.

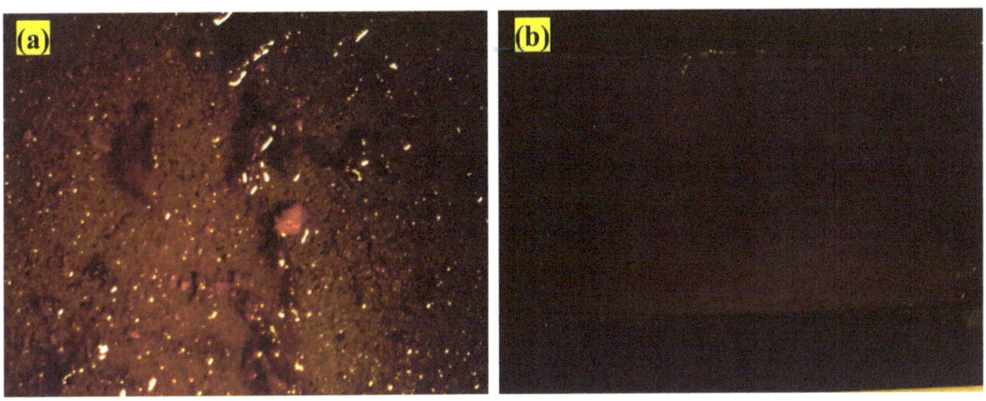

Figure 13. Photos of sample (**a**) without polyaniline, (**b**) containing 10 wt.% polyaniline after 3 months.

4. Environmental Challenges and Toxicity Effects of Biocides

Biocides are chemical substances that control or eliminate harmful organisms such as bacteria, viruses, fungi, and algae. While these compounds play a vital role in safeguarding public health and preventing the spread of diseases, their use poses significant environmental challenges and toxic effects. This summary will delve into two major aspects of these concerns: environmental impacts and the potential toxicity of biocides. The utilization of biocides can lead to a range of environmental challenges. One major issue is the persistence of these substances in ecosystems. Biocides, particularly those with high chemical stability, can accumulate in the environment over time, resulting in long-term contamination. This persistence poses risks to various organisms within the ecosystem, including aquatic organisms, birds, and mammals. Furthermore, biocides can undergo bioaccumulation, where they accumulate in the tissues of organisms at higher trophic levels. This bioaccumulation can magnify toxic effects, potentially disrupting entire food chains and ecological balances [94–96].

Another significant environmental challenge associated with biocides is the potential for non-target effects. Biocides are designed to target specific organisms, but their application may also affect non-target species, including beneficial insects, pollinators, and other organisms critical to ecosystem functioning. These unintended consequences can disrupt ecological interactions, reduce biodiversity, and impact ecosystems' overall health and resilience. Additionally, the release of biocides into the environment via different pathways, such as runoff or leaching, can result in contamination of water, soil, and air, further exacerbating environmental challenges. In terms of toxicity, biocides have the potential to cause adverse effects on both human health and the environment. Some biocides are classified as hazardous substances due to their toxic properties. These substances may pose acute or chronic health risks to humans, ranging from skin and eye irritation to respiratory problems and even carcinogenic effects. Moreover, exposure to biocides has been linked to adverse effects on aquatic organisms, such as fish and amphibians, as well as on soil organisms and beneficial microorganisms critical for soil fertility and nutrient cycling.

In conclusion, the use of biocides presents significant environmental challenges and toxic effects. The persistence of these substances in ecosystems, potential non-target effects, and the toxicity risks they pose to both human health and the environment underscores the need for responsible and judicious use of biocides. Implementing appropriate risk assessment and management strategies, exploring alternative pest control methods, and promoting sustainable practices can help mitigate the negative impacts associated with biocides while ensuring effective disease prevention and control.

5. Conclusions

This study investigated the critical role of biocide release rates in the efficiency of antifouling paints. It examined the release rates of Cu_2O and ZnO biocides in paints with varying resin compositions, particle sizes, and immersion times. Acrylic resin-based paints exhibited higher release rates due to the hydrolysis of acrylic copolymers, which weakened bonds and facilitated biocide release. Conversely, epoxy resin-based paints had lower release rates due to their insolubility and high crosslinking density, limiting water penetration and biocide dissolution. Vinyl/rosin-based paints showed intermediate release rates due to the combined hydrophobic properties of their constituent resins. Immersion time was found to be significant, with initial high release rates followed by a steady state, attributed to the formation of penetration barriers and biocide depletion. Biocide content in the formulation influenced release rates, showing a linear increase until a threshold was reached. Particle size also played a role, with smaller particles having higher release rates, especially in epoxy and vinyl/rosin resins. Interestingly, ZnO biocides exhibited release rates comparable to those of Cu_2O despite their lower concentration. The study emphasized the importance of maintaining optimal release rates for effective antifouling performance, with fouling growth observed in samples with rates below recommendations. Curing time and hardness varied among different resin types while meeting standard requirements for physical properties.

Supplementary Materials: The following supporting information can be downloaded at: https://www.mdpi.com/article/10.3390/polym15193948/s1, Table S1: Specification of artificial seawater; Figure S1: Real photo of the designed laboratory system; Figure S2: Schematic of the designed laboratory system; Figure S3: Real photo of built setup for algae growth; Table S2: Chemical composition of used water in the algae growth environment.

Author Contributions: A.J.: Investigation, Conceptualization, Methodology. A.A.: Conceptualization, Methodology, Supervision. S.B.M.: Formal analysis, Writing original draft. M.H.: Formal analysis, Writing original draft. All authors have read and agreed to the published version of the manuscript.

Funding: This research received no external funding.

Institutional Review Board Statement: Not applicable.

Data Availability Statement: All data generated or analyzed during this study are included in this published article.

Conflicts of Interest: The authors declare that they have no known competing financial interests or personal relationships that could have appeared to influence the work reported in this paper.

References

1. Chasse, K.; Scardino, A.; Swain, G. Corrosion and fouling study of copper-based antifouling coatings on 5083 aluminum alloy. *Prog. Org. Coat.* **2020**, *141*, 105555. [CrossRef]
2. Tesler, A.B.; Kim, P.; Kolle, S.; Howell, C.; Ahanotu, O.; Aizenberg, J. Extremely durable biofouling-resistant metallic surfaces based on electrodeposited nanoporous tungstite films on steel. *Nat. Commun.* **2015**, *6*, 8649. [CrossRef] [PubMed]
3. Han, X.; Wu, J.; Zhang, X.; Shi, J.; Wei, J.; Yang, Y.; Wu, B.; Feng, Y. Special issue on advanced corrosion-resistance materials and emerging applications. The progress on antifouling organic coating: From biocide to biomimetic surface. *J. Mater. Sci. Technol.* **2021**, *61*, 46–62. [CrossRef]
4. Ali, A.; Jamil, M.I.; Jiang, J.; Shoaib, M.; Amin, B.U.; Luo, S.; Zhan, X.; Chen, F.; Zhang, Q. An overview of controlled-biocide-release coating based on polymer resin for marine antifouling applications. *J. Polym. Res.* **2020**, *27*, 85. [CrossRef]
5. Chen, C.-L.; Maki, J.S.; Rittschof, D.; Teo, S.L.-M. Early marine bacterial biofilm on a copper-based antifouling paint. *Int. Biodeterior. Biodegrad.* **2013**, *83*, 71–76. [CrossRef]
6. Shenashen, M.; El-Safty, S.A.; Higazy, S.; Selim, M.; Isago, H.; Elmarakbi, A. Recent progress in marine foul-release polymeric nanocomposite coatings. *Prog. Mater. Sci.* **2017**, *87*, 1–32.
7. Long, W.; Li, H.; Yang, B.; Huang, N.; Liu, L.; Gai, Z.; Jiang, X. Superhydrophobic diamond-coated Si nanowires for application of anti-biofouling'. *J. Mater. Sci. Technol.* **2020**, *48*, 1–8. [CrossRef]
8. Callow, J.A.; Callow, M.E. Trends in the development of environmentally friendly fouling-resistant marine coatings. *Nat. Commun.* **2011**, *2*, 244. [CrossRef]
9. Tian, J.; Xu, K.; Hu, J.; Zhang, S.; Cao, G.; Shao, G. Durable self-polishing antifouling Cu-Ti coating by a micron-scale Cu/Ti laminated microstructure design. *J. Mater. Sci. Technol.* **2021**, *79*, 62–74. [CrossRef]
10. Hadfield, M.G. Biofilms and Marine Invertebrate Larvae: What Bacteria Produce That Larvae Use to Choose Settlement Sites. *Annu. Rev. Mar. Sci.* **2011**, *3*, 453–470. [CrossRef]
11. Qian, P.-Y.; Lau, S.C.K.; Dahms, H.-U.; Dobretsov, S.; Harder, T. Marine Biofilms as Mediators of Colonization by Marine Macroorganisms: Implications for Antifouling and Aquaculture. *Mar. Biotechnol.* **2007**, *9*, 399–410. [CrossRef] [PubMed]
12. Shoar, Z.K.; Pourpasha, H.; Heris, S.Z.; Mousavi, S.B.; Mohammadpourfard, M. The effect of heat transfer characteristics of macromolecule fouling on heat exchanger surface: A dynamic simulation study. *Can. J. Chem. Eng.* **2023**, *101*, 5802–5817. [CrossRef]
13. Yu, Y.; Xu, N.; Zhu, S.; Qiao, Z.; Zhang, J.; Yang, J.; Liu, W. A novel Cu-doped high entropy alloy with excellent comprehensive performances for marine application. *J. Mater. Sci. Technol.* **2021**, *69*, 48–59. [CrossRef]
14. Bai, X.; Li, J.; Zhu, L.; Wang, L. Effect of Cu content on microstructure, mechanical and anti-fouling properties of TiSiN-Cu coating deposited by multi-arc ion plating. *Appl. Surf. Sci.* **2018**, *427*, 444–451. [CrossRef]
15. Ma, F.; Li, J.; Zeng, Z.; Gao, Y. Tribocorrosion behavior in artificial seawater and anti-microbiologically influenced corrosion properties of TiSiN-Cu coating on F690 steel. *J. Mater. Sci. Technol.* **2019**, *35*, 448–459. [CrossRef]
16. Yebra, D.M.; Kiil, S.; Weinell, C.E.; Dam-Johansen, K. Effects of marine microbial biofilms on the biocide release rate from antifouling paints—A model-based analysis. *Prog. Org. Coat.* **2006**, *57*, 56–66. [CrossRef]
17. Sharma, B.K.; Gupta, A.K.; Khare, N.; Dhawan, S.; Gupta, H. Synthesis and characterization of polyaniline–ZnO composite and its dielectric behavior. *Synth. Met.* **2009**, *159*, 391–395. [CrossRef]
18. Prasad, G.; Takei, T.; Yonesaki, Y.; Kumada, N.; Kinomura, N. Hybrid nanocomposite based on NbWO6 nanosheets and polyaniline. *Mater. Lett.* **2006**, *60*, 3727–3730. [CrossRef]
19. Deshpande, N.; Gudage, Y.; Sharma, R.; Vyas, J.; Kim, J.; Lee, Y. Studies on tin oxide-intercalated polyaniline nanocomposite for ammonia gas sensing applications. *Sens. Actuators B Chem.* **2009**, *138*, 76–84. [CrossRef]
20. Gurunathan, K.; Amalnerkar, D.; Trivedi, D. Synthesis and characterization of conducting polymer composite (PAn/TiO_2) for cathode material in rechargeable battery. *Mater. Lett.* **2003**, *57*, 1642–1648. [CrossRef]
21. Mostafaei, A.; Nasirpouri, F. Epoxy/polyaniline–ZnO nanorods hybrid nanocomposite coatings: Synthesis, characterization and corrosion protection performance of conducting paints. *Prog. Org. Coat.* **2014**, *77*, 146–159. [CrossRef]
22. Nami, S.H.; Mousavi, S.B. Nitrate Removal Performance of Different Granular Adsorbents Using a Novel Fe-Exchanged Nanoporous Clinoptilolite. *Ind. Eng. Chem. Res.* **2023**, *62*, 3659–3671. [CrossRef]
23. Hosseini, M.; Jafari, M.; Najjar, R. Effect of polyaniline–montmorillonite nanocomposite powders addition on corrosion performance of epoxy coatings on Al 5000. *Surf. Coat. Technol.* **2011**, *206*, 280–286. [CrossRef]
24. Heris, S.Z.; Etemadi, M.; Mousavi, S.B.; Mohammadpourfard, M.; Ramavandi, B. Preparation and characterizations of TiO_2/ZnO nanohybrid and its application in photocatalytic degradation of tetracycline in wastewater. *J. Photochem. Photobiol. A Chem.* **2023**, *443*, 114893.
25. Yousefi, F.; Mousavi, S.B.; Heris, S.Z.; Naghash-Hamed, S. UV-shielding properties of a cost-effective hybrid PMMA-based thin film coatings using TiO_2 and ZnO nanoparticles: A comprehensive evaluation. *Sci. Rep.* **2023**, *13*, 7116. [CrossRef] [PubMed]
26. Mousavi, S.B.; Heris, S.Z.; Hosseini, M.G. Experimental investigation of MoS_2/diesel oil nanofluid thermophysical and rheological properties. *Int. Commun. Heat Mass Transf.* **2019**, *108*, 104298. [CrossRef]
27. Naghash-Hamed, S.; Arsalani, N.; Mousavi, S.B. Facile fabrication of $CuFe_2O_4$ coated with carbon quantum dots nanocomposite as an efficient heterogeneous catalyst toward the reduction of nitroaniline compounds for management of aquatic resources. *J. Photochem. Photobiol. A Chem.* **2023**, *443*, 114822. [CrossRef]

28. Naghash-Hamed, S.; Arsalani, N.; Mousavi, S.B. The catalytic performance of $CuFe_2O_4$@CQD nanocomposite as a high-perform heterogeneous nanocatalyst in nitroaniline group reduction. *Sci. Rep.* **2023**, *13*, 3329. [CrossRef]
29. Mousavi, S.B.; Heidari, M.; Rahmani, F.; Sene, R.A.; Clough, P.T.; Ozmen, S. Highly robust ZrO_2-stabilized CaO nanoadsorbent prepared via a facile one-pot MWCNT-template method for CO_2 capture under realistic calcium looping conditions. *J. Clean. Prod.* **2023**, *384*, 135579. [CrossRef]
30. Athari, M.-J.; Tahmasebpoor, M.; Azimi, B.; Heidari, M.; Pevida, C. Waste oleaster seed-derived activated carbon mixed with coarse particles of fluid catalytic cracking as a highly-efficient CO2 adsorbent at low temperatures. *Process. Saf. Environ. Prot.* **2023**, *178*, 580–594. [CrossRef]
31. Tahmasebpoor, M.; Iranvandi, M.; Heidari, M.; Azimi, B.; Pevida, C. Development of novel waste tea-derived activated carbon promoted with SiO2 nanoparticles as highly robust and easily fluidizable sorbent for low-temperature CO_2 capture. *J. Environ. Chem. Eng.* **2023**, *11*, 110437. [CrossRef]
32. Kadri, Y.; Srasra, E.; Bekri-Abbes, I.; Herrasti, P. Facile and eco-friendly synthesis of polyaniline/ZnO composites for corrosion protection of AA-2024 aluminium alloy. *J. Electroanal. Chem.* **2021**, *893*, 115335. [CrossRef]
33. Rui, M.; Zhu, A. The synthesis and corrosion protection mechanisms of PANI/CNT nanocomposite doped with organic phosphoric acid. *Prog. Org. Coat.* **2021**, *153*, 106134. [CrossRef]
34. Najjar, R.; Katourani, S.A.; Hosseini, M.G. Self-healing and corrosion protection performance of organic polysulfide@urea-formaldehyde resin core-shell nanoparticles in epoxy/PANI/ZnO nanocomposite coatings on anodized aluminum alloy. *Prog. Org. Coat.* **2018**, *124*, 110–121. [CrossRef]
35. Pagotto, J.F.; Recio, F.J.; Motheo AD, J.; Herrasti, P. Multilayers of PAni/n-TiO_2 and PAni on carbon steel and welded carbon steel for corrosion protection. *Surf. Coat. Technol.* **2016**, *289*, 23–28. [CrossRef]
36. Shetty, K.; Jayadev; Raj, K. Synthesized conducting Polyaniline—TiO_2 based nanocomposite for corrosion control on steel 316. *Mater. Today Proc.* **2021**, *38*, 2493–2498. [CrossRef]
37. Sumi, V.; Arunima, S.; Deepa, M.; Sha, M.A.; Riyas, A.; Meera, M.; Saji, V.S.; Shibli, S. PANI-Fe_2O_3 composite for enhancement of active life of alkyd resin coating for corrosion protection of steel. *Mater. Chem. Phys.* **2020**, *247*, 122881. [CrossRef]
38. Bandgar, D.K.; Navale, S.T.; Vanalkar, S.A.; Kim, J.H.; Harale, N.S.; Patil, P.S.; Patil, V.B. Synthesis, structural, morphological, compositional and electrical transport properties of polyaniline/α-Fe_2O_3 hybrid nanocomposites. *Synth. Met.* **2014**, *195*, 350–358. [CrossRef]
39. Gholami, A.; Mousavi, S.B.; Heris, S.Z.; Mohammadpourfard, M. Highly efficient treatment of petrochemical spent caustic effluent via electro-Fenton process for COD and TOC removal: Optimization and experimental. *Biomass Convers. Biorefin.* **2023**, 1–17. [CrossRef]
40. Ebratkhahan, M.; Naghash Hamed, S.; Zarei, M.; Jafarizad, A.; Rostamizadeh, M. Removal of neutral red dye via electro-fenton process: A response surface methodology modeling. *Electrocatalysis* **2021**, *12*, 579–594. [CrossRef]
41. Khouri, O.; Goshayeshi, H.R.; Nami, S.H.; Mousavi, S.B.; Heris, S.Z.; Chaer, I. Evaluating the impact of graphene oxide nanoparticles on enhancing heat transfer characteristics within the indirect heat exchanger of a natural city gate station. *Res. Sq.* **2023**. preprint.
42. Yong, H.E.; Krishnamoorthy, K.; Hyun, K.T.; Kim, S.J. Preparation of ZnO nanopaint for marine antifouling applications. *J. Ind. Eng. Chem.* **2015**, *29*, 39–42. [CrossRef]
43. Seyedi, S.S.; Shabgard, M.R.; Mousavi, S.B.; Heris, S.Z. The impact of SiC, Al_2O_3, and B_2O_3 abrasive particles and temperature on wear characteristics of 18Ni (300) maraging steel in abrasive flow machining (AFM). *Int. J. Hydrogen Energy* **2021**, *46*, 33991–34001. [CrossRef]
44. Al-Belushi, M.A.; Myint MT, Z.; Kyaw, H.H.; Al-Naamani, L.; Al-Mamari, R.; Al-Abri, M.; Dobretsov, S. ZnO nanorod-chitosan composite coatings with enhanced antifouling properties. *Int. J. Biol. Macromol.* **2020**, *162*, 1743–1751. [CrossRef] [PubMed]
45. Rajan, R.; Selvaraj, M.; Palraj, S.; Subramanian, G. Studies on the anticorrosive & antifouling properties of the Gracilaria edulis extract incorporated epoxy paint in the Gulf of Mannar Coast, Mandapam, India. *Prog. Org. Coat.* **2016**, *90*, 448–454.
46. Ni, C.; Feng, K.; Li, X.; Zhao, H.; Yu, L. Study on the preparation and properties of new environmentally friendly antifouling acrylic metal salt resins containing indole derivative group. *Prog. Org. Coat.* **2020**, *148*, 105824. [CrossRef]
47. Adeleye, A.S.; Oranu, E.A.; Tao, M.; Keller, A.A. Release and detection of nanosized copper from a commercial antifouling paint. *Water Res.* **2016**, *102*, 374–382. [CrossRef] [PubMed]
48. Palanichamy, S.; Subramanian, G. Antifouling properties of marine bacteriocin incorporated epoxy based paint. *Prog. Org. Coat.* **2017**, *103*, 33–39. [CrossRef]
49. Quan, X.; Wang, J.; Zhao, S.; Cai, W.; Wang, Z.; Wang, S.; Cui, X. Improved antibacterial, antifouling and corrosion protective performance of epoxy coatings with poly (m-aminophenol). *Prog. Org. Coat.* **2018**, *115*, 9–17. [CrossRef]
50. Quan, X.; Wang, J.; Souleyman, T.; Cai, W.; Zhao, S.; Wang, Z. Antibacterial and antifouling performance of bisphenol-A/Poly (ethylene glycol) binary epoxy coatings containing bromine-benzyl-disubstituted polyaniline. *Prog. Org. Coat.* **2018**, *124*, 61–70. [CrossRef]
51. Tonelli, M.; Perini, I.; Ridi, F.; Baglioni, P. Improving the properties of antifouling hybrid composites: The use of Halloysites as nano-containers in epoxy coatings. *Colloids Surf. A Physicochem. Eng. Asp.* **2021**, *623*, 126779. [CrossRef]
52. Ma, J.; Ma, C.; Yang, Y.; Xu, W.; Zhang, G. Biodegradable polyurethane carrying antifoulants for inhibition of marine biofouling. *Ind. Eng. Chem. Res.* **2014**, *53*, 12753–12759. [CrossRef]

53. Shtykova, L.; Fant, C.; Handa, P.; Larsson, A.; Berntsson, K.; Blanck, H.; Simonsson, R.; Nydén, M.; Härelind, H.I. Adsorption of antifouling booster biocides on metal oxide nanoparticles: Effect of different metal oxides and solvents. *Prog. Org. Coat.* **2009**, *64*, 20–26. [CrossRef]
54. Razmjoo, F.; Sadeghi, E.; Alizadeh-Sani, M.; Noroozi, R.; Azizi-Lalabadi, M. Fabrication and application of functional active packaging material based on carbohydrate biopolymers formulated with Lemon verbena/Ferulago angulata extracts for the preservation of raw chicken meat. *Food Process. Preserv.* **2022**, *46*, e16830. [CrossRef]
55. *ISIRI6456*; Paints and Varnishes; Surface-Drying Test; Ballotini Method. Institute of Standards and Industrial Research of Iran: Tehran, Iran, 2014.
56. Takahashi, K. Release rate of biocides from antifouling paints. *Ecotoxicol. Antifouling Biocides* **2009**, 3–22.
57. *ASTM D6442-06*; Standard Test Method for Determination of Copper Release Rate From Antifouling Coatings in Substitute Ocean Water. ASTM International: West Conshohocken, PA, USA, 2020.
58. Singh, S.; Singh, R.; Sharma, N.R.; Singh, A. Ethanolic Extract of Cockroach Wing Powder as Corrosion Inhibitor for N80 Steel in an ASTM D1141-98(2013) Standard Artificial Seawater Solution. *Int. J. Electrochem. Sci.* **2021**, *16*, 210841. [CrossRef]
59. *ASTM D1141*; Standard Practice for Preparation of Substitute Ocean Water. ASTM International: West Conshohocken, PA, USA, 2021.
60. Yebra, D.M.; Kiil, S.; Dam-Johansen, K. Antifouling technology—Past, present and future steps towards efficient and environmentally friendly antifouling coatings. *Prog. Org. Coat.* **2004**, *50*, 75–104. [CrossRef]
61. Brady, R.F. Clean hulls without poisons: Devising and testing nontoxic marine coatings. *J. Coat. Technol.* **2000**, *72*, 45–56. [CrossRef]
62. Sapozhnikova, Y.; Wirth, E.; Schiff, K.; Fulton, M. Antifouling biocides in water and sediments from California marinas. *Mar. Pollut. Bull.* **2013**, *69*, 189–194. [CrossRef]
63. Beitelman, A.D.; Marsh, C.P.; Carlson, T.A. Low heat-transfer coatings in heat distribution systems. In Proceedings of the CORROSION 2009, Atlanta, GA, USA, 22–26 March 2009. NACE–09496 NACE CORROSION.
64. *ASTM D1186*; Standard Test Methods for Nondestructive Measurement of Dry Film Thickness of Nonmagnetic Coatings Applied to a Ferrous Base. ASTM International: West Conshohocken, PA, USA, 2021.
65. Patel, C.J.; Mannari, V. Air-drying bio-based polyurethane dispersion from cardanol: Synthesis and characterization of coatings. *Prog. Org. Coat.* **2014**, *77*, 997–1006. [CrossRef]
66. *ASTM D562*; Standard Test Method for Consistency of Paints Measuring Krebs Unit (KU) Viscosity Using a Stormer-Type Viscometer. ASTM International: West Conshohocken, PA, USA, 2010.
67. *ASTM D3363*; Standard Test Method for Film Hardness by Pencil Test. ASTM International: West Conshohocken, PA, USA, 2022.
68. *ASTM D4366*; Standard Test Methods for Hardness of Organic Coatings by Pendulum Damping Tests. ASTM International: West Conshohocken, PA, USA, 2021.
69. Azadi, M.; Bahrololoom, M.E.; Heidari, F. Enhancing the mechanical properties of an epoxy coating with rice husk ash, a green product. *J. Coat. Technol. Res.* **2011**, *8*, 117–123. [CrossRef]
70. Kurt, Ş.; Özçifçi, A. Effect of various fire retardants on Brinell hardness of some wood. *BioResources* **2009**, *4*, 960–969. [CrossRef]
71. *ASTM D522*; Standard Test Methods for Mandrel Bend Test of Attached Organic Coatings. ASTM International: West Conshohocken, PA, USA, 2008.
72. *ASTM D4541*; Standard Test Method for Pull-Off Strength of Coatings Using Portable Adhesion Testers. ASTM International: West Conshohocken, PA, USA, 2022.
73. Schilling, M. *Coating Adhesion Testing in Accordance with ASTM D4541 Sticky Business*; Corrosion Probe, Inc.: Centerbrook, CT, USA, 2004.
74. Leloup, F.B.; Audenaert, J.; Obein, G.; Ged, G.; Hanselaer, P. Repeatability and reproducibility of specular gloss meters in theory and practice. *J. Coat. Technol. Res.* **2016**, *13*, 941–951. [CrossRef]
75. *ASTM D523*; Standard Test Method for Specular Gloss. ASTM International: West Conshohocken, PA, USA, 1999.
76. *ASTM D1794*; Standard Test Method for Alcohol-Benzene Soluble Matter in Cellulose. ASTM International: West Conshohocken, PA, USA, 1991.
77. Gore, P.M.; Balakrishnan, S.; Kandasubramanian, B. Superhydrophobic corrosion inhibition polymer coatings. In *Superhydrophobic Polymer Coatings*; Elsevier: Amsterdam, The Netherlands, 2019; pp. 223–243.
78. *ASTM D4060*; Standard Test Method for Abrasion Resistance of Organic Coatings by the Taber Abraser. ASTM International: West Conshohocken, PA, USA, 2019.
79. Browning, R.; Lim, G.-T.; Moyse, A.; Sue, H.-J.; Chen, H.; Earls, J. Quantitative evaluation of scratch resistance of polymeric coatings based on a standardized progressive load scratch test. *Surf. Coat. Technol.* **2006**, *201*, 2970–2976. [CrossRef]
80. *ASTM D7027*; Standard Test Method for Evaluation of Scratch Resistance of Polymeric Coatings and Plastics Using an Instrumented Scratch Machine. ASTM International: West Conshohocken, PA, USA, 2020.
81. *ASTM G102-89*; Standard Practice for Calculation of Corrosion Rates and Related Information from Electrochemical Measurements. ASTM International: West Conshohocken, PA, USA, 2010.
82. Vitelaru, C.; Ghiban, N.; Parau, A.C.; Balaceanu, M.; Miculescu, F.; Vladescu, A. Corrosion behaviour of Ti6Al4V alloy in artificial saliva solution with fluoride content and low pH value: Korrosionsverhalten der Legierung Ti6Al4V in künstlicher Speichelflüssigkeit mit Fluoridanteilen und geringem pH-Wert. *Mater. Werkst.* **2014**, *45*, 91–98. [CrossRef]

83. Krishnan, R.; Manivannan, S.; Patel, P.B.; Patil, P.P.; Venkatesh, R.; Thanappan, S. Studies on Corrosion Behavior of Mg-Al-Zn-RE Cast Alloy with Powder-Coated Al and CED Mg by Salt Spray Test, Immersion Test, and Electrochemical Test. *Int. J. Chem. Eng.* **2022**, *2022*, 1891419. [CrossRef]
84. *ASTM B117*; Standard Practice for Operating Salt Spray (Fog) Apparatus. ASTM International: West Conshohocken, PA, USA, 2019.
85. Wang, X.-H.; Li, J.; Zhang, J.-Y.; Sun, Z.-C.; Yu, L.; Jing, X.-B.; Sun, Z.-X.; Ye, Z.-J. Polyaniline as marine antifouling and corrosion-prevention agent. *Synth. Met.* **1999**, *102*, 1377–1380. [CrossRef]
86. Mostafaei, A.; Nasirpouri, F. Preparation and characterization of a novel conducting nanocomposite blended with epoxy coating for antifouling and antibacterial applications. *J. Coat. Technol. Res.* **2013**, *10*, 679–694. [CrossRef]
87. Chen, S.; Zhu, J.; Zhou, T.; He, B.; Huang, W.; Wang, B. Preparation and properties study of polyaniline conductive anti-fouling coatings. *Int. J. Electrochem. Sci.* **2012**, *7*, 8170. [CrossRef]
88. Gitlitz, M.H.; Leiner, H.H. Erodible Ship-Bottom Paints for Control of Marine Fouling. U.S. Patent 4,593,055, 3 June 1986.
89. Yebra, D.M.; Kiil, S.; Weinell, C.E.; Dam-Johansen, K. Dissolution rate measurements of sea water soluble pigments for antifouling paints: ZnO. *Prog. Org. Coat.* **2006**, *56*, 327–337. [CrossRef]
90. Candries, M. Drag, Boundary-Layer and Roughness Characteristics of Marine Surfaces Coated with Antifoulings. Ph.D. Thesis, University of Newcastle upon Tyne, Newcastle upon Tyne, UK, 2001.
91. Almeida, E.; Diamantino, T.C.; de Sousa, O. Marine paints: The particular case of antifouling paints. *Prog. Org. Coat.* **2007**, *59*, 2–20. [CrossRef]
92. Kiil, S.; Dam-Johansen, K.; Weinell, C.E.; Pedersen, M.S. Seawater-soluble pigments and their potential use in self-polishing antifouling paints: Simulation-based screening tool. *Prog. Org. Coat.* **2002**, *45*, 423–434. [CrossRef]
93. Sobhani, E.; Heris, S.Z.; Mousavi, S.B. The Synergistic Effect of Intumescent Fire-Resistive Paint Containing TiO_2 Nanoparticles and Chlorinated Paraffin onto Atmospheric-Metallic Substrates. *ChemistrySelect* **2022**, *7*, e202203513. [CrossRef]
94. Tahmasebpoor, M.; Hosseini Nami, S.; Khatamian, M.; Sanaei, L. Arsenate removal from contaminated water using Fe_2O_3-clinoptilolite powder and granule. *Environ. Technol.* **2022**, *43*, 116–130. [CrossRef] [PubMed]
95. Khatamian, M.; Nami, S.H.; Mosayeb, S.G.H.; Divband, B. Preparation and characterization of nano- $Fe(OH)_3$, its composites with two natural zeolites, and granulation of them for nitrate removal from polluted water. *Phys. Scr.* **2023**, *98*, 035024. [CrossRef]
96. Khatamian, M.; Afshar No, N.; Hosseini Nami, S.; Fazli-Shokouhi, S. Synthesis and characterization of zeolite A, Fe_3O_4/zeolite A, and Fe2O3/zeolite A nanocomposites and investigation of their arsenic removal performance. *J. Iran. Chem. Soc.* **2023**, *20*, 1657–1670. [CrossRef]

polymers
MDPI

Article

Highly Stable, Mechanically Enhanced, and Easy-to-Collect Sodium Alginate/NZVI-rGO Gel Beads for Efficient Removal of Cr(VI)

Qi Jing *, Yuheng Ma, Jingwen He and Zhongyu Ren

Faculty of Architecture, Civil and Transportation Engineering, Beijing University of Technology, Beijing 100124, China; mayuheng@emails.bjut.edu.cn (Y.M.); a_wjbzd@163.com (J.H.); renzhy@bjut.edu.cn (Z.R.)
* Correspondence: jingqi@bjut.edu.cn; Tel.: +86-185-0195-00178

Abstract: Nanoscale zero-valent iron (NZVI) is a material that is extensively applied for water pollution treatment, but its poor dispersibility, easy oxidation, and inconvenient collection limit its application. To overcome these drawbacks and limit secondary contamination of nanomaterials, we confine NZVI supported by reduced graphene oxide (rGO) in the scaffold of sodium alginate (SA) gel beads (SA/NZVI-rGO). Scanning electron microscopy showed that the NZVI was uniformly dispersed in the gel beads. Fourier transform infrared spectroscopy demonstrated that the hydrogen bonding and conjugation between SA and rGO allowed the NZVI-rGO to be successfully embedded in SA. Furthermore, the mechanical strength, swelling resistance, and Cr(VI) removal capacity of SA/NZVI-rGO were enhanced by optimizing the ratio of NZVI and rGO. Interestingly, cation exchange may drive Cr(VI) removal above 82% over a wide pH range. In the complex environment of actual Cr(VI) wastewater, Cr(VI) removal efficiency still reached 70.25%. Pseudo-first-order kinetics and Langmuir adsorption isotherm are preferred to explain the removal process. The mechanism of Cr(VI) removal by SA/NZVI-rGO is dominated by reduction and adsorption. The sustainable removal of Cr(VI) by packed columns could be well fitted by the Thomas, Adams–Bohart, and Yoon–Nelson models, and importantly, the gel beads maintained integrity during the prolonged removal. These results will contribute significant insights into the practical application of SA/NZVI-rGO beads for the Cr(VI) removal in aqueous environments.

Keywords: gel bead; nanoscale zero-valent iron; reduced graphene oxide; sodium alginate; Cr(VI) removal

1. Introduction

Water pollution caused by heavy metals is a critical issue since their detrimental effects on the environment and living organisms. Chromium is a typical heavy metal element commonly occurring in industrial wastewater, such as electroplating, textile printing and dyeing, steel and leather manufacturing, and accumulating in soil and water [1]. The toxicity of chromium is highly correlated with its valence, and hexavalent chromium (Cr(VI)), which is highly toxic relative to trivalent chromium (Cr(III)), is even more hazardous [2]. Since Cr(VI) is strongly oxidizing and water-soluble, long periods of exposure to high levels of Cr(VI) in humans may cause cancer and cellular tissue malformations [3]; therefore, there is a necessity to remove Cr(VI) from the aqueous environment.

Nano-zero-valent iron (NZVI) has been extensively applied to remove pollutants from groundwater, including heavy metals [4–7] and organics [7–9], because of its large specific surface area, high reactivity [10], and the possibility of injection into aquifers [11]. However, if NZVI is present at higher concentrations, NZVI tends to aggregate into larger particles owing to magnetic and van der Waals forces, thus reducing the specific surface area and mobility in water [12]. Moreover, under natural conditions, the structure and chemical properties of NZVI change due to its high reactivity, gradually oxidizing and corroding with time [13]. To enhance the dispersion and stability of NZVI, loading NZVI on porous materials is an effective approach. Currently, NZVI has been successfully loaded onto graphene [14], silica [15], carbon nanotubes [16], activated carbon [17], bentonite [18], and other materials. Loading NZVI onto graphene can effectively reduce particle agglomeration and show efficient treatment efficiency for organic matter [19] and heavy metals [20].

However, the practical applications of such nanocomposites are still restricted for reasons of economy and safety. Since most of these composites are nano-powder materials, they are difficult to separate and collect after application in water pollution remediation. From an economic point of view, the fact that inconvenient collection and reuse will undoubtedly cause an increase in treatment costs. What is more difficult is that from the perspective of environment and biomedicine, these nanomaterials that are inconvenient to collect are likely to cause secondary pollution to the environment. It has been reported that nano-scale materials, such as NZVI and GO are harmful to plants, animals, and even humans [21,22]. Therefore, the development of a remediation agent that is easy to collect, safe, and economical has become a hot topic of current research.

Alginate, a natural polysaccharide derived from seaweed, is a biodegradable, non-toxic, and water-insoluble gel gent [23]. Alginate can be chelated with divalent cations to form hydrogels, and alginate–calcium hydrogel/bead embedding is commonly used in the food industry and biomedical field for purposes such as the immobilization of living cells [24] and transportation of drugs [25]. In addition, the alginate-encapsulated NZVI technology has been applied in water treatment technology and the encapsulation does not affect the reactivity of NZVI. Bezbaruah et al. found that the reactivity of alginate-encapsulated NZVI for nitrate removal was comparable to that of bare NZVI and that entrapped NZVI could overcome the fluidity and sedimentation challenges related to bare NZVI [26]. The encapsulation process improves the co-removal efficiency of Cu(II) and MCB compared to Ni/Fe nanoparticles alone, and the gel beads can be reused multiple times with little effect on the removal efficiency, indicating the positive effect of encapsulation on degradation and reduction as well as improved economics [27]. However, alginate gels inevitably swell under certain conditions, which reduces their mechanical properties and limits the use of hydrogels [28]. It has been shown that the mechanical properties of sodium alginate (SA) beads can be improved by encapsulating graphene oxide (GO) in SA beads and then reducing them to form a reduced graphene oxide/sodium alginate (rGO/SA) bead [29]. Hence, the formation of gel beads by coating NZVI and rGO with sodium alginate not only avoids the disadvantages of nano-powder materials but also sufficiently utilizes the advantage that carbon materials can overcome the poor mechanical properties of sodium alginate.

Considering that rGO can not only improve the dispersibility and lifetime of NZVI but also enhance the mechanical strength of sodium alginate. Therefore, in this study, pre-synthesized NZVI supported by rGO (NZVI-rGO) was mixed with SA, and then the mixture was pumped into calcium chloride ($CaCl_2$) solution to form the final product SA/NZVI-rGO gel beads. We demonstrate that SA/NZVI-rGO gel beads improve durability, mechanical strength, and NZVI dispersibility. Further, the ratio of NZVI-rGO was optimized to boost the anti-swelling and mechanical strength of SA/NZVI-rGO gel beads. Considering Cr(VI) as the target pollutant, the effects of bead composition, dose ratio of NZVI and rGO, pH, initial concentration of Cr(VI), and solid–liquid ratio on the removal effect were discussed. The kinetics, adsorption isotherms, and possible removal mechanisms of Cr(VI) removal by SA/NZVI-rGO gel beads were investigated. Finally, the performance of the gel beads under continuous removal was investigated by fixed-bed column experiments and simulated using three different models at different flow rates, concentrations, and column heights.

2. Materials and Methods

2.1. Materials

Sodium alginate (SA) and potassium dichromate ($K_2Cr_2O_7$) were obtained from Tianjin Fuchen Chemical Reagent Co. Ltd., Tianjin, China. Sodium borohydride ($NaBH_4$), diphenyl carbamide, anhydrous calcium chloride ($CaCl_2$), and ferrous sulfate heptahydrate ($FeSO_4 \cdot 7H_2O$) were bought from the Sinopharm Group Chemical Reagent Co. Ltd., Shanghai, China. Graphene oxide powder was received from Suzhou Hengqiu Technology Co. Ltd., Suzhou, China. None of the chemicals involved in the experiment were further purified. All solutions were made using deionized water.

2.2. Procedures for SA/NZVI-rGO Beads Synthesis

To load NZVI onto rGO (Scheme 1a), GO (5 mg/mL) dissolved in water was pulverized by an ultrasonic cell pulverizer (Biosafer, 900-92, Biosafer Co., Nanjing, China) for 10 min, and then sonicated for 1 h in a water bath. Then, the GO suspension mixed with a solution of $FeSO_4 \cdot 7H_2O$ (2.482 g/ 20 mL) was transferred to a three-necked flask, and $NaBH_4$ solution (2.5133 g/ 50 mL) was then dripped into the three-necked flask using a peristaltic pump at a rate of 1.5 mL/min. The mixture was constantly stirred and introduced with nitrogen gas. With the intense reductivity of $NaBH_4$, Fe^{2+} and GO were reduced to NZVI-rGO. The final synthesized material was filtered through a 0.45 μm filter membrane and freeze-dried for 24 h. The NZVI-rGO at a ratio of NZVI: rGO = 1:1 was made. Considering the influence of different dosing ratios of NZVI and rGO on the final product, NZVI-rGO were also produced in ratios of 2:1, 3:1, and 5:1. The composite materials were named as xNZVI-rGO (1NZVI-rGO, 2NZVI-rGO, 3NZVI-rGO, and 5NZVI-rGO), where the x was the ratio of NZVI and rGO.

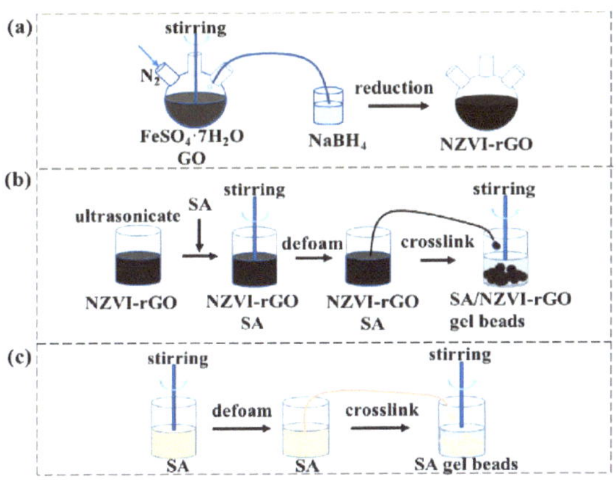

Scheme 1. Preparation process of the (**a**) NZVI-rGO, (**b**) SA/NZVI-rGO gel beads, and (**c**) SA gel beads.

Next, to encapsulate the NZVI-rGO in SA (Scheme 1b), 3 g of SA was dispersed in 50 mL of purified water and stirred for 3 h, and then left to defoam. NZVI-rGO (0.8 g/40 mL) dissolved in purified water was treated with an ultrasonic cell crusher for 10 min. The mixed SA and NZVI-rGO solutions were subsequently stirred under nitrogen for 1 h and then left to defoam. To obtain SA/NZVI-rGO gel beads, the mixed SA and NZVI-rGO were slowly dropped into $CaCl_2$ solution (500 mL 10 mg/mL) using a peristaltic pump (Scheme 1c). SA/NZVI-rGO gel beads rinsed repeatedly with purified water were freeze-dried. The beads were washed several times with deionized water and then freeze-dried. Unless otherwise stated, SA/NZVI-rGO beads below refer to SA/2NZVI-rGO beads.

To compare the performance of different gel beads, SA, SA mixed NZVI, and SA mixed rGO solutions were, respectively, dropped into 500 mL of 10 mg/mL $CaCl_2$ solution with mechanical stirring to obtain SA gel beads, SA/NZVI gel beads, and SA/rGO gel beads.

To fit the mechanical test apparatus, SA/xNZVI-rGO mixture was poured into PTFE molds and injected with $CaCl_2$ to gelate for 48 h to make rectangles (10 mm × 10 mm × 15 mm).

2.3. Characterization and Analysis Methods

Scanning electron microscopy (SEM, JSM-7900 F, JEOL Ltd., Tokyo, Japan) and Fourier transform infrared spectroscopy (FTIR, IRAffinity-1S, Shimadzu, Tokyo, Japan) were applied to observe the surface morphology and analyze the functional groups. An X-ray photoelectron spectroscopy (XPS, ESCALAB 250Xi, Thermo Fisher, Warrington, UK) was applied to determine elemental composition and valence state of gel beads. The zeta potential of products is measured with zeta potential analyzer (90 Plus, Brookhaven, NY, USA) at 25 °C. The concentration of Cr(VI) was quantified by UV-Vis spectrophotometer (2802S UV/VIS, Shanghai Unicosh Instruments Co., Shanghai, China). The compressive strength was tested using an electronic universal testing machine (CMT6103, MTS, Akron, OH, USA) with a compression rate of 0.5 mm min^{-1}. The swelling ratio (SR%) and water content (WC%) are illustrated in Text S1.

2.4. Batch Experiments and Column Experiments

The Cr(VI) removal batch experiments were conducted by different gel beads (gel beads prepared by coating different components with SA as well as SA/NZVI-rGO beads by NZVI and rGO in different dosing ratios) to compare the removal effect and further optimize the SA/NZVI-rGO gel beads. The removal of Cr(VI) (100 mL) by beads at differing conditions was compared by batch experiments. The removal experiments were all performed using a water bath thermostatic oscillator at 150 rad/min. Aliquots were obtained at specified intervals and filtered with a 0.22 μm filter membrane. The effects of pHs (from 3 to 11), initial Cr(VI) concentrations (from 10 to 400 mg/L), and solid-to-liquid ratios (from 0.5 to 5 mg/mL) on Cr(VI) removal were considered separately. The calculation of removal capacity and removal efficiency is shown in Text S2.

The practical applicability of SA/NZVI-rGO gel beads was verified by treating real Cr(VI)-containing wastewater, which was taken from a chemical factory in Beijing, China. The composition of the actual Cr(VI)-containing wastewater is shown in Table S2. Cation and anion contents were detected using inductively coupled plasma mass spectrometry (ICP-MS 7800, Agilent, Beijing, China) and ion chromatography (ICS-1100, Thermo Dionex, Warrington, UK), respectively.

Column experiments were conducted in Plexiglas columns (2 cm inner diameter × 14 cm height), which were tightly packed with SA/NZVI-rGO beads in the middle and packed with 2 cm thick glass beads at the top and bottom to ensure uniform water inflow and to prevent leakage of SA/NZVI-rGO beads from the bottom of the column, with the effluent flowing through the column in an upward direction. The effluent was obtained from the sampling port at regular intervals and sample collection was stopped when effluent concentration was almost equivalent to influent concentration. The effects of different flow rates (0.7, 2, 4 mL/min), bed heights (4.5, 6.5, 8.5 cm), and influent concentrations of Cr(VI) (10, 30, 40 mg/L) at a temperature of 25 °C and pH 3 of the Cr(VI) solution were considered. The breakthrough curve was obtained by plotting C_t/C_0 versus time.

3. Results and Discussion

3.1. Characterization of SA/NZVI-rGO Beads

The digital photographs of SA/NZVI beads, SA/NZVI-rGO beads, SA beads, and freeze-dried SA/NZVI-rGO beads, respectively, were shown in Figures 1 and S1. All the beads showed a sphere-like shape with a diameter of about 3 mm and the color ranged from transparent (SA beads) to black (SA/NZVI beads and SA/NZVI-rGO beads). The encapsulated structure facilitates an easy separation of the gel beads from the treated wastewater. A certain mass of SA/NZVI-rGO beads was freeze-dried and then weighed, and the water content was 96.61%, as calculated by Equation (S2). In addition, SA/NZVI beads' color turned from black to brownish red due to the rapid oxidation of the highly reactive NZVI in a short period (Figure 1a,c). However, SA/NZVI-rGO beads did not change their color after 22 h of Cr(VI) removal (Figure 1d), and the color of the SA/NZVI-rGO beads remained unchanged after 226 days of sealed storage (Figure S1b), which indicates that the mere SA coating of NZVI did not reduce the oxidation rate of NZVI, but the pre-loading of NZVI on rGO can reduce the oxidation rate of SA/NZVI-rGO beads, which is beneficial for preservation in practical applications. And previous studies have also shown that NZVI can be coupled with carbon materials to build galvanic cells, which could expedite the electron release of NZVI and prevent the formation of oxidation films [30,31].

Figure 1. Photographs of just-prepared (**a**) SA/NZVI beads and (**b**) SA/NZVI-rGO beads, of (**c**) SA/NZVI beads exposed to oxygen for 2 h, of (**d**) SA/NZVI-rGO beads after removal of Cr(VI) for 22 h.

The SEM images in Figure 2 and Figure S2 show the morphology and structure of NZVI, NZVI-rGO, and SA/NZVI-rGO. Figure S2 reveals that massive amounts of NZVI particles are tightly connected to form large chain-like aggregates with poor dispersion. Figure 2a reveals the NZVI particles loaded on rGO nanosheets with folded lamellar structures are well dispersed. Our previous studies have demonstrated that well-dispersed NZVI could improve antioxidant properties [32]. These observations imply that NZVI with enhanced dispersion by rGO will contribute to increased reactive sites for the removal of contaminants. Figure 2b shows that SA/NZVI-rGO beads have a continuous three-dimensional staggered fibrous structure, and NZVI is well dispersed in beads, which indicates that SA/NZVI-rGO beads are more abundant in folds and channels than NZVI-rGO and greatly enhance the adsorption capacity of heavy metal contaminants.

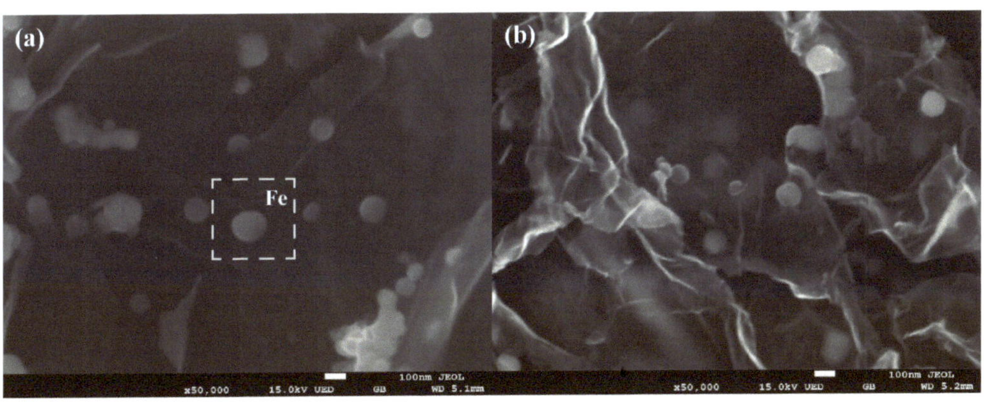

Figure 2. SEM of (**a**) NZVI-rGO and (**b**) SA/ NZVI-rGO beads.

FTIR spectra of SA beads, NZVI-rGO powder, and SA/ NZVI-rGO beads before and after Cr(VI) removal are compared in Figure 3. For SA beads, the wide band around 3400 cm^{-1} represented the telescoping vibrations of O-H, C-H at 2922 cm^{-1}, O=C-O at 1430 cm^{-1}, C=O at 1625 cm^{-1}, and O-H at 1029 cm^{-1} represent the typical structure of alginate [33]. In NZVI-rGO, there is also a strong and wide peak near 3400 cm^{-1} associated with the telescoping vibration of the O-H group, with representative absorption peaks at 1648, 1398, 1128, and 1029 cm^{-1} ascribed to the deformation vibration peak of aromatic C=C, carboxy O=C-O, epoxy C-O, and hydroxyl O-H, respectively [34–36]. The weak peak at 692 cm^{-1} may be the Fe-O telescoping vibration belonging to Fe_3O_4 nanoparticles [37], which is consistent with the XPS results. Additionally, the spectra of NZVI-rGO and SA/NZVI-rGO remain almost identical, which indicates that SA successfully encapsulates NZVI-rGO. Specifically, the O=C-O of the SA/NZVI-rGO spectrum shifted significantly from 1398 cm^{-1} to 1428 cm^{-1} compared with the NZVI-rGO spectrum, indicating that some interfacial hydrogen bonds were formed between rGO and SA, which could cause a tight adhesion between the SA and rGO surfaces [38]. Meanwhile, the carbon atoms in rGO are bonded with sp^2 hybrid orbitals, and the remaining electrons in the p orbitals can form delocalized π bonds. The O-H on the SA can contribute electrons to the free-moving π-electrons, thereby enhancing the π-electron cloud densities of the conjugated systems, which in turn promotes the tight adhesion of rGO on the SA [39]. These results indicate that NZVI-rGO can be uniformly and stably embedded in SA. The FTIR spectra of SA/NZVI-rGO before and after the removal of Cr(VI) (Figure 3) were utilized to identify the functional groups participating in the Cr(VI) removal. After Cr(VI) removal, a broad peak of O-H at 3400 cm^{-1} was observed to be narrowed, suggesting that O-H participated in the removal of Cr(VI). Also, an enhancement of the vibrational peaks of carboxyl groups at 1428 cm^{-1} (O=C-O) and 1618 cm^{-1} (C=O) was noted, which is congruous with prior studies that hydroxyl groups could serve as an electron donator to reduce Cr(VI) by being oxidized to carboxyl groups [40,41].

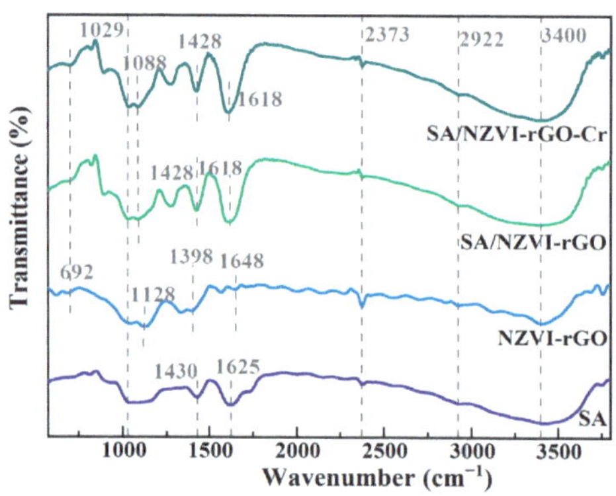

Figure 3. FTIR spectrums of SA, NZVI-rGO, and SA/NZVI-rGO before and after Cr(VI) removal.

XPS spectra identified the elemental formation and chemical valence of the SA/NZVI-rGO beads before and after the removal of Cr(VI). Previous research has demonstrated NZVI has a kernel-shell system with an Fe(0) kernel and an iron oxide shell [42]. As shown in the Fe 2p spectra of SA/NZVI-rGO beads (Figure 4b), Fe_2O_3 at 711.95 eV (Fe 2p3/2) and 725.52 eV (Fe 2p1/2), Fe_3O_4 at 710.31 eV (Fe 2p3/2) and 723.11 eV (Fe 2p1/2), and Fe(0) at 705.79 eV (Fe 2p3/2), which proves the successful loading of NZVI particles on rGO nanosheets [43]. In Figure 4b, after Cr(VI) removal, the peak area of Fe(III) was elevated by 21.86% and that of Fe(II) was lessened by 2.22%, and Fe(0) disappeared, which confirmed that oxidative conversion of Fe(0) and Fe(II) to Fe(III) during Cr(VI) reduction by SA/NZVI-rGO gel beads. From the O1s spectrum (Figure 4c), the peaks observed at 531.06 eV, 529.51 eV, and 529.43 eV are attributed to O-H, O=C/O=C-O, and Fe-O, separately [44]. The mol ratio of Fe-O went up from 15.49% to 19.75% after the removal of Cr(VI), further demonstrating that the Fe(0) that provided electrons to Cr(VI) was being oxidized. Additionally, the mol ratio of O-H lessened from 66.43% to 52.84% and the molar ratio of O=C/O=C-O increased from 20.07% to 27.41%, which indicates that the hydroxyl groups donated electrons and were oxidized to carboxyl groups for the process of reducing Cr(VI) [45,46], which is consistent with the FTIR results. Cr 2p binding energy peak of Cr atoms was observed at 578.08 eV for the SA/NZVI-rGO beads after the removal in Figure 4a, and as shown in the Cr 2p split-peak spectrum (Figure 4d), Cr(III) corresponded at 576.03 eV (Cr 2p3/2) and 585.39 eV (Cr 2p1/2), while Cr(VI) at 578.43 eV (Cr 2p3/2) and 587.79 eV (Cr 2p1/2), indicating that both Cr(VI) and Cr(III) being reduced were adsorbed on gel beads. The results show that Cr(III) accounts for 85.69% of the total Cr on SA/NZVI-rGO, which means that the reduction is the primary cause for Cr(VI) removal. We further found that the intensity of the Ca 2p peak decreased (Figure S3), which implies that part of the Cr(III) ions possibly displaces Ca^{2+} to form carboxylated metal complexes within the gel beads instead of releasing them directly into the aqueous solution, which is consistent with the previous findings [47].

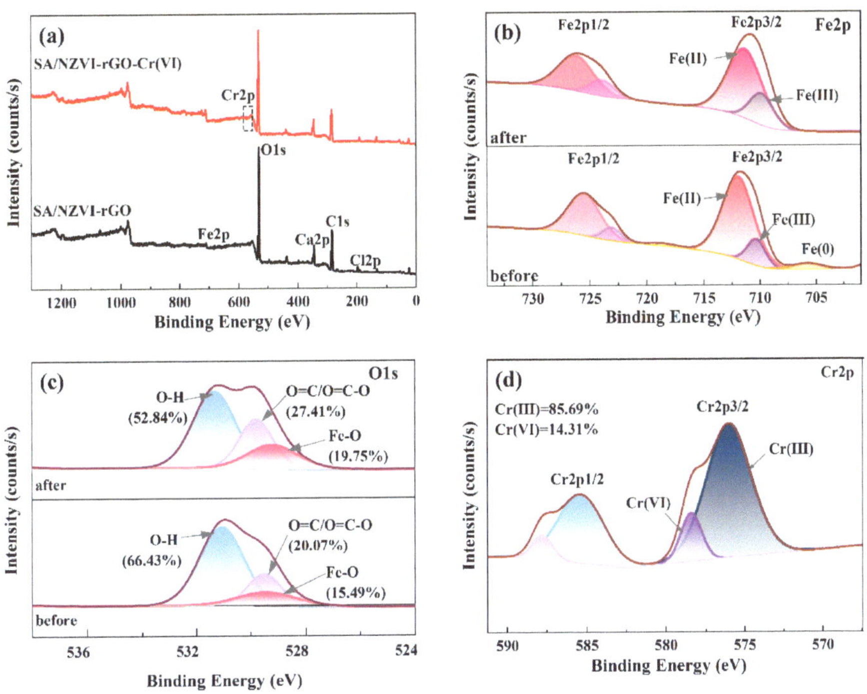

Figure 4. (**a**) Wide XPS spectrums, (**b**) Fe 2p and (**c**) O 1s XPS spectrums of SA/NZVI-rGO beads before and after Cr(VI) removal, and (**d**) Cr 2p XPS spectra of SA/NZVI-rGO beads after removal.

The zeta potential is applied to analyze the electrostatic characteristics of the material. To further explore the removal mechanism, Figure 5a demonstrates the zeta potential of SA/NZVI-rGO at different pHs. It is observed that SA/NZVI-rGO is positively charged at a low pH and has a strong negative charge when the pH was higher. At a low pH, Cr(VI) exists primarily as $HCrO_4^-$ [43], and as the pH < zero point charge (pHzpc), the surface of SA/NZVI-rGO is electropositive because of the carboxylate protonation, thus benefiting SA/NZVI-rGO in an acidic environment for Cr(VI) retention. However, when the pH > pHzpc, the gel beads are electronegative on account of the deprotonation of functional groups, which causes an electrostatic repulsion against the negatively charged $Cr_2O_4^{2-}$, thereby being unfavorable to the adsorption. These prove that electrostatic interactions have a major role in Cr(VI) adsorption on SA/NZVI-rGO.

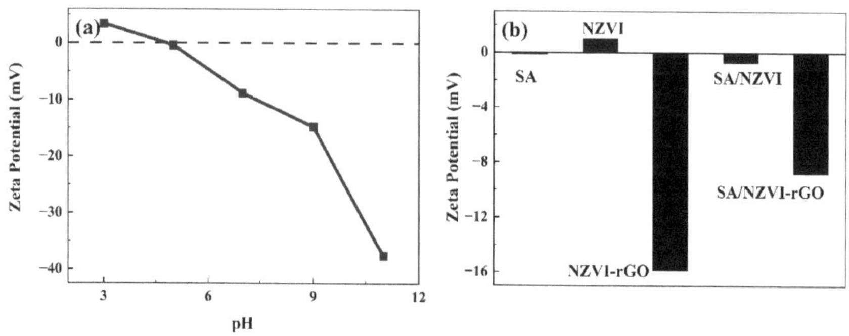

Figure 5. (**a**) Zeta potential of SA/NZVI-rGO gel under different pH. (**b**) Zeta potential of SA gel, NZVI powder, NZVI-rGO powder, SA/NZVI gel, and SA/ NZVI-rGO gel at pH = 7.

To investigate the effect of the surface charge on the stability of gels, Figure 5b shows the zeta potential values of SA gel, NZVI powder, NZVI-rGO powder, SA/NZVI gel, and SA/ NZVI-rGO gel. The zeta potential of SA gel is −0.16 mV, the NZVI powder is 1 mV, and the NZVI-rGO powder is −15.90 mV; therefore, the NZVI-rGO powder contains more charge than the first two. And SA/ NZVI-rGO gel has a higher zeta potential value than SA/NZVI gel, probably because the stronger electrostatic repulsion between SA and NZVI-rGO makes the gel enhanced and forms a more stable gel [29].

3.2. Swelling and Mechanical Properties

For the mechanical tests, all the gels were made into rectangles, as shown in Figure S4. The picture of the gel before, during, and after compression show that the gel is elastic. The stress–strain curves and Young's modulus of gels with different rGO contents are shown in Figure 6a and Figure 6b, respectively. For SA/NZVI-rGO gels, no yielding was exhibited in any of the stress–strain curves, indicating neither fracture strain nor fracture strength is generated during compression [48]. The gels with a higher rGO content exhibited a higher Young's modulus, indicating better compressive properties. For SA/NZVI gels without rGO addition, yielding occurred at a strain of 19.05% (Figure 6a), when the stress was 10 KPa, indicating that the addition of rGO and the higher rGO content resulted in better mechanical properties of the gels. Other studies have also found improved mechanical properties after adding rGO to alginate-based gels [29,49]. In Figure 6c, different gel beads were immersed in deionized water at different pHs, and the experimental procedure revealed that the SA/5NZVI-rGO gel beads in deionized water at pH = 13 were the first to dissolve and break compared to other pHs. At the same pH, the lower the rGO content, the higher the swelling ratio, which is consistent with the change in the decrease in the mechanical strength, further demonstrating that the swelling behavior decreases the mechanical strength. The reason for this phenomenon may be that rGO nanosheets dispersed uniformly in SA can restrict the movement of alginate chains [50]. Meanwhile, FTIR demonstrated that the interfacial hydrogen bonding and conjugated electron interaction between rGO and SA could enhance the compact structure within the gel beads, thus hindering and weakening the diffusion of water molecules within the gel beads. Thus, SA/NZVI-rGO gel beads, which improve the mechanical strength by adding the appropriate amount of rGO and maintain excellent anti-swelling properties in a broad pH range, are ideal for practical applications in the treatment of Cr(VI) contaminated wastewater.

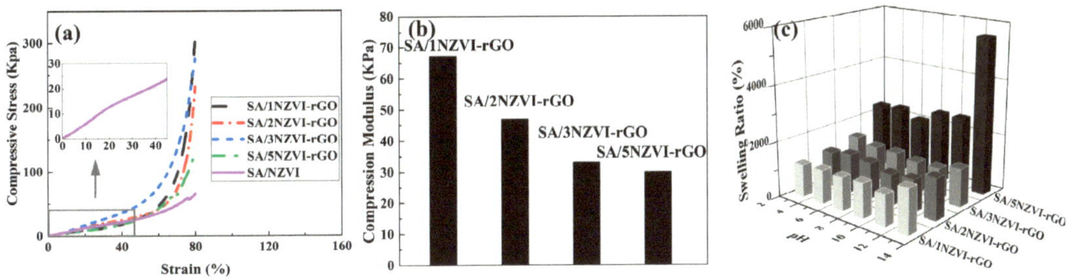

Figure 6. (**a**) Compressive stress–strain curves and (**b**) Young's modulus of gels with different rGO contents and (**c**) swelling ratio of SA/xNZVI-rGO gels at different pH.

3.3. Bath Experiments

3.3.1. Effect of Different Bead Components

The prepared SA/NZVI-rGO gel beads showed a strong performance in removing Cr(VI), which is shown in Figure 7a. Cr(VI) at an initial concentration of 10 mg/L was removed with 0.5 g of SA/NZVI-rGO gel beads. The Cr(VI) removal reached equilibrium at the 80th minute, in which 98% of Cr(VI) was removed. To compare the contribution of the three components (SA, NZVI, and rGO) within the beads to the Cr(VI) removal, SA beads, SA/NZVI beads, and SA/rGO beads were made, respectively. As shown in Figure 7a, SA beads removed only about 1% of Cr(VI), SA/rGO beads improved the removal efficiency by about 14% over SA beads, whereas SA/NZVI beads improved the removal of Cr(VI) by about 95% over SA beads and by about 81% over SA/rGO beads. The SA/NZVI beads showed a relatively lower removal efficiency but a relatively higher removal rate than the SA/NZVI-rGO beads. In the gel beads, the NZVI acts as the active component to reduce Cr(VI) to Cr(III), and its high reactivity leads to a quicker and larger removal of Cr(VI). Nevertheless, compared to SA/NZVI-rGO gel beads, SA/NZVI gel beads are more prone to oxidation and show poor mechanical properties, which would limit the practical application of SA/NZVI beads. Therefore, we chose a SA/NZVI-rGO gel bead with oxidation resistance, enhanced mechanical properties, and effective Cr(VI) removal as the remediation agent.

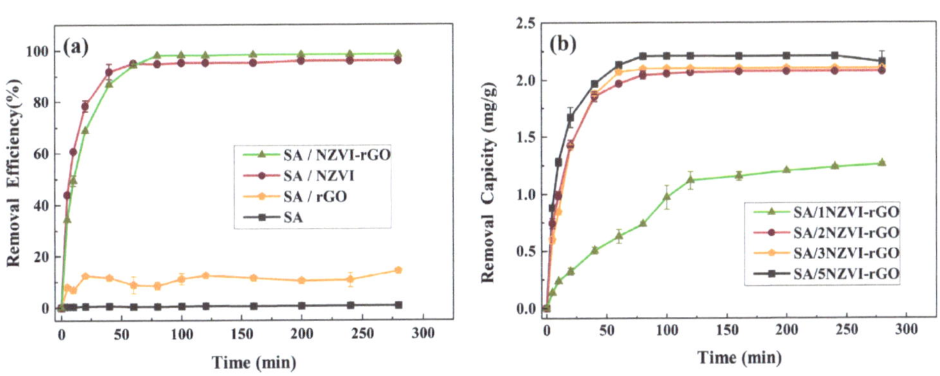

Figure 7. The effects of (**a**) different beads components and (**b**) different dosing ratios of NZVI and rGO on the Cr(VI) removal (initial concentration 10 mg/L, pH 3, dosage 0.5 g, 298 K).

3.3.2. Effect of Different Dosing Ratios of NZVI and rGO

NZVI, which has strong reducibility, is the active ingredient in the reduction of Cr(VI) to Cr(III), and rGO is the component that can improve the mechanical properties of beads. To consider the influence of the dose ratio of NZVI and rGO on the removal capacity, 0.5 g SA/xNZVI-rGO gel beads were used to remove the Cr(VI) solution (100 mL, 10 mg/L). The removal capacity of the SA/NZVI-rGO beads for Cr(VI) increased with the dosage ratio of NZVI and rGO increasing (Figure 7b). The removal capacity of SA/2NZVI-rGO beads for Cr(VI) (2.07 mg/g) was almost twice that of SA/1NZVI-rGO beads (1.24 mg/g), which may be accounted for the more available active sites. However, as the dosing ratios of NZVI and rGO were further increased, SA/3NZVI-rGO beads and SA/5NZVI-rGO beads no longer showed a significant increase in the Cr(VI) removal capacity compared with SA/2NZVI-rGO beads, and the reaction rates of the three beads were similar in the early stage of Cr(VI) removal. The cause may be that the excess NZVI particles are severely agglomerated during the loading of rGO, thereby reducing the active sites, resulting in a no longer evident improvement in the removal capacity of Cr(VI) [51]. Also, considering the effect of rGO dosage on the mechanical properties and swelling ratio of the gel beads, SA/2NZVI-rGO was finally selected as the gel beads used in the subsequent experiments. The SA/2NZVI-rGO gel beads were used for reusability experiments, details and results are available in Text S3.

3.3.3. Effect of Initial Solution pH

The different initial pH of solution also influences the Cr(VI) removal. As demonstrated in Figure 8a, with an initial pH of 3, the removal efficiency of Cr(VI) at the 80th minute could reach over 98%. As the pH increases to 5 and 7, the removal of Cr(VI) dropped to 91.74% and 89.67%, separately. When the solution was alkaline, the removal efficiency of Cr(VI) further dropped to 88.41% (pH 9) and 82.09% (pH 11). The decrease in the Cr(VI) removal efficiency with an increasing pH is mainly related to the charge on the gel beads and the species of Cr(VI) ions. On the one hand, SA/NZVI-rGO gel beads are susceptible to protonation in acidic solutions and have a strong electrostatic attraction to Cr(VI) anions. This can be confirmed by zeta potential of SA/NZVI-rGO gel beads at a different pH (Figure 5a). On the other hand, a different pH affects the Cr(VI) species, which influence the adsorption. The Cr(VI) species at different pH values were analyzed using Visual MINTEQ software (Figure 8b). At a low pH range, Cr(VI) was primarily in the form of $HCrO_4^-$ and when the pH increases it mainly appears as CrO_4^{2-}. And the reduction of Cr(VI) to Cr(III) is dependent on H^+ ions in solution, as shown in Equations (1) and (2). $HCrO_4^-$ has less sorption-free energy compared to CrO_4^{2-}, which facilitates the adsorption [52,53].

$$HCrO_4^- + 7H^+ + 3e^- \rightarrow Cr^{3+} + 4H_2O \quad (1)$$

$$CrO_4^{2-} + 8H^+ + 3e^- \rightarrow Cr^{3+} + 4H_2O \quad (2)$$

However, the removal of Cr(VI) in the present study was almost pH-independent (even at pH = 11, Cr(VI) removal reached over 80%), which may be explained by the synergy of adsorption and reduction. Specifically, the reduction of Cr(VI) to Cr(III) is accompanied by the consumption of protons; however, it is noteworthy that with the increase in Cr(III), the cation exchange between Cr(III) and Ca(II) occurs on the three-dimensional structure of alginate, which is presented in the XPS results (Figure S3), and Ca(II) is re-released into environment, resulting in a localized pH decrease (hydrolysis process), which would again facilitate the adsorption and reduction. The minimum removal efficiency of SA/NZVI-rGO gel beads for Cr(VI) at a broad pH range was compared with other NZVI-based materials, and the results are presented in Table S1. The results show that the SA/NZVI-rGO gel beads can remove Cr(VI) efficiently in a broad pH range.

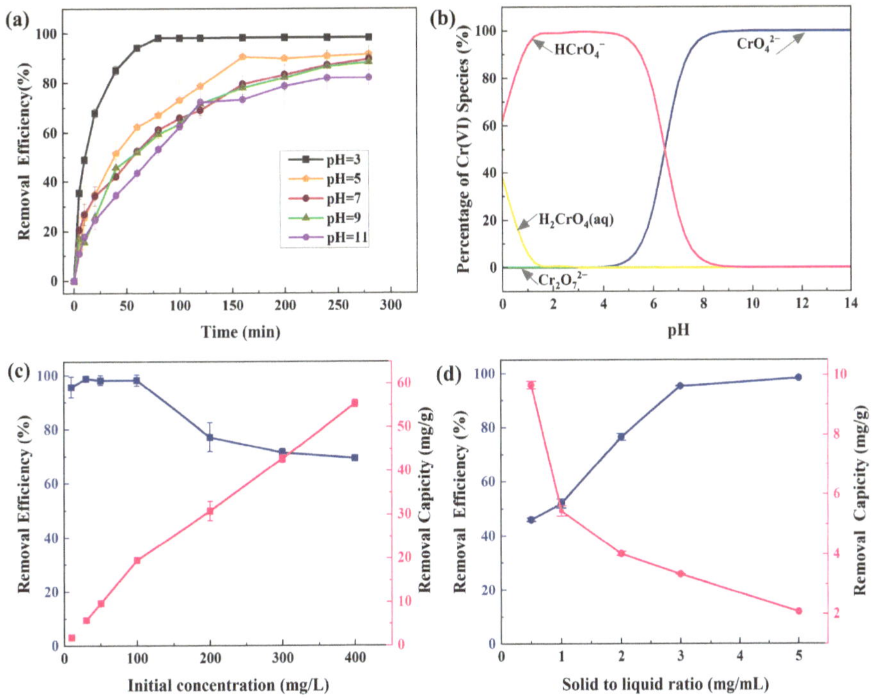

Figure 8. The effects of (**a**) different pH (initial concentration 10 mg/L, dosage 0.5 g, 298 K), (**c**) different initial concentrations (dosage 0.5 g, pH 3, 298 K), and (**d**) different solid-to-liquid ratios (initial concentration 10 mg/L, pH 3, 298 K) on the Cr(VI) removal. And (**b**) species distribution of Cr(VI) at different pHs.

3.3.4. Effect of Initial Cr(VI) Concentration

Figure 8c shows the removal capacity and removal efficiency of Cr(VI) by SA/NZVI-rGO beads under different initial concentrations (10 mg/L–400 mg/L). The removal capacity of the SA/NZVI-rGO beads exhibited an increasing tendency with an increasing initial concentration, and the removal capacity of SA/NZVI-rGO beads reached a maximum of 55.42 mg/g at the initial concentration of Cr(VI) of 400 mg/L. When the removal reached equilibrium, the removal efficiency of Cr(VI) reached about 99% (at the initial concentration of 10 mg/L–100 mg/L), but as the initial concentration was further increased, the removal efficiency of Cr(VI) showed a decreasing trend, and the removal efficiency dropped to 70% at the initial concentration of 400 mg/L. When the concentration of Cr(VI) is high, a Fe-Cr hydroxide layer forms on gel beads to surround NZVI, thus preventing the reduction of Cr(VI) by electron transfer from NZVI [54].

3.3.5. Effect of Solid-to-Liquid Ratio

The solid-to-liquid ratio between gel beads and Cr(VI) waste solution is also a key factor affecting the removal of Cr(VI). The Cr(VI) (100 mL) removal increased continuously (from 45% to 98%) with an increasing solid-to-liquid ratio (from 0.5 mg/mL to 5 mg/mL), while the removal capacity showed the opposite trend, decreasing from 9.57 mg/g to 2.07 mg/g, as shown in Figure 8d. It is attributed to a large amount of Cr(VI) competing for the available sites of gel beads at low dosing levels, resulting in a low removal efficiency but high removal capacity. As the dosage of SA/NZVI-rGO beads was appropriately increasing (from 0.5 mg/mL to 3 mg/mL), the increase in the Cr(VI) removal efficiency showed a rapid upward trend; however, as the dosage of gel beads was further increased (from

3 mg/mL to 5 mg/mL), the removal efficiency of Cr(VI) no longer increased significantly. The reason for this is that the higher the solid–liquid ratio, the more overlapping active sites there are [55].

3.3.6. Cr(VI) Removal from Real Wastewater

The performance of the SA/NZVI-rGO gel beads for Cr(VI) removal from real wastewater was investigated. As shown in Figure 9, under the same operating conditions, when the removal reached equilibrium, the removal efficiency of the simulated Cr(VI)-containing wastewater reached 91.5%, while the removal efficiency of Cr(VI) in the actual wastewater reached 70.25%. As shown in Table S2, the coexisting ions in real wastewater are usually quite complex. After the removal experiments, Ca^{2+} and Na^{+} in the wastewater increased due to the ion exchange of SA/NZVI-rGO gel beads, and most of the remaining anion and cation concentrations decreased. This proves that substantial coexisting anions and cations in the actual Cr(VI)-containing wastewater will vie with Cr(VI) ions for available sorption sites, leading to a decreased Cr(VI) removal. However, despite the complexity of the actual wastewater environment, the SA/NZVI-rGO beads still showed a nice Cr(VI) removal efficiency.

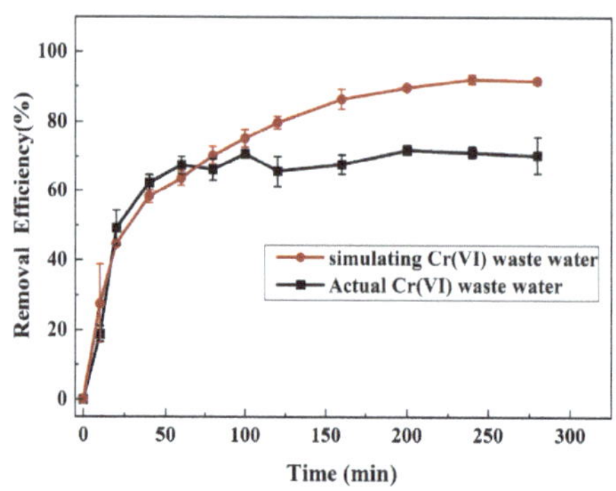

Figure 9. The removal effect of SA/NZVI-rGO beads on real and simulated wastewater with equal Cr(VI) concentration (initial concentration 13.75 mg/L, 100 mL, pH 5.58, dosage 0.4 g, 298 K).

3.4. Kinetics Study of the Reaction

Based on the above results, the Cr(VI) removal by the SA/NZVI-rGO gel beads involves both adsorption and reduction. The adsorption kinetics models and reduction kinetics models were fitted to the experimental data for clarifying detailed kinetics of Cr(VI) removal.

3.4.1. Adsorption Kinetics

Pseudo-first-order adsorption kinetics model and pseudo-second-order adsorption kinetics model are usually fitted for adsorption kinetics [56]. The pseudo-first-order adsorption kinetics equation and pseudo-second-order adsorption kinetics equation are below, respectively:

$$Q_t = Q_e(1 - e^{-k_1 t}) \tag{3}$$

$$Q_t = \frac{k_2 Q_e^2 t}{1 + k_2 Q_e t} \tag{4}$$

where Q_e is the adsorption amount at equilibrium (mg/g); Q_t is the adsorption amount at time t (mg/g); k_1 is the rate constant of the pseudo-first-order adsorption kinetics model; k_2 is the rate constant of the pseudo-second-order adsorption kinetics model.

According to the experimental data of Cr(VI) removal by the SA/xNZVI-rGO beads, the fitting results of the adsorption kinetics are presented in Table 1. As a result, correlation coefficients (R^2) of pseudo-first-order adsorption kinetics are larger than that of pseudo-second-order adsorption kinetics, and the fitted Q_e is closer to the measured q_e, indicating that the Cr(VI) adsorption by SA/NZVI-rGO gel beads is more consistent with the pseudo-first-order adsorption kinetics, which demonstrates that physical sorption is the main rate control step in adsorption process of SA/NZVI-rGO gel beads [57].

Table 1. Pseudo-first-order/pseudo-second-order adsorption kinetics parameters for Cr(VI) adsorption.

	Parameter	SA/1NZVI-rGO	SA/2NZVI-rGO	SA/3NZVI-rGO	SA/5NZVI-rGO
	q_e	1.243	2.073	2.103	2.219
pseudo-first-order	k_1 (min^{-1})	0.012	0.066	0.056	0.080
	Q_e (mg g^{-1})	1.299	2.057	2.110	2.188
	R^2	0.986	0.994	0.998	0.988
pseudo-second-order	k_2 ($g^{-1}mg^{-1}min^{-1}$)	0.007	0.043	0.034	0.051
	Q_e (mg g^{-1})	1.692	2.23	2.31	2.348
	R^2	0.981	0.984	0.965	0.987

3.4.2. Reduction Kinetics

According to the experiments on the Cr(VI) removal by the SA/xNZVI-rGO beads, using the pseudo-first-order reduction kinetics model and the pseudo-second-order reduction kinetics model to fit the above experimental data [58]. The models of pseudo-first-order reduction kinetics and pseudo-second-order reduction kinetics can be described below:

$$\ln \frac{C_t}{C_0} = -k_{obs1}t \tag{5}$$

$$\ln\left(\frac{1}{C_t} - \frac{1}{C_0}\right) = k_{obs2}t \tag{6}$$

where C_0 is the initial contaminant concentration (mg/L); C_t is the contaminant concentration at time t (mg/L); k_{obs1} is the pseudo-first-order rate constant (min^{-1}); k_{obs2} is the pseudo-second-order rate constant (L/mg/min).

The pseudo-first-order/pseudo-second-order reduction kinetics parameters are listed in Table 2. The R^2 of the pseudo-first-order kinetics fits ranged from 0.937 to 0.992; however, the R^2 of pseudo-second-order kinetics fits were all less than 0.9, indicating that the Cr(VI) reduction by SA/NZVI-rGO gel beads is consistent with the pseudo-first-order reduction kinetics. It is in agreement with the statement that the pseudo-first-order kinetics model can correctly characterize the kinetics of NZVI particles and modified NZVI [59,60]. Additionally, when the ratio of NZVI to rGO was increased to 2:1, an abrupt increase in k_{obs1} was observed, indicating that SA/2NZVI-rGO has fast reduction kinetics. A comparison of k_1 for the pseudo-first-order adsorption kinetics (Table 1) and k_{obs1} for the pseudo-first-order reduction kinetics (Table 2) revealed that adsorption served as the velocity-limiting step since the reduction process is far faster than that of adsorption [58].

Table 2. Pseudo-first-order/pseudo-second-order reduction kinetics parameters for Cr(VI) reduction.

	Pseudo-First-Order		Pseudo-Second-Order	
	k_{obs1} ($10^{-2}min^{-1}$)	R^2	k_{obs2} (10^{-2} L/mg/min)	R^2
SA/1NZVI-rGO	0.385	0.937	0.979	0.773
SA/2NZVI-rGO	5.829	0.992	1.531	0.615
SA/3NZVI-rGO	4.351	0.969	0.985	0.490
SA/5NZVI-rGO	6.792	0.984	1.341	0.575

3.5. Adsorption Isotherm Analysis

Further analysis of the adsorption mechanism was based on different initial concentration experiments. The experimental data were fitted using the Langmuir adsorption isotherm model and the Freundlich adsorption isotherm model. The Langmuir isotherm model hypothesizes that adsorption takes place on a monolayer where all adsorption sites are of equal energy, and the adsorption capacity is dependent on the number of adsorption sites on the adsorbent [61]. The linear form of Langmuir model could be written as follows:

$$\frac{C_e}{Q_e} = \frac{1}{k_L Q_m} + \frac{C_e}{Q_m} \tag{7}$$

where C_e (mg/L) and Q_e (mg/g) are the equilibrium concentration and equilibrium adsorption amount of Cr(VI) after removal; k_L (L/mg) is a constant correlated with the adsorption energy; Q_m (mg/g) is theoretical maximum adsorption amount.

Langmuir separation factor (R_L) can reflect basic features of Langmuir isotherm, and R_L determines whether the adsorption process is advantageous or not, and the equation is as follows:

$$R_L = \frac{1}{1 + k_L C_0} \tag{8}$$

where C_0 is initial concentration of adsorbent.

Freundlich isotherm model is empirical. It assumes that adsorbent has a non-homogeneous sorption surface with different binding forces at each adsorption site, and is multilayer adsorption [62]. The linear equation for the Langmuir model is represented below:

$$\ln Q_e = \ln k_F + \frac{\ln C_e}{n} \tag{9}$$

where k_F is a constant relevant to the capacity of the adsorbent; n is a constant that responds to the sorption strength of the adsorbent.

The fitted isothermal adsorption parameters from the experimental data of different initial concentrations are shown in Table 3. The relevant coefficient of the Langmuir isothermal model (R^2 = 0.968) is larger than that of the Freundlich isothermal model (R^2 = 0.924), so the Langmuir isothermal adsorption model provides a more favorable fitting of the measured data. The R_L was between 0 and 1 for this experiment, indicating that adsorption is easy to perform [63]. Furthermore, the maximum theorized capacity for Cr(VI) removal by SA/NZVI-rGO gel beads using the Langmuir isothermal adsorption model is 53.42 mg/g, which is approaching the measured maximum sorption capacity (55.42 mg/g). The removal capacity of SA/NZVI-rGO gel beads for Cr(VI) was contrasted with that of other NZVI-based materials, and the results are shown in Table S3. It can be observed that the SA/NZVI-rGO gel beads have a comparable removal capacity. Although this capacity does not remarkably increase to other published NZVI-based materials, the amount of NZVI used in this study is relatively low (per gram of SA/NZVI-rGO beads contains only 0.14 g NZVI).

Table 3. Langmuir and Freundlich isotherms parameters of Cr(VI) removal by SA/NZVI-rGO gel beads.

Langmuir Model				Freundlich Model		
k_L	Q_m	R^2	R_L	k_F	n	R^2
0.111	53.42	0.968	0.474–0.022	6.932	2.238	0.924

3.6. Removal Mechanism

Reasonable removal mechanisms, including reduction, electrostatic interaction, and ion exchange, are proposed in Figure 10. The Cr 2p XPS spectra and kinetic results confirmed Cr(VI) removal by gel beads was achieved by adsorption and reduction, which was the main rate control step. Specifically, the high water content (96.61%) of the SA/NZVI-rGO gel beads promoted the penetration of the Cr(VI) solution, and the abundant folds and channels within the gel beads shown by SEM favored the attachment of Cr(VI). Under different pHs, the zeta potential of SA/NZVI-rGO beads (Figure 5a) and Cr(VI) removal experiments (Figure 8a) showed that the electrostatic attraction between positively charged SA/NZVI-rGO gel beads and negatively charged Cr(VI) favored the adsorption of Cr(VI) by SA/NZVI-rGO in acidic environments. Furthermore, the XPS spectra of SA/NZVI-rGO before and after Cr(VI) removal demonstrated that Fe(0) was oxidized to Fe(II) and Fe(III), as well Cr(VI) was reduced to Cr(III) through obtaining electrons given by Fe(0), and Cr(III) reacts with OH^- in an aqueous solution to form $Cr(OH)_3$ deposited on the gel beads. The presence of rGO led to a faster reduction of Cr(VI) to Cr(III) due to the formation of massive primary batteries by NZVI and rGO to accelerate the release of electrons [64,65]. Then, Fe(II), which is the result of the oxidation of Fe(0) losing electrons, can further reduce Cr(VI) to Cr(III), and Fe(II) was eventually oxidized to Fe(III). Additionally, FTIR (Figure 3) and O 1s XPS spectra (Figure 4c) of SA/NZVI-rGO before and after Cr(VI) removal demonstrate the hydroxyl group is an electron donor that can reduce Cr(VI) to Cr(III) and then to be oxidized to carboxyl groups. It is interesting to note that in the alginate matrix, Cr(III) can undergo an ion exchange with -COO...Ca...OOC- (Figure S3), thus releasing Ca(II) to further decrease the solution's pH. This means that more H^+ is available in the aqueous solution for electrostatic adsorption due to the protonation of functional groups and for reduction, which may indirectly improve the removal [66]. Thus, the synergistic combination of the proposed removal mechanisms can provide the SA/NZVI-rGO gel beads with a great removal capacity for the effective Cr(VI) removal from aqueous solutions.

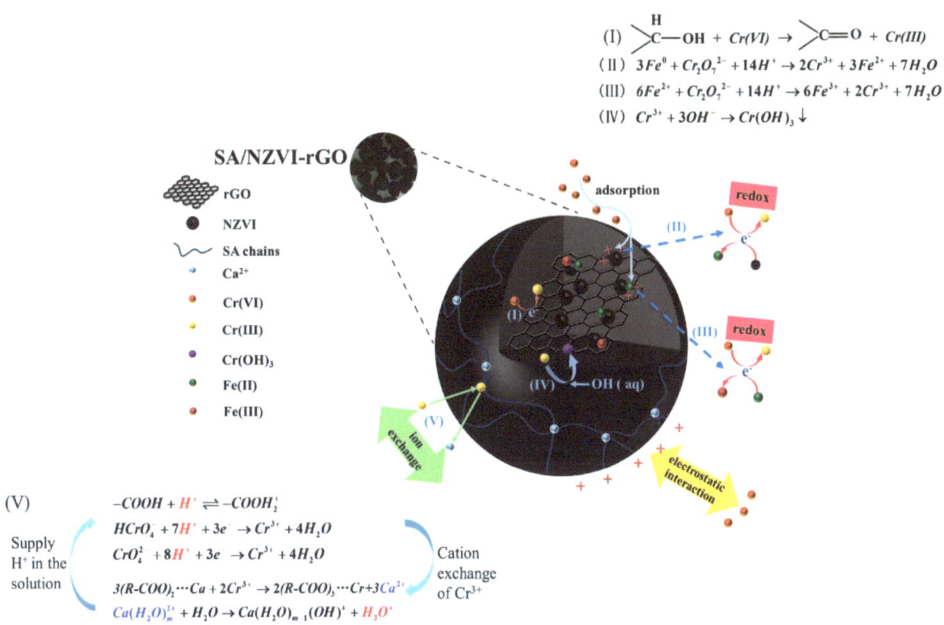

Figure 10. Schematic diagram of the mechanism of Cr(VI) removal by SA/NZVI-rGO.

3.7. Column Experiments and Model Fitting

Column experiments were applied to research the continuous Cr(VI) removal by SA/NZVI-rGO beads. Theoretical column models are available to simulate the breakthrough performance of columns packed with SA/NZVI-rGO beads, and the Thomas, Adams—Bohart, and Yoon—Nelson models were fitted and analyzed to the measured data, and the above models are described in detail in Text S4. As can be seen from Figure 11a,d,g, when the bed height rose from 4.5 cm to 8.5 cm (flow rate 0.7 mL/min, influent concentration 10 mg/L), the breakthrough time and removal performance increased due to the increase in the adsorption binding sites and the prolonged exposure time of Cr(VI) to the gel beads, resulting in an improved removal efficiency of Cr(VI) [47]. As a result, the higher filling of the column would slowly decrease the removal performance of the column, which is ideal for operation, and a similar finding has been published in the literature [67]. When the influent concentration of Cr(VI) was increased from 10 mg/mL to 40 mg/mL (flow rate 0.7 mL/min, bed height 8.5 cm) (Figure 11b,e,h), the breakthrough time and depletion time of gel beads bed will be advanced because the binding sites would saturate more quickly at higher concentrations [68]. When the flow rate was accelerated from 0.7 mL/min to 4.0 mL/min (bed height 4.5 cm, influent concentration 10 mg/L) (Figure 11c,f,i), both the exhaustion time and removal performance decreased, and the faster flow rate reduced the contact time of Cr(VI) with SA/NZVI-rGO beads, resulting in a steeper penetration curve. This is because a too fast flow rate increases the mass transfer rate, which leads to the increase in the Cr(VI) adsorbed per unit bed height and faster saturation of the bed column, which is consistent with the literature reports [69,70]. Furthermore, due to the excellent mechanical properties and enhanced swelling resistance of the SA/NZVI-rGO gel beads, the gel beads in fixed-bed column keep their integrity after continuous removal. NZVI-rGO confinement in SA reduces nanomaterial contamination due to uncontrolled migration of nanoparticles. The SA/NZVI-rGO gel beads are available for treatment of actual Cr(VI) wastewater.

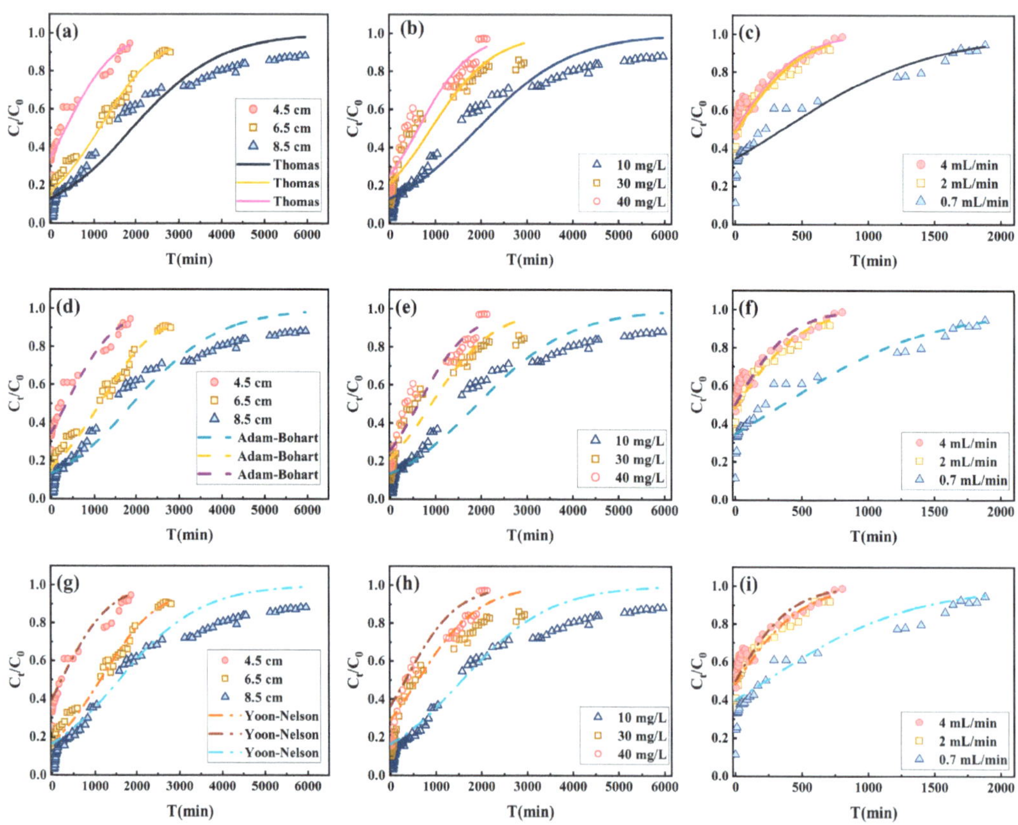

Figure 11. Effect of (**a**) bed height, (**b**) influent concentration, and (**c**) flow rate on the breakthrough curve of continuous Cr(VI) removal by SA/NZVI-rGO. The lines represent the curves fitted with (**a**–**c**) Thomas, (**d**–**f**) Adam–Bohart, and (**g**–**i**) Yoon–Nelson models.

As presented in Figure 11, these models were fitted to the measured data from column experiments under different conditions. The parameters of each model were acquired from the nonlinear form of the model equation and summarized in Tables S4–S6. When the bed height increased, the flow rate increased and the influent concentration decreased, respectively, K_T, K_{AB}, and K_{YN} increased. The well-fitted Thomas model suggested external diffusion and internal diffusion are not restrictive steps, and the well-fitted Adam–Bohart model demonstrates that external mass transfer governs dynamics of the whole system [67]. It can be concluded from the Yoon–Nelson model' fit that τ increases with a higher bed height and a lower influent concentration and flow rate, respectively, and τ increases indicating a slower fixed bed exhaustion, which is ideal for removal processes. All models have high correlation coefficients ($R^2 > 0.823$), indicating that all three models have a good fit, there is evidence that the above models, which are mathematically verified to be equivalent, all exhibit high levels of fit performance [71].

4. Conclusions

In this study, the NZVI that was pre-loaded on rGO was embedded into SA to synthesize SA/NZVI-rGO gel beads for removing Cr(VI) from the aqueous solution. The research proved the following: (1) The optimized gel beads simultaneously improved oxidation resistance, swell resistance, and compression strength. (2) The SA/NZVI-rGO gel beads could achieve a substantial Cr(VI) removal in a broad pH range (3–11) or a wide Cr(VI) concentration range (10–100 mg/L). The Langmuir adsorption isotherm demonstrated that the adsorption of Cr(VI) by SA/NZVI-rGO occurred easily, and the theoretical maximum removal capacity was 53.42 mg/g. (3) The removal mechanism of Cr(VI) is a synergy of adsorption, reduction, and ion exchange, where the reduction is dominant. (4) The Adams–Bohart, Thomas, and Yoon–Nelson models showed a good fit for the continuous Cr(VI) removal under different conditions in packed columns. Importantly, the gel beads in the column maintained integrity all the time, owing to their excellent mechanical strength, not only facilitating collection but also limiting the uncontrolled migration of NZVI-rGO, which provides a reliable basis for reducing secondary environmental contamination by nanomaterials in practical applications. Overall, by demonstrating improved oxidation resistance, enhanced mechanical properties and swelling resistance, easy collection, and efficient Cr(VI) removal, the SA/NZVI-rGO beads are a promising remediation agent in Cr(VI) wastewater treatment.

Supplementary Materials: The following supporting information can be downloaded at: https://www.mdpi.com/article/10.3390/polym15183764/s1, Figure S1: Photographs of (a) SA/NZVI beads and (b) SA/NZVI-rGO beads stored under seal for 226 days; Photographs of (c) SA beads and (d) freeze-dried SA/NZVI-rGO beads; Figure S2: SEM of NZVI; Figure S3: Ca 2p XPS spectra of SA/NZVI-rGO beads before and after removal Cr(VI); Figure S4: Photographs of hydrogels before, during, and after compression; Figure S5: The reuse of SA/NZVI-rGO gel beads (initial concentration 20 mg/L, pH3, dosage 0.5 g, 298 K); Table S1: Comparison of the removal efficiency of Cr(VI) by NZVI-based materials under alkaline conditions; Table S2: Composition of the actual Cr(VI) wastewater before and after treatment; Table S3: Comparison of maximum adsorption capacity of Cr(VI) by NZVI-based materials; Table S4: Effect of bed height on column adsorption model constants; Table S5: Effect of initial concentration on column adsorption model constants; Table S6: Effect of flow rate on column adsorption model constants. References [72–82] are cited in the supplementary materials.

Author Contributions: Q.J.: conceptualization, methodology, resources, supervision, writing—review and editing, project administration, funding acquisition. Y.M.: methodology, investigation, data curation, writing original draft, formal analysis. J.H.: investigation. Z.R.: writing—review and editing. All authors have read and agreed to the published version of the manuscript.

Funding: This research was funded by the National Natural Science Foundation of China, Grant No. 41907176.

Data Availability Statement: The data that have been used are confidential.

Acknowledgments: We would like to thank all the participants. This work was financially supported by the National Natural Science Foundation of China (Grant No. 41907176).

Conflicts of Interest: The authors declare that they have no known competing financial interest or personal relationships that could have appeared to influence the work reported in this paper.

References

1. Barrera-Díaz, C.E.; Lugo-Lugo, V.; Bilyeu, B. A review of chemical, electrochemical and biological methods for aqueous Cr(VI) reduction. *J. Hazard. Mater.* **2012**, *223–224*, 1–12. [CrossRef] [PubMed]
2. Kimbrough, D.E.; Cohen, Y.; Winer, A.M.; Creelman, L.; Mabuni, C. A Critical Assessment of Chromium in the Environment. *Crit. Rev. Environ. Sci. Technol.* **1999**, *29*, 1–46. [CrossRef]
3. Saha, R.; Nandi, R.; Saha, B. Sources and toxicity of hexavalent chromium. *J. Coord. Chem.* **2011**, *64*, 1782–1806. [CrossRef]
4. Azzam, A.M.; El-Wakeel, S.T.; Mostafa, B.B.; El-Shahat, M. Removal of Pb, Cd, Cu and Ni from aqueous solution using nano scale zero valent iron particles. *J. Environ. Chem. Eng.* **2016**, *4*, 2196–2206. [CrossRef]
5. Dongsheng, Z.; Wenqiang, G.; Guozhang, C.; Shuai, L.; Weizhou, J.; Youzhi, L. Removal of heavy metal lead(II) using nanoscale zero-valent iron with different preservation methods. *Adv. Powder Technol.* **2018**, *30*, 581–589. [CrossRef]
6. Li, S.; Wang, W.; Liang, F.; Zhang, W.-X. Heavy metal removal using nanoscale zero-valent iron (nZVI): Theory and application. *J. Hazard. Mater.* **2017**, *322*, 163–171. [CrossRef]
7. Stefaniuk, M.; Oleszczuk, P.; Ok, Y.S. Review on nano zerovalent iron (nZVI): From synthesis to environmental applications. *Chem. Eng. J.* **2016**, *287*, 618–632. [CrossRef]
8. Chen, J.; Qiu, X.; Fang, Z.; Yang, M.; Pokeung, T.; Gu, F.; Cheng, W.; Lan, B. Removal mechanism of antibiotic metronidazole from aquatic solutions by using nanoscale zero-valent iron particles. *Chem. Eng. J.* **2012**, *181–182*, 113–119. [CrossRef]
9. Li, Q.; Chen, Z.; Wang, H.; Yang, H.; Wen, T.; Wang, S.; Hu, B.; Wang, X. Removal of organic compounds by nanoscale zero-valent iron and its composites. *Sci. Total. Environ.* **2021**, *792*, 148546. [CrossRef] [PubMed]
10. Morgada, M.E.; Levy, I.K.; Salomone, V.; Farías, S.S.; López, G.; Litter, M.I. Arsenic (V) removal with nanoparticulate zerovalent iron: Effect of UV light and humic acids. *Catal. Today* **2009**, *143*, 261–268. [CrossRef]
11. Zhang, W.-X. Nanoscale Iron Particles for Environmental Remediation: An Overview. *J. Nanoparticle Res.* **2003**, *5*, 323–332. [CrossRef]
12. Zhao, X.; Liu, W.; Cai, Z.; Han, B.; Qian, T.; Zhao, D. An overview of preparation and applications of stabilized zero-valent iron nanoparticles for soil and groundwater remediation. *Water Res.* **2016**, *100*, 245–266. [CrossRef]
13. Calderon, B.; Fullana, A. Heavy metal release due to aging effect during zero valent iron nanoparticles remediation. *Water Res.* **2015**, *83*, 1–9. [CrossRef]
14. Jabeen, H.; Kemp, K.C.; Chandra, V. Synthesis of nano zerovalent iron nanoparticles—Graphene composite for the treatment of lead contaminated water. *J. Environ. Manag.* **2013**, *130*, 429–435. [CrossRef] [PubMed]
15. Wu, M.; Ma, Y.; Wan, J.; Wang, Y.; Guan, Z.; Yan, Z. Investigation of factors affecting the physicochemical properties and degradation performance of nZVI@mesoSiO2 nanocomposites. *J. Mater. Sci.* **2019**, *54*, 7483–7502. [CrossRef]
16. Xue, X.; Lv, X.; Jiang, G.; Baig, S.A.; Xu, X. Influence of Environmental Factors on Hexavalent Chromium Removal from Aqueous Solutions by Nano-Adsorbent Composites. *CLEAN-Soil Air Water* **2016**, *44*, 162–168. [CrossRef]
17. Xu, C.-H.; Zhu, L.-J.; Wang, X.-H.; Lin, S.; Chen, Y.-M. Fast and Highly Efficient Removal of Chromate from Aqueous Solution Using Nanoscale Zero-Valent Iron/Activated Carbon (NZVI/AC). *Water Air Soil Pollut.* **2014**, *225*, 1–13. [CrossRef]
18. Shi, L.-N.; Zhang, X.; Chen, Z.-L. Removal of Chromium (VI) from wastewater using bentonite-supported nanoscale zero-valent iron. *Water Res.* **2011**, *45*, 886–892. [CrossRef]
19. Jing, Q.; You, W.; Tong, L.; Xiao, W.; Kang, S.; Ren, Z. Response surface design for removal of Cr(VI) by hydrogel-supported sulfidated nano zero-valent iron (S-nZVI@H). *Water Sci. Technol. J. Int. Assoc. Water Pollut. Res.* **2021**, *84*, 1190–1205. [CrossRef]
20. Ren, L.; Dong, J.; Chi, Z.; Huang, H. Reduced graphene oxide-nano zero value iron (rGO-nZVI) micro-electrolysis accelerating Cr(VI) removal in aquifer. *J. Environ. Sci.* **2018**, *73*, 96–106. [CrossRef] [PubMed]
21. Xu, M.; Zhu, J.; Wang, F.; Xiong, Y.; Wu, Y.; Wang, Q.; Weng, J.; Zhang, Z.; Chen, W.; Liu, S. Improved In Vitro and In Vivo Biocompatibility of Graphene Oxide through Surface Modification: Poly(Acrylic Acid)-Functionalization is Superior to PEGylation. *ACS Nano* **2016**, *10*, 3267–3281. [CrossRef] [PubMed]
22. Xue, W.; Huang, D.; Zeng, G.; Wan, J.; Cheng, M.; Zhang, C.; Hu, C.; Li, J. Performance and toxicity assessment of nanoscale zero valent iron particles in the remediation of contaminated soil: A review. *Chemosphere* **2018**, *210*, 1145–1156. [CrossRef] [PubMed]
23. Li, Y.; Liu, F.; Xia, B.; Du, Q.; Zhang, P.; Wang, D.; Wang, Z.; Xia, Y. Removal of copper from aqueous solution by carbon nanotube/calcium alginate composites. *J. Hazard. Mater.* **2010**, *177*, 876–880. [CrossRef] [PubMed]
24. Olivas, G.I.; Barbosa-Cánovas, G.V. Alginate–calcium films: Water vapor permeability and mechanical properties as affected by plasticizer and relative humidity. *Lwt-Food Sci. Technol.* **2008**, *41*, 359–366. [CrossRef]
25. Lee, K.Y.; Mooney, D.J. Alginate: Properties and biomedical applications. *Prog. Polym. Sci.* **2012**, *37*, 106–126. [CrossRef]
26. Bezbaruah, A.N.; Krajangpan, S.; Chisholm, B.J.; Khan, E.; Bermudez, J.J.E. Entrapment of iron nanoparticles in calcium alginate beads for groundwater remediation applications. *J. Hazard. Mater.* **2009**, *166*, 1339–1343. [CrossRef]

27. Kuang, Y.; Du, J.; Zhou, R.; Chen, Z.; Megharaj, M.; Naidu, R. Calcium alginate encapsulated Ni/Fe nanoparticles beads for simultaneous removal of Cu (II) and monochlorobenzene. *J. Colloid Interface Sci.* **2015**, *447*, 85–91. [CrossRef]
28. Kamata, H.; Akagi, Y.; Kayasuga-Kariya, Y.; Chung, U.-I.; Sakai, T. "Nonswellable" Hydrogel Without Mechanical Hysteresis. *Science* **2014**, *343*, 873–875. [CrossRef]
29. Zhuang, Y.; Yu, F.; Chen, H.; Zheng, J.; Ma, J.; Chen, J. Alginate/graphene double-network nanocomposite hydrogel beads with low-swelling, enhanced mechanical properties, and enhanced adsorption capacity. *J. Mater. Chem. A* **2016**, *4*, 10885–10892. [CrossRef]
30. Dou, X.; Li, R.; Zhao, B.; Liang, W. Arsenate removal from water by zero-valent iron/activated carbon galvanic couples. *J. Hazard. Mater.* **2010**, *182*, 108–114. [CrossRef]
31. Lv, X.; Xue, X.; Jiang, G.; Wu, D.; Sheng, T.; Zhou, H.; Xu, X. Nanoscale Zero-Valent Iron (nZVI) assembled on magnetic Fe_3O_4/graphene for Chromium (VI) removal from aqueous solution. *J. Colloid Interface Sci.* **2014**, *417*, 51–59. [CrossRef]
32. Jing, Q.; You, W.; Qiao, S.; Ma, Y.; Ren, Z. Comprehensive understanding of adsorption and reduction on 2,4-DCP and Cr(VI) removal process by NZVI-rGO: Performance and mechanism. *J. Water Process. Eng.* **2023**, *51*. [CrossRef]
33. Roosen, J.; Pype, J.; Binnemans, K.; Mullens, S. Shaping of Alginate–Silica Hybrid Materials into Microspheres through Vibrating-Nozzle Technology and Their Use for the Recovery of Neodymium from Aqueous Solutions. *Ind. Eng. Chem. Res.* **2015**, *54*, 12836–12846. [CrossRef]
34. Bharti; Khurana, I.; Shaw, A.K.; Saxena, A.; Khurana, J.M.; Rai, P.K. Removal of Trinitrotoluene with Nano Zerovalent Iron Impregnated Graphene Oxide. *Water Air Soil Pollut.* **2018**, *229*, 17. [CrossRef]
35. Li, J.; Chen, C.; Zhu, K.; Wang, X. Nanoscale zero-valent iron particles modified on reduced graphene oxides using a plasma technique for Cd(II) removal. *J. Taiwan Inst. Chem. Eng.* **2016**, *59*, 389–394. [CrossRef]
36. Sahoo, S.K.; Padhiari, S.; Biswal, S.; Panda, B.; Hota, G. Fe_3O_4 nanoparticles functionalized GO/g-C3N4 nanocomposite: An efficient magnetic nanoadsorbent for adsorptive removal of organic pollutants. *Mater. Chem. Phys.* **2020**, *244*, 122710. [CrossRef]
37. Gu, M.; Sui, Q.; Farooq, U.; Zhang, X.; Qiu, Z.; Lyu, S. Enhanced degradation of trichloroethylene in oxidative environment by nZVI/PDA functionalized rGO catalyst. *J. Hazard. Mater.* **2018**, *359*, 157–165. [CrossRef]
38. Modrogan, C.; Pandele, A.M.; Bobirică, C.; Dobrotă, D.; Dăncilă, A.M.; Gârleanu, G.; Orbuleţ, O.D.; Borda, C.; Gârleanu, D.; Orbeci, C. Synthesis, Characterization and Sorption Capacity Examination for a Novel Hydrogel Composite Based on Gellan Gum and Graphene Oxide (GG/GO). *Polymers* **2020**, *12*, 1182. [CrossRef]
39. Zhao, W.; Qi, Y.; Wang, Y.; Xue, Y.; Xu, P.; Li, Z.; Li, Q. Morphology and Thermal Properties of Calcium Alginate/Reduced Graphene Oxide Composites. *Polymers* **2018**, *10*, 990. [CrossRef]
40. Song, L.; Liu, F.; Zhu, C.; Li, A. Facile one-step fabrication of carboxymethyl cellulose based hydrogel for highly efficient removal of Cr(VI) under mild acidic condition. *Chem. Eng. J.* **2019**, *369*, 641–651. [CrossRef]
41. Zhao, C.; Hu, L.; Zhang, C.; Wang, S.; Wang, X.; Huo, Z. Preparation of biochar-interpenetrated iron-alginate hydrogel as a pH-independent sorbent for removal of Cr(VI) and Pb(II). *Environ. Pollut.* **2021**, *287*, 117303. [CrossRef] [PubMed]
42. Yan, W.; Herzing, A.A.; Kiely, C.J.; Zhang, W.-X. Nanoscale zero-valent iron (nZVI): Aspects of the core-shell structure and reactions with inorganic species in water. *J. Contam. Hydrol.* **2010**, *118*, 96–104. [CrossRef] [PubMed]
43. Zhao, M.; Zhang, C.; Yang, X.; Liu, L.; Wang, X.; Yin, W.; Li, Y.C.; Wang, S.; Fu, W. Preparation of highly-conductive pyrogenic carbon-supported zero-valent iron for enhanced Cr(VI) reduction. *J. Hazard. Mater.* **2020**, *396*, 122712. [CrossRef] [PubMed]
44. Majumder, S.; Sardar, M.; Satpati, B.; Kumar, S.; Banerjee, S. Magnetization Enhancement of Fe_3O_4 by Attaching onto Graphene Oxide: An Interfacial Effect. *J. Phys. Chem. C* **2018**, *122*, 21356–21365. [CrossRef]
45. Song, L.; Feng, Y.; Zhu, C.; Liu, F.; Li, A. Enhanced synergistic removal of Cr(VI) and Cd(II) with bi-functional biomass-based composites. *J. Hazard. Mater.* **2020**, *388*, 121776. [CrossRef]
46. Zhu, C.; Liu, F.; Zhang, Y.; Wei, M.; Zhang, X.; Ling, C.; Li, A. Nitrogen-doped chitosan-Fe(III) composite as a dual-functional material for synergistically enhanced co-removal of Cu(II) and Cr(VI) based on adsorption and redox. *Chem. Eng. J.* **2016**, *306*, 579–587. [CrossRef]
47. Patel, H. Fixed-bed column adsorption study: A comprehensive review. *Appl. Water Sci.* **2019**, *9*, 45. [CrossRef]
48. Yao, W.; Geng, C.; Han, D.; Chen, F.; Fu, Q. Strong and conductive double-network graphene/PVA gel. *RSC Adv.* **2014**, *4*, 39588–39595. [CrossRef]
49. Zhuang, J.; Dai, J.; Ghaffar, S.H.; Yu, Y.; Tian, Q.; Fan, M. Development of highly efficient, renewable and durable alginate composite aerogels for oil/water separation. *Surf. Coatings Technol.* **2020**, *388*, 125551. [CrossRef]
50. Li, J.; Ma, J.; Chen, S.; Huang, Y.; He, J. Adsorption of lysozyme by alginate/graphene oxide composite beads with enhanced stability and mechanical property. *Mater. Sci. Eng. C* **2018**, *89*, 25–32. [CrossRef]
51. Deng, J.; Dong, H.; Zhang, C.; Jiang, Z.; Cheng, Y.; Hou, K.; Zhang, L.; Fan, C. Nanoscale zero-valent iron/biochar composite as an activator for Fenton-like removal of sulfamethazine. *Sep. Purif. Technol.* **2018**, *202*, 130–137. [CrossRef]
52. Pérez-Candela, M.; Martín-Martínez, J.M.; Torregrosa-Maciá, R. Chromium(VI) removal with activated carbons. *Water Res.* **1995**, *29*, 2174–2180. [CrossRef]
53. Yi, Y.; Tu, G.; Zhao, D.; Tsang, P.E.; Fang, Z. Key role of FeO in the reduction of Cr(VI) by magnetic biochar synthesised using steel pickling waste liquor and sugarcane bagasse. *J. Clean. Prod.* **2020**, *245*, 118886. [CrossRef]

54. Huang, X.-Y.; Ling, L.; Zhang, W.-X. Nanoencapsulation of hexavalent chromium with nanoscale zero-valent iron: High resolution chemical mapping of the passivation layer. *J. Environ. Sci.* **2018**, *67*, 4–13. [CrossRef]
55. Zhang, Y.; Liu, N.; Yang, Y.; Li, J.; Wang, S.; Lv, J.; Tang, R. Novel carbothermal synthesis of Fe, N co-doped oak wood biochar (Fe/N-OB) for fast and effective Cr(VI) removal. *Colloids Surfaces A Physicochem. Eng. Asp.* **2020**, *600*, 124926. [CrossRef]
56. Özcan, A.S.; Gök, Ö.; Özcan, A. Adsorption of lead(II) ions onto 8-hydroxy quinoline-immobilized bentonite. *J. Hazard. Mater.* **2009**, *161*, 499–509. [CrossRef]
57. Wang, Z.; Wu, X.; Luo, S.; Wang, Y.; Tong, Z.; Deng, Q. Shell biomass material supported nano-zero valent iron to remove Pb^{2+} and Cd $^{2+}$ in water. *R. Soc. Open Sci.* **2020**, *7*, 201192. [CrossRef]
58. Chen, Z.; Wang, T.; Jin, X.; Chen, Z.; Megharaj, M.; Naidu, R. Multifunctional kaolinite-supported nanoscale zero-valent iron used for the adsorption and degradation of crystal violet in aqueous solution. *J. Colloid Interface Sci.* **2013**, *398*, 59–66. [CrossRef]
59. Liu, X.; Zhang, S.; Zhang, X.; Guo, H.; Cao, X.; Lou, Z.; Zhang, W.; Wang, C. A novel lignin hydrogel supported nZVI for efficient removal of Cr(VI). *Chemosphere* **2022**, *301*, 134781. [CrossRef]
60. Shi, L.-N.; Lin, Y.-M.; Zhang, X.; Chen, Z.-L. Synthesis, characterization and kinetics of bentonite supported nZVI for the removal of Cr(VI) from aqueous solution. *Chem. Eng. J.* **2011**, *171*, 612–617. [CrossRef]
61. Langmuir, I. The adsorption of gases on plane surfaces of glass, mica and platinum. *J. Am. Chem. Soc.* **1918**, *40*, 1361–1403. [CrossRef]
62. Freundlich, H.M.F. Über die Adsorption in Lösungen. *Z. Phys. Chem.* **1907**, *57U*, 385–470. [CrossRef]
63. Liu, J.; Wang, Y.; Fang, Y.; Mwamulima, T.; Song, S.; Peng, C. Removal of crystal violet and methylene blue from aqueous solutions using the fly ash-based adsorbent material-supported zero-valent iron. *J. Mol. Liq.* **2018**, *250*, 468–476. [CrossRef]
64. Gu, H.; Gao, Y.; Xiong, M.; Zhang, D.; Chen, W.; Xu, Z. Removal of nitrobenzene from aqueous solution by graphene/biochar supported nanoscale zero-valent-iron: Reduction enhancement behavior and mechanism. *Sep. Purif. Technol.* **2021**, *275*, 119146. [CrossRef]
65. Mon, P.P.; Cho, P.P.; Chanadana, L.; Kumar, K.A.; Dobhal, S.; Shashidhar, T.; Madras, G.; Subrahmanyam, C. Bio-waste assisted phase transformation of Fe_3O_4/carbon to nZVI/graphene composites and its application in reductive elimination of Cr(VI) removal from aquifer. *Sep. Purif. Technol.* **2023**, *306*. [CrossRef]
66. Zhang, W.; Wang, H.; Hu, X.; Feng, H.; Xiong, W.; Guo, W.; Zhou, J.; Mosa, A.; Peng, Y. Multicavity triethylenetetramine-chitosan/alginate composite beads for enhanced Cr(VI) removal. *J. Clean. Prod.* **2019**, *231*, 733–745. [CrossRef]
67. Omitola, O.B.; Abonyi, M.N.; Akpomie, K.G.; Dawodu, F.A.; Adams-Bohart, Y.N. Thomas modeling of the fix-bed continuous column adsorption of amoxicillin onto silver nanoparticle-maize leaf composite. *Appl. Water Sci.* **2022**, *12*, 94. [CrossRef]
68. Jain, M.; Garg, V.; Kadirvelu, K. Cadmium(II) sorption and desorption in a fixed bed column using sunflower waste carbon calcium–alginate beads. *Bioresour. Technol.* **2012**, *129*, 242–248. [CrossRef]
69. Ko, D.C.; Porter, J.F.; McKay, G. Optimised correlations for the fixed-bed adsorption of metal ions on bone char. *Chem. Eng. Sci.* **2000**, *55*, 5819–5829. [CrossRef]
70. Njaramba, L.; Kim, S.; Kim, Y.; Cha, B.; Kim, N.; Yoon, Y.; Park, C. Remarkable adsorption for hazardous organic and in-organic contaminants by multifunctional amorphous core–shell structures of metal–organic framework-alginate compo-sites. *Chem. Eng. J.* **2022**, *431*, 133415. [CrossRef]
71. Chu, K.H. Breakthrough curve analysis by simplistic models of fixed bed adsorption: In defense of the century-old Bohart-Adams model. *Chem. Eng. J.* **2019**, *380*, 122513. [CrossRef]
72. Bohart, G.S.; Adams, E.Q. Some Aspects of the Behavior of Charcoal with Respect to Chlorine. *J. Am. Chem. Soc.* **1920**, *42*, 523–544. [CrossRef]
73. Cheng, Y.; Dong, H.; Hao, T. CaCO3 coated nanoscale zero-valent iron (nZVI) for the removal of chromium(VI) in aqueous solution. *Sep. Purif. Technol.* **2020**, *257*, 117967. [CrossRef]
74. Lv, X.; Qin, X.; Wang, K.; Peng, Y.; Wang, P.; Jiang, G. Nanoscale zero valent iron supported on MgAl-LDH-decorated reduced graphene oxide: Enhanced performance in Cr(VI) removal, mechanism and regeneration. *J. Hazard. Mater.* **2019**, *373*, 176–186. [CrossRef]
75. Sarojini, G.; Babu, S.V.; Rajamohan, N.; Kumar, P.S.; Rajasimman, M. Surface modified polymer-magnetic-algae nanocomposite for the removal of chromium- equilibrium and mechanism studies. *Environ. Res.* **2021**, *201*, 111626. [CrossRef]
76. Tian, Z.; Jing, Q.; Qiao, S.; You, W. Encapsulation/capture of S-nZVI particles by PANa-PAM hydrogel limits their leakage and improves Cr(VI) removal. *Process. Saf. Environ. Prot.* **2023**, *172*, 124–135. [CrossRef]
77. Wang, T.; Sun, Y.; Bai, L.; Han, C.; Sun, X. Ultrafast removal of Cr(VI) by chitosan coated biochar-supported nano zero-valent iron aerogel from aqueous solution: Application performance and reaction mechanism. *Sep. Purif. Technol.* **2023**, *306*, 122631. [CrossRef]
78. Wen, R.; Tu, B.; Guo, X.; Hao, X.; Wu, X.; Tao, H. An ion release controlled Cr(VI) treatment agent: Nano zero-valent iron/carbon/alginate composite gel. *Int. J. Biol. Macromol.* **2019**, *146*, 692–704. [CrossRef]
79. Yi, Y.; Wang, X.; Zhang, Y.; Yang, K.; Ma, J.; Ning, P. Formation and mechanism of nanoscale zerovalent iron supported by phosphoric acid modified biochar for highly efficient removal of Cr(VI). *Adv. Powder Technol.* **2023**, *34*, 103826. [CrossRef]
80. Yoon, Y.H.; Nelson, J.H. Application of Gas Adsorption Kinetics I. A Theoretical Model for Respirator Cartridge Service Life. *Am. Ind. Hyg. Assoc. J.* **1984**, *45*, 509–516. [CrossRef] [PubMed]

81. Zhao, R.; Zhou, Z.; Zhao, X.; Jing, G. Enhanced Cr(VI) removal from simulated electroplating rinse wastewater by amino-functionalized vermiculite-supported nanoscale zero-valent iron. *Chemosphere* **2018**, *218*, 458–467. [CrossRef] [PubMed]
82. Zhou, H.; Ma, M.; Zhao, Y.; Baig, S.A.; Hu, S.; Ye, M.; Wang, J. Integrated green complexing agent and biochar modified nano zero-valent iron for hexavalent chromium removal: A characterisation and performance study. *Sci. Total. Environ.* **2022**, *834*, 155080. [CrossRef] [PubMed]

MDPI AG
Grosspeteranlage 5
4052 Basel
Switzerland
Tel.: +41 61 683 77 34

Polymers Editorial Office
E-mail: polymers@mdpi.com
www.mdpi.com/journal/polymers

www.ingramcontent.com/pod-product-compliance
Lightning Source LLC
Chambersburg PA
CBHW041346210926
51850CB00013B/3001
* 9 7 8 3 7 2 5 8 4 4 9 1 3 *